# Green Energy and Infrastructure

# Green Energy and Infrastructure

## Securing a Sustainable Future

Edited by

Jacqueline A. Stagner and David S-K. Ting

CRC Press
Taylor & Francis Group
Boca Raton London New York

CRC Press is an imprint of the
Taylor & Francis Group, an informa business

First edition published 2021
by CRC Press
6000 Broken Sound Parkway NW, Suite 300, Boca Raton, FL 33487-2742

and by CRC Press
2 Park Square, Milton Park, Abingdon, Oxon, OX14 4RN



CRC Press is an imprint of Taylor & Francis Group, LLC



*Library of Congress Cataloging-in-Publication Data*
Names: Stagner, Jacqueline A., editor. | Ting, David S-K., editor.
Title: Green energy and infrastructure : securing a sustainable future /
edited by Jacqueline A. Stagner and David S-K. Ting.
Description: First edition. | Boca Raton, FL : CRC Press, 2021. |
Includes index.
Identifiers: LCCN 2020020229 (print) | LCCN 2020020230 (ebook) |
ISBN 9780367559496 (hardback) | ISBN 9781003095811 (ebook)
Subjects: LCSH: Renewable energy sources. | Sustainable development.
Classification: LCC TJ808 .G688 2021 (print) | LCC TJ808 (ebook) |
DDC 333.79/4—dc23
LC record available at https://lccn.loc.gov/2020020229
LC ebook record available at https://lccn.loc.gov/2020020230

ISBN: 978-0-367-55949-6 (hbk)
ISBN: 978-1-003-09581-1 (ebk)

Typeset in Times
by codeMantra

*To the individuals who practice and foster sustainability in their everyday living, securing a sustainable future for those succeeding them.*

# Contents

# Preface

More than half a century ago, C. S. Lewis wrote, "The task of the modern educator is not to cut down jungles but to irrigate deserts." This is ever more relevant today. To secure a sustainable future, we need to enhance the "irrigating of the deserts" and teach young minds to do likewise. Reader breaks the ice with the first chapter, "Energy for Buildings: Practices, Policies, and Prospects", by highlighting the fact that building energy is a dominant contributor to global energy usage and greenhouse gas emissions. Within this context, he points out that the energy sector plays an important role in realizing sustainable development goals embedded in the United Nations plan to eliminate global poverty. To advance building energy, Balo and Polat illustrate how sustainable building materials can simultaneously improve energy, economic, and environmental bottom lines using a mini automotive-repair facility as an example in Chapter 2, "Green Design Effectiveness for a Mini Automotive-Repair Facility." A companion house for a research hospital is used to demonstrate the substantial energy improvement that can be realized via proper greening by Balo et al. in Chapter 3, "Green Hospitals and Sustainability: Case of Companion House of a Research Hospital."

It would defeat the purpose if energy savings are realized at the expense of occupant health and comfort. Seduikyte et al. caution that this undesirable consequence happens in their case study in Chapter 4, "Indoor Environment and Well-being: the Case of Academic Workplace in Historic Building."

Biomass is a renewable energy which can be further "greened." Zhang et al. disclose the underlying properties of biomass and their influence on furthering the energy conversion technologies in Chapter 5, "Properties and Conversion Technologies of Biomass." Wind energy will continue to play a vital role in greening tomorrow, and its progress leans on accurate assessment and forecasting.

Xydis and Nanaki timely convey "Wind Resource Forecasting Error in Flat and Complex Terrains" in Chapter 6. One potent means to improve wind forecasting is via deep learning methods as delineated by Gökgöz and Filiz in Chapter 7, "Wind Power Forecasting via Deep Learning Methods." Whether it is wind, solar, geothermal, or tidal, storage can substantially improve the integration of these largely intermittent energy sources.

Santhakumari discusses energy storage in terms of power balancing, grid stability, and reliability in Chapter 8, "Green Energy: Solar, Wind, Geothermal, Tidal Storage." One cannot talk about energy storage without giving an account on compressed air energy storage. In Chapter 9, Llamas et al. study abandoned coal mines for their suitability for this purpose, "New Energy Mining: Compressed Air Energy Storage in Abandoned Mines." Also promising and arising on the horizon is "Hydrostatically Compensated Energy Storage Technology," as expounded by Ebrahimi et al. in Chapter 10.

Green infrastructures encompasses the notion that they are in harmony and integration with both the spatial and temporal domains. Barbeta and Abdallah enlighten us on the need to respect nature in our intervention in Chapter 11, "Bioconstruction and Harmonic Complexity of Biomimicry Organisms." Yes, the message is "Back to the Basics: Return to the Origin, Gaudí and Nature," as proclaimed by Estévez and Urbano in Chapter 12.

This book ends with "Triple Bottom Line Analysis, Methodology and Its Implementation" as Chapter 13. In this chapter, Patel et al. elucidate the three pillars of sustainability, environmental sustainability, social sustainability, and economic sustainability. All three pillars must be sustained as we strive in securing a sustainable future.

# Acknowledgments

This book would have remained but a vision without the helping hands from many paladins. The superlative Taylor & Francis team, Joseph Clements and Lisa Wilford, as well as Assunta Petrone et al. at codeMantra, deserve a big shout out. Also, without the experts who formulated the individual chapters, the state-of-the-art volume would not have been possible. Finally, a sincere "thank you" goes to the reviewers who improved the quality of the manuscript through their constructive remarks.

# Editors

**Dr. Jacqueline A. Stagner** is the Undergraduate Programs Coordinator of Engineering at the University of Windsor. She has a PhD in Materials Science and Engineering, a Master of Business Administration, and a Bachelor's Degree in Mechanical Engineering. She worked as a release engineer in the automotive industry prior to joining the University of Windsor. As an adjunct graduate faculty member in the Department of Mechanical, Automotive, and Materials Engineering, she co-advises students in the sustainability and renewable energy areas, in the Turbulence & Energy Laboratory.

**Dr. David S-K. Ting** is the founder of the Turbulence & Energy Laboratory, University of Windsor. As a professor in the Department of Mechanical, Automotive, and Materials Engineering, he supervises students on a wide range of research projects in the thermofluids, flow turbulence, energy conversion and conservation, and renewable energy areas. To date, he has co/supervised over 75 graduate students and co-authored more than 140 journal papers.

# Contributors

**Yomna Abdallah**
IBAG-UIC
Universitat Internacional de Catalunya
Barcelona, Spain

**Figen Balo**
Department of Industrial Engineering
Firat University
Elazığ, Turkey

**Gabriel Barbeta**
Ecoarquitectura
Universitat de Girona
Girona, Spain

**Rupp Carriveau**
Department of Civil and Environmental Engineering
University of Windsor
Windsor, Canada

**Mehdi Ebrahimi**
Turbulence and Energy Laboratory
University of Windsor
Windsor, Canada

**Alberto T. Estévez**
Institute for BioDigital Architecture & Genetics
Universitat Internacional de Catalunya
Barcelona, Spain

**Fahrettin Filiz**
Doctoral Program in Management, Social Sciences Institute
Ankara University
Ankara, Turkey

**Paris Fokaides**
Faculty of Civil Engineering and Architecture
Kaunas University of Technology
Kaunas, Lithuania
School of Engineering
Frederick University
Nicosia, Cyprus

**Wenming Fu**
School of Energy Science and Engineering
Harbin Institute of Technology
Harbin, China

**Fazıl Gökgöz**
Quantitative Methods Division, Department of Management, Faculty of Political Sciences
Ankara University
Ankara, Turkey

**Indre Grazuleviciute-Vileniske**
Faculty of Civil Engineering and Architecture
Kaunas University of Technology
Kaunas, Lithuania

**Baocheng Jiang**
School of Energy Science and Engineering
Harbin Institute of Technology
Harbin, China

**Bingxi Li**
School of Energy Science and Engineering
Harbin Institute of Technology
Harbin, China

**Bernardo Llamas**
ETSI Minas y Energía
Universidad Politecnica de Madrid
Madrid, Spain

**Andrew McGillis**
Hydrotor
Windsor, Canada

**Pedro Mora**
ETSI Minas y Energía
Universidad Politecnica de Madrid
Madrid, Spain

**Evanthia A. Nanaki**
Department of Business Development and Technology
Aarhus University
Herning, Denmark

**Marcelo F. Ortega**
ETSI Minas y Energía
Universidad Politecnica de Madrid
Madrid, Spain

**Amit R Patel**
Mechanical Engineering Department
Maharaja Sayajirao University of Baroda
Vadodara, India

**Hasan Polat**
Department of Architecture
Firat University
Elazığ, Turkey

**Graham T. Reader**
Department of Mechanical, Automotive and Materials Engineering
University of Windsor
Windsor, Canada

**Manju Santhakumari**
CSIR—National Geophysical Research Institute
Hyderabad, India

**Prabir Sarkar**
Department of Mechanical Engineering
Indian Institute of Technology Ropar
Ropar, India

**Lina Seduikyte**
Faculty of Civil Engineering and Architecture
Kaunas University of Technology
Kaunas, Lithuania

**Harpreet Singh**
School of Mechanical, Materials & Energy Engineering
Indian Institute of Technology Ropar
Ropar, India

**Tomas Sorius**
Faculty of Civil Engineering and Architecture
Kaunas University of Technology
Kaunas, Lithuania

**Egle Staniskiene**
School of Economics and Business
Kaunas University of Technology
Kaunas, Lithuania

**Zivile Stankeviciute**
School of Economics and Business
Kaunas University of Technology
Kaunas, Lithuania

**Laura Stasiuliene**
Faculty of Civil Engineering and Architecture
Kaunas University of Technology
Kaunas, Lithuania

**Lutfu S. Sua**
Industrial Engineering
Firat University
Elazığ, Turkey

**David S-K. Ting**
Turbulence and Energy Laboratory
University of Windsor
Windsor, Canada

**Himanshu Tyagi**
Department of Mechanical Engineering
Indian Institute of Technology Ropar
Ropar, India

**Judith Urbano**
Department of Humanities
Universitat Internacional de Catalunya
Barcelona, Spain

**Belén Vallespir**
ETSI Minas y Energía
Universidad Politecnica de Madrid
Madrid, Spain

**Pingfei Xu**
School of Energy Science and Engineering
Harbin Institute of Technology
Harbin, China

**George Xydis**
Department of Business Development and Technology
Aarhus University
Herning, Denmark

**Yaning Zhang**
School of Energy Science and Engineering
Harbin Institute of Technology
Harbin, China

# 1 Energy for Buildings
## *Practices, Policies, and Prospects*

*G. T. Reader*
University of Windsor

## CONTENTS



## 1.1 INTRODUCTION: BACKGROUND AND DRIVING FORCES

As the global human population continues to increase, so too does its consumption of energy. The ability to provide and supply increasing amounts of affordable and clean energy is fast becoming one of the major political and technologically challenges of the 21st century. These challenges will only intensify as the global population, over the next 81 years, could be between 22% and 65% higher than it is now. Such rises are realistic according to the latest United Nations (UN) Population Division report (2019), in which it is projected that the population will grow to 10.9 billion by 2100, within a range as *low* as 9.4 or as *high* as 12.7 billion.

A major user of energy is the buildings and construction sectors, whether commercial or residential. Although the energy used in buildings as a sector does not attract as much media attention as the other categorized energy-usage sectors, it is proportionally just as significant, if not more so, than the transportation and industrial sectors. According to the 2018 United Nations Environmental Programme (UN EP)—International Energy Agency (IEA) report, the building construction and operation sector accounted for, *36% of global final energy use and nearly 40% of energy related carbon dioxide ($CO_2$) emissions in 2017* (UNEP 2018). These percentages include the manufacture of materials and products for construction, which are not always included by other agencies under the "building" sector category. Consequently, there can be some published variations from one national or international agency to another depending on the precise definitions used to describe the building sector and its sub-components. Regardless of these usually minor differences, there is no doubt that building-related energy consumption is a significant proportion of the global energy use, and while the IEA data perhaps paints a slightly exaggerated picture, most of the literature indicates that one-third of global energy is used in building services.

The energy consumed in buildings is used for a variety of human needs and wants. The main uses, since humans first learned how to harness, control, and start *fire*, being heating (space and water), cooking, and lighting. More recently, energy has been used for space and food cooling, manifest in air-conditioning and refrigeration systems, respectively, and for powering a multitude of electrical equipment such as computers, communications (radio, television, telephones), labor-saving devices, and appliances. In addition, in the commercial sector, energy is also required for machines and services such as sewage (wastewater) treatment.

The amount and type of Building energy demand varies by geographical location, national, regional, and local, the inherent prevailing weather and climate, occupant behavior, and the underlying wealth of the populace. How wealthy a country is, based on measures such as Gross Domestic Product (GDP), Gross National Income (GNI), or Purchasing power parity (PPP) per capita, has a marked impact on energy use in general, and particularly Building energy (Lafrance and Schembri 2002, 27–33). Nevertheless, even in some *rich*, developed countries, there are those in the population who can ill-afford Building energy services and so experience living conditions termed or defined as *energy poverty* (Enbridge 2019). Some countries and their constituent parts, states, provinces, and so on, rely on *imported* energy. Depending upon the proportion of imported-to-indigenous energies, a nation's *energy security* could be severely impacted by an interruption, temporary or long term, of the imported supplies.

But what exactly is energy security? There are several definitions provided in the open literature, but the one promoted by the IEA, namely that energy security is *the uninterrupted availability of energy sources at an affordable price*, has much to recommend it when discussing global energy issues (IEA accessed 2019). For example, over the next two decades, it has been forecast that in the developing countries of the Asia-Pacific region, energy growth demand will account for ⅔ of global growth, and that much of this demand will have to be met by energy imports into these countries (IEA report June 2019). Subsequently, energy *crises*, akin to the 20th-century oil crises which economically distressed many western countries, could have similar impacts in many parts of the Asia-Pacific region. Energy self-sufficiency could overcome such socio-economic risks within the 2020–2040 timeframe and most likely for many decades beyond; targeted efforts to improve *energy efficiency* appear to offer some hope of countering rising energy security concerns.

### 1.1.1 Urbanization and Sustainability

In terms of future energy challenges for buildings, there are other key elements that need to be tackled such as, *urbanization* and *sustainability*. Urbanization, i.e., the movement of the population from rural to urban areas (cities), is a relatively new phenomenon within human history, whose modern origins can be traced to the industrial age and which has accelerated since the mid-20th century. *This is illustrated in Figure 1.1 adapted from the article by Ritchie and Roser* (2019). According to the UN's (2019, World Urbanization) latest revision to the *World Urbanization Prospects*, 56% of today's global population live in urban areas compared to just over 29% seven decades ago, and the proportion is expected to rise to 68% by 2050. This increase has obvious ramifications for buildings, not least of which is the so-called *Heat Island* impact whereby, depending upon the time of day, urban areas air temperatures can be between 1°C and 12°C higher than the surrounding rural areas, especially in summer seasons (US EPA 2019). This effect usually leads to increasing energy demands and, in some cases, increased greenhouse gas (GHG) emissions. Consequently, improving Building energy efficiency is not only an intuitive pathway for the Asia-Pacific region but also globally.

As the population increases throughout the century and the proven reserves, especially involving oil deposits, of the currently dominant fossil fuel energy sources decrease, significant transitions to other forms of energy will be required to be included in the global energy mix. For permanent effect, these other forms should be *sustainable*, which requires them to also be *renewable* (Reader 2019). Moreover, sustainability will need to apply not only to energy sources and energy conversion devices but also to the buildings themselves. But what is building sustainability? The term *Green* building is usually used interchangeably with *sustainable* building, but different organizations, regional, and national governments have differing definitions and interpretations of exactly what constitutes a green or sustainable building.

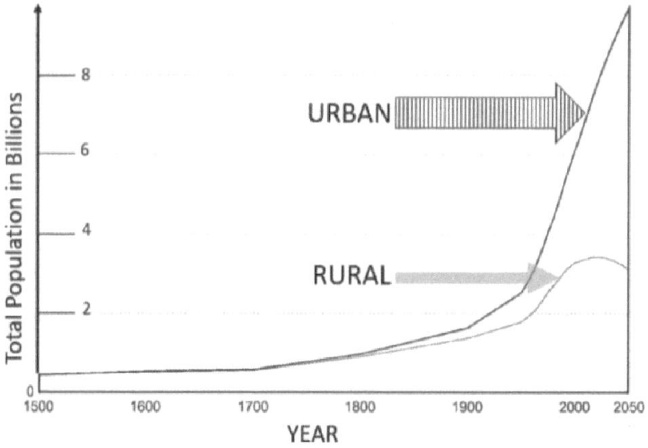


**FIGURE 1.1** Total urban and rural population, given as estimates to 2016 and UN Projections to 2050. (Author redrawn from original of H. Ritchie and M. Roser. 2019. Published online at OurWorldInData.org.)

The definition promoted by *Green Building Canada*, which is fairly typical states, *Green building is a resource-efficient method of construction that produces healthier buildings which have less impact on the environment and cost less to maintain. This sustainable approach to construction accounts for a building's entire life cycle* (Green Building Canada 2019). The United States Environmental Protection Agency (EPA) had proposed a similar definition, but with the arrival of a new Federal Administration, the definitions and explanations were archived as of 2017 (US EPA 2016).

Nevertheless, there are several non-governmental or quasi-government organizations in the United States, Canada, and worldwide, who are involved in certifying *Green or Sustainable* buildings using the constantly evolving Leadership in Energy and Environmental Design (LEED) system since its origins three decades ago.[1] More recently, there is an increasing trend to advocate for and specify *Net Zero-Energy Buildings* (NZEB) whereby, over a period of say a year, the building infrastructure generates more electricity than it consumes (Torcellini et al. 2006, 12). On a global basis, this could be a challenging concept, particularly since over a billion people, at least 11% of the current world population, do not have access to electricity (World Bank 2019). A concomitant trend is the intent to develop *Zero-Carbon* buildings, which means that no form of carbonaceous fuel should be used to directly or indirectly provide energy services to a building or, presumably, any carbon dioxide generated by the energy consumed by the building, must be sequestrated and eventually stored (Canada Green Building Council 2017). Another building concept is the voluntary *Passive House* or the Adamson–Feist Passivhaus standard whose aim is to reduce the ecological footprint of residential buildings through energy efficiency improvements (The Passive House Institute 2019). Since the founding of the Passive House Institute (PHI) in 1996, the Passivhaus standard has also been used for new and retrofitted residential buildings as well as some commercial and local government buildings. The use of the PHI standard originally started in Germany and Austria but has expanded to many other countries including the United Kingdom (UK), Canada, and the United States, the latter having its own version of the Passive House standard—PHIUS+ (The UK Passive House Organization 2019; Passive House Institute US 2019).

All the factors involved in the provision and operation of Building energy services, i.e., energy poverty and security, urbanization and sustainability, Net Zero Energy, zero carbon, and Passivhaus, have a modern backdrop of the deep concerns embedded in the United Nation's Framework Convention on Climate Change (UNFCC). The UNFCC is an international environmental treaty which became effective over 25 years ago in March 1994 and evolved into the 2015 *Paris Agreement*, described as, *a landmark agreement to combat climate change and to accelerate and intensify the actions and investments needed for a sustainable low carbon future,* which, as of 21 November 2019, has been ratified by 187 *parties* (member states plus the European Union) (Kuyper, Schroeder, and Linnér 2018, 343–368; United Nations Climate Change). The scientific background related to the

[1] Developed by the US environmental advocacy group the National Resources Defence Council, https://en.wikipedia.org/wiki/Natural_Resources_Defense_Council (accessed November 21, 2019).

UNFCC agreements on climate change is provided by another United Nations body, the Intergovernmental Panel on Climate Change (IPCC) through numerous reports and regular updates, dealing with anthropogenic climate change and its potential impacts on global society.[2]

The IPCC is not a policy-making body, but its investigative research findings are used by the UNFCC policy makers in formulating global environmental agreements and climate change mitigation action plans. The IPCC and UNFCC both play roles in the establishment of the United Nations *Sustainable Development Goals* (SDGs) whose overall objective is *to achieve a better and more sustainable future for all* (United Nations 2015, SDG). Of the 17 current SDGs, two are particularly pertinent to energy for buildings, SDG 7: *Affordable and Clean Energy*, and SDG 11: *Sustainable Cities and Communities*. In both cases, in addition to all the factors highlighted at the start of the previous paragraph, another key factor is the *energy efficiency* of buildings.

If the portentous forecasts of the IPCC, the UNFCC, and much of the learned scientific literature published by climate researchers regarding anthropogenic climate change and its causes are correct, then in the necessary transition to a clean and sustainable energy future, buildings will play a vital role (IEA 2019). Consequently, one of the ramifications of the attempts to mitigate, or in a utopian scenario reverse, the negative aspects of anthropogenic climate change will likely be the need to have enforceable global statutory building regulations to ensure that the Building energy factors are thoroughly addressed and SDG 7 and 11 are achieved.

The obligations and constraints to be placed on new building design, construction, and operation are then somewhat daunting. Although many of these requirements, such as energy efficiency, are starting to be defined in increasingly stringent local, regional, and national building regulations, much more research and analysis is required before the vagaries of anthropogenic climate change and energy transitions can be satisfactorily embedded in building codes and regulations (United Kingdom Government 2019). Akin to the formulation and development of anthropogenic climate change models, Building energy models have been under development since the start of the 21st century, and in the United States, comprehensive building models are being developed to benchmark the energy performance and characteristics of common types of commercial buildings (Underwood and Yik 2004; Deru et al. 2011, 1–118). No doubt these models could prove as useful as anthropogenic climate change models in providing insights into the complexities of energy flows and transitions in the context of building design, construction, and operation especially in the increasingly urbanized environments.

To validate models and hypotheses, they must be tested against measured empirical data. While models may correlate well with past data, this does not necessarily guarantee that they will accurately predict or, in modern parlance, *project* future measured outcomes, but it is usual for insights to be gained into the likely effects of changes to key model parameters. For instance, the key input parameters to climate change models are the external factors that alter the amount of the sun's energy that

[2] The IPCC was jointly established by the World Meteorological Organization (WMO) and the United Nations Environmental Program (UNEP): https://www.ipcc.ch/.

is absorbed by the earth or trapped in the surrounding atmosphere. These external factors are called *forcings*, and their individual values can be changed by modelers to project the effect of model outputs, for example, the amount of $CO_2$ in the atmosphere can be increased to ascertain how this could increase the surface temperature. The model projections will eventually be compared to actual data, and if there are differences between the forecast parameter values and the measure data, the model will need to be refined. However, as emphasized by Richard Lorch (2017) in a recent publication, *the failure of buildings to meet their hypothecated energy targets; as well poor predictive software*, has given rise to a so-called *performance gap* in adequately estimating current Building energy requirements and consumption, yet alone future projections. Evidently better models and more accurate datasets are necessary.

In the absence of validated models, perhaps the next best approach is to benchmark the energy requirements and consumption for a defined set of building types, probably identified by size (floor area/volume), use, and location (Underwood and Yik 2004; Deru et al. 2011). Energy reduction targets could then be set for actual buildings and consumption measurements made to determine the degree of compliance. Associated with the energy targets could be emission reduction targets, specifically for GHG generation levels. In this way, some of the anthropogenic climate change anxieties of the *Paris Agreement* signatories could be addressed, because of the large amounts of energy consumed and the emissions produced by buildings. At the very least, the approach could provide policy makers with options for addressing their environmental concerns.

To address all these issues, researchers need to determine the contribution of each element of Building energy consumption and consequent emissions, especially to estimate future needs as the global population rises. The quality of such data, especially historical, is paramount. Moreover, the level of detail required, if all the elements are to be considered, requires national collection, collation, and analyses agencies. Although many such organizations now exist in the *developed world*, the differences in definitions and terminologies between the countries are not always helpful in obtaining an overarching picture. In this necessarily succinct chapter, an attempt will be made to minimize the vagaries of this somewhat asymmetrical dataset situation by extensive, but not exclusive, use of the authoritative and comprehensive energy datasets produced by the federal agencies of the United States and Canada, and the multinational IEA. This stratagem, hopefully, should provide a dependable basis for further discussion of the energy for building issues raised so far.

The US Energy Information Administration (USEIA) publishes monthly, and annual, energy consumption data for the United States, international users, and separate national and global regions.[3] It is an authoritative source on energy consumption in general and valuably categorizes Building energy consumption by what constitutes the *residential* and *commercial* building sectors. To paraphrase the USEIA definitions (USEIA 2019 How much energy):

---

[3] USEIA, https://www.eia.gov/.

> Residential Sector - An energy-consuming sector that consists of living quarters for private households . . . but excludes institutional living quarters.
> Commercial Sector - An energy-consuming sector that consists of service-providing facilities and equipment of businesses; Federal, State, and local governments; and other private and public organizations . . . [including] institutional living quarters.

The Federal Canadian Government, through their *Natural Resources Canada* (NRCan) agency, produces a myriad of publications and webpages on energy matters (National Resources Canada home page). This includes a National Energy Use Database (NEUD) (NRCan 2019 Energy Use Data Handbook), of which the 2019, 16th edition, *Energy Use Data Handbook* covering the period 1990–2016 (NRCan 2019 Energy Use 16th Edition), will be used in this chapter, in addition to the relevant USEIA data. The IEA produces annual reports on the *World Energy Outlook* in which *medium to long-term energy projections* are described using possible scenarios based on energy databases and a *large-scale simulation* termed the *World Energy Model* (WEM) (IEA-WEM 2019). The scenarios may change from year to year and the simulation updated, but the databases used are normally the most currently available. For the building sector, the WEM takes account of the uses of Building energy services as well as some of the factors already mentioned, *this is illustrated in Figure 1.2 taken from the 2019 version of WEM.* It should be noted that the *Brown Goods* identified on the figure refer to televisions and computers and that building construction materials are largely dealt with under the industrial sector analyses with the model.

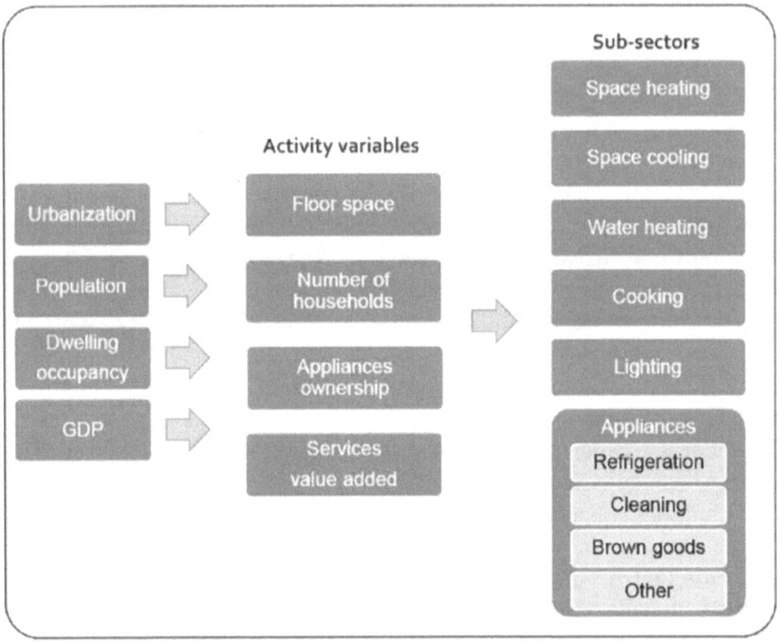


**FIGURE 1.2** Factors in the WEM for buildings. (IEA. 2019. *World Energy Model Documentation*, https://www.iea.org/weo/weomodel/docs/.)

## 1.2 CLIMATE AND BUILDING ENERGY SERVICES

The amount of energy consumed globally by the building sector is a significant percentage, 30% to 40%, of total world energy consumption, but the specific uses of Building energy vary by geographical location, the prevailing climate, and the GDP. Typically, the climate variations are noticeable in regions and countries where certain portions of their land area are close to the polar ice caps, while others are closer to the equator. Such variations have implications for the energy needs of buildings. It can be anticipated that buildings in *cold* low temperature climate zones will use more energy for heating purposes rather than for cooling, whereas in *hot* high-temperature climate zones, more energy will be consumed for cooling as opposed to heating. However, it is not just temperature which decides a defined *climate zone*, other criteria are also included in the descriptions to provide more precise classifications.

Two of the more modern climate zone classifications systems are known as the Köppen–Geiger (Peel, Finlayson, and McMahon 2007) and the Köppen–Trewartha modification (Belda et al. 2014). Both are extensions or modifications of Wladimir Köppen's original system published in the 19th century (Arnfield 2019). The basic Köppen system has five main climate groups ranging from tropical through to polar and are given a letter A to E. Each group is further divided to include a temperature and precipitation pattern with each given a lowercase letter. So, for example, within the *C* (temperate) group, a zone labeled *Cfb* means an oceanic climate with warm summers.[4] Some countries now use six, seven, and sometimes eight climate groups and use slightly different terminology to describe their various zones but largely remain in harmony with the Köppen system. All are very useful in identifying the types of vegetation that grow, or can be grown, in specific region, but how does this help with characterizing Building energy needs?

In the latter part of the 20th century, using an enhanced Köppen approach, a non-profit organization, the International Code Council (ICC), was established with the aim to develop a set of national building codes (International Code Council *Driving Growth*). Out of their deliberations and consultations, the International Energy Conservation Code (IECC) was formulated in 2000 and has now been adopted by many US states and local governments. The main driver of the code is *Building energy efficiency*, which continues to be a *hot (sic) topic* globally. Subsequently, there has been significant collaboration between the IECC and the US Department of Energy (DOE) Building Energy Codes Program to establish building practices in construction and operation to achieve more energy savings as part of the DOE's Office of Energy Efficiency and Renewable Energy (EERE) *Building America* programs, which are based on US climate zones.

In the Building America approach, to more precisely define climate zones, two additional criteria have been introduced to augment the current Köppen systems, namely *heating degree days* (HDDs) (Chen 2019) and *cooling degree days* (CDDs) (Scott 2019). To calculate an HDD, a baseline temperature is defined, which is the

---

[4] On-Line https://en.wikipedia.org/wiki/K%C3%B6ppen_climate_classification, retrieved December 1, 2019.

average of a day's high and low temperature, say 65°F, then if the actual day's average is 45°F, the day's HDD is (65–45), i.e., 20. If every day in a 30-day month had an average daily temperature of 45°F, then the month's HDD value would be 30 × 20 or 600 based on the Fahrenheit scale. Some countries use the Celsius scale to calculate their HDDs. However, in the United States, the scale of choice is Fahrenheit, and on an annual basis, a *cold climate is generally defined as a region with approximately 5,400 HDDs or more and fewer than approximately 9,000 HDDs* (USDOE 2019) *as illustrated on the Building America Climate Zone classification map shown in Figure 1.3* (USDOE 2010). CDDs are calculated in a similar manner to HDDs, but in this case, the number of degrees and days *above* the benchmark or daily average are used. CDDs describe the likely overall cooling needs as part of building planning processes but, at least in the United States, HDDs are more commonly used in climate zone classifications. The DOE's EERE then publishes guidelines combining the IECC[5] and professional agency codes (e.g., American Society of Heating, Refrigeration and Air-Conditioning Engineers[6] (ASHRAE)) for builders and occupants to achieve the most energy savings in buildings, especially residential in the identified climate zones. While such guidelines are extremely helpful, they are only useful if their realization is affordable to consumers, and the involved technologies are available. Thus, while the code guidelines may be applicable globally, not all jurisdictions are able to use them.

Unforeseen shifts in weather conditions, temperatures, wind, and precipitation, can alter the historically projected HDDs and CCDs in a climate zone, and these

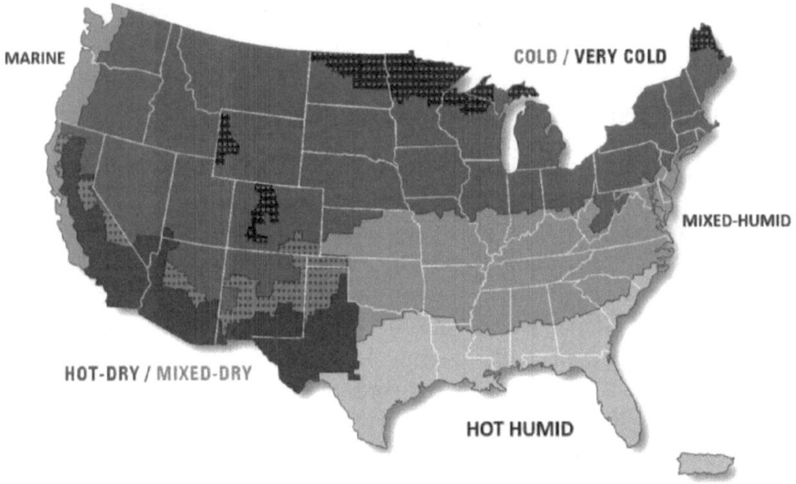


**FIGURE 1.3** Building America climate zone classification maps. (USDOE. 2010. *Building Technologies Program, Vol 7.1: Guide to Determining Climate Regions by County*, https://www1.eere.energy.gov/buildings/publications/pdfs/building_america/ba_climateguide_7_1.pdf.)

[5] The current 2018 IECE codes are to be updated in 2021.

[6] https://www.ashrae.org/home.

can lead to economic losses caused, for example, by crop losses and transportation disruption. According to Ganti (2019), at least 20% of the US economy is susceptible to the adverse effects of weather conditions. Although the impacts of some dramatic and catastrophic weather events can be covered by conventional insurance, an increasingly popular method of offsetting some weather losses are the so-called *Weather Derivatives* (Chen 2018). These have been available since the 1990s, and they can be purchased by companies and individuals through stock exchanges to *hedge against risk of weather-related losses* and the derivatives are usually based on agreed and defined climate parameters such as precipitation and HDDs (Chen 2018). Departures from the predetermined parameter values can result in the derivative seller either paying the holder or receiving payments from the holder. These *financial securities* are appealing to farmers and energy companies and represent a multibillion dollar industry.

Climate zone classifications, such as defined by the US DOE, can provide planners and constructors with precise information on the energy requirements of new buildings, but also, when older buildings are to be retrofitted or renovated, their existing energy consumption levels can be benchmarked against future obligations. The zone specifications also indicate how occupants can be expected to use the energy supplied, in a cold zone, for example, it is to be anticipated that the demand for energy for space heating will be higher than the need for space cooling. Of course, it is essential that the identification and specifications of zones be regularly updated as the effects of urbanization, population increases, regulated carbon mitigation requirements, and the inevitable energy source transitions become more pronounced. To fully assess these issues, it is crucial to know exactly how building occupants use energy and from what sources the energy is obtained.

## 1.3 USES AND SOURCES OF BUILDING ENERGY

In most cases, the amount of energy supplied to buildings, residential and commercial, can be measured by grid metering. The measurement of the precise uses of this energy is more problematic and so national energy agencies rely on consumer surveys to ascertain how Building energy is consumed. Models are then applied to determine the financial expenditures of each energy consumption element, which can provide some insights into individual affordability of these elements and the choices made, or which must be made, by building occupants. Moreover, energy use costs do not fully correlate with actual energy consumption, because of regional tariffs and municipal pricing variations, regardless of the climate zone location. However, despite the uncertainties associated with surveys of Building energy uses, they are closely related to inherent weather patterns and prevailing climates, *as can be seen on Figure 1.4* for the US climate zones (Berry, Lawson, and Woodward 2018).

### 1.3.1 Heating, Cooling, and Climate Zones

In all these zones, except the *hot-humid* zone which covers a geographic area from Florida to southeastern Texas (USEIA 2017, 3), more energy is expended on space heating compared to cooling. Nevertheless, in the United States by 2018 between

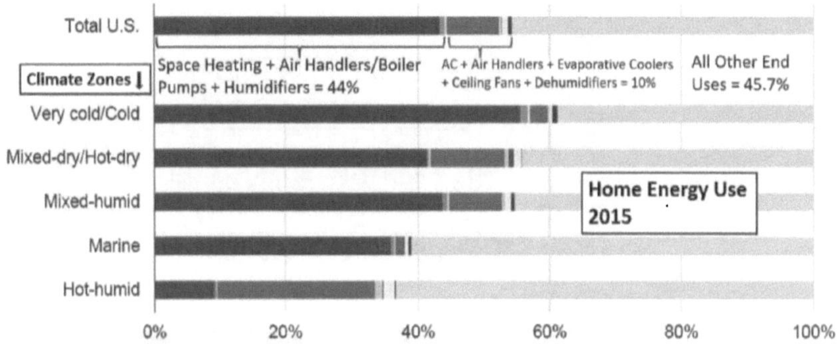


**FIGURE 1.4** EIA 2015 Residential Energy Consumption Survey. (Berry, C., Lawson, G., and Woodward, M. 2018. Highlights from the 2015 RECS, webinar 31 July, https://www.eia.gov/consumption/residential/webinar_slides/highlights_from_the_2015_RECS.pdf; Author Redrawn.)

85% and 87% (Lester 2015) of the 118.2 million households (USEIA 2017, 2) had "air-conditioning."[7] Almost without exception, cooling units use electrical power, so regardless of what type of energy source is provided to buildings, electricity will be required and this is also invariably the case for lighting.

Home air-conditioning units became affordable, and therefore popular, in the United States from 1947 onward. This popularity was not replicated for several decades by its colder northern neighbor, Canada, but by 2011, 58% of Canadian households had such units (NRCan 2014). Nevertheless, in 2016, Canada's residential sector used only 2.5% of total Building energy for cooling, compared with 60% for space heating (NRCan 2019 Energy Use 16th Edition). Even so, since 1990, the energy used for residential heating has fallen by 7.6% while that for cooling has risen by 270%, and although commercial Building energy use still lags, the amount used in residential applications *both heating and cooling energy demands have increased as shown in Figure 1.5* (adapted from NRCan 2019 Energy Use 16th Edition). Could the measurable increase in energy for cooling be a sign of the effects of anthropogenic climate change, or increased personal affluence, or a mix of both? Canada has four Köppen climate zones all of which are classified as *cold,* and several subgroups are identified within the four zones, but only one of the subgroups is defined as having a *hot summer*, that being the Windsor–Essex area in the south-western region of the Province of Ontario. The population of this province is almost 40% of the total Canadian population and is largely concentrated in the southern and central regions around the Greater Toronto Area (GTA).[8] Nevertheless, many areas of Canada have *warm* summers, *as shown in Figure 1.6 the Köppen climate types of Canada* (Peterson 2016), with over 70% of Canadians living within 100 km of the US border and in highly urbanized environments.

[7] A term used in S.W. Cramer's US patent application of 1906 and first used in a textbook by G. B. Wilson two years later in 1908—see Lester, 2015.

[8] *Greater Toronto Area*, https://en.wikipedia.org/wiki/Greater_Toronto_Area (accessed December 4, 2019).

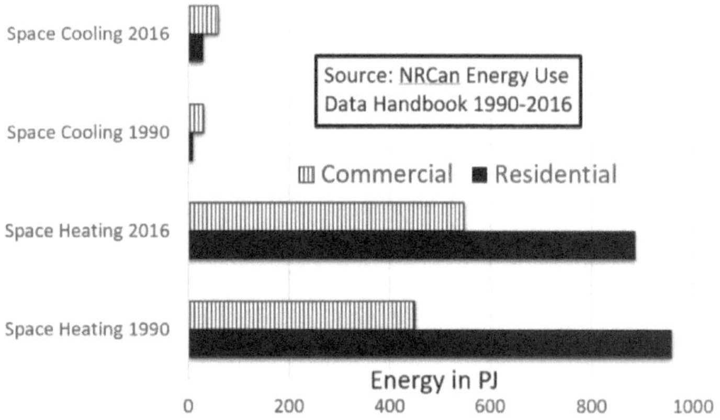


**FIGURE 1.5** Changes in energy use for heating and cooling in Canada 1990–2016. (Data Source: NRCan, *Energy Use Data Handbook*, 16th edition, 1990 to 2016, retrieved 28 November 2019, http://oee.nrcan.gc.ca/corporate/statistics/neud/dpa/menus/trends/handbook/tables.cfm. Author Constructed.)

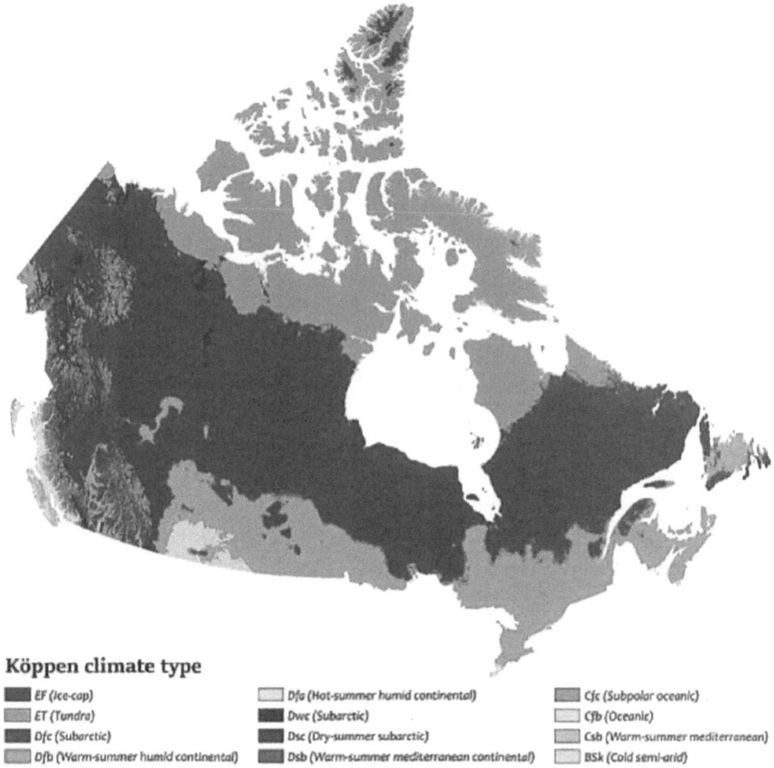


**FIGURE 1.6** Canada's Köppen climate types. (Peterson, A. 2016. https://commons.wikimedia.org/wiki/File:Canada_K%C3%B6ppen.svg.)

In contrast to Canada's four, and the US eight, climate zones, the UK has only one climate zone, *temperate*, and like many other European countries has only relatively recently embraced the use of the air-conditioning technology in the *service* building sectors. In the UK, about 6% of the Building energy in the services sector is used for cooling, about 1% less than in Canada and half the proportionate level used in the United States (*OpenLearn*® 2016). Even so, the use of air-conditioners in *domestic* and *residential* settings is still rarely encountered in the UK with probably less than 0.5% homes possessing such cooling appliances (Barford 2013). This situation could change quickly over the next three decades as heatwaves in the UK are expected to become more prevalent and the number of heat-related deaths forecast to rise from 2,000 annually to as many as 7,000 by 2050 according to the Environmental Audit Committee of the UK parliament, an increase of 257%! (Wharton 2018). This is perhaps another example of a climate change impact, but if so, can this need for more cooling be attributed to anthropogenic causes alone? In recent years, the UK has made great efforts to replace fossil fuels with renewable and non-polluting energy sources, and in 2019, these sources will likely be responsible for almost 50% of electricity generation surpassing the amount produced by fossil fuels (Thomas 2019). These concerted efforts have resulted in $CO_2$ emissions in the UK being reduced by 43.5% from 1990 to 2018, but despite this achievement, heatwaves have not been prevented (Penistone 2019, 2). This is perhaps not too surprising since it has been projected that *regional* heatwave characteristics are impacted by *global* mean warming (Perkins-Kirkpatrick and Gibson, 2017). It could be argued that without the UK's efforts, their local heatwave occurrences may have increased, or it could be concluded that the efforts of one small individual country alone have little impact on the global climate. The UK's Building energy use profiles are *illustrated in Figure 1.7 showing domestic and service building sector energy from one of their most recent surveys,*

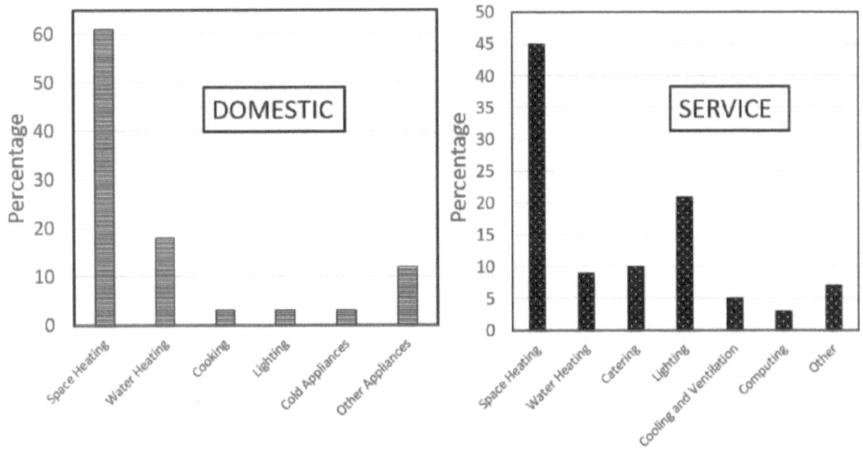


**FIGURE 1.7** UK Building Sector Energy Use Profiles. (*OpenLearn*® 2016. Energy in buildings courseware, retrieved 14 Nov 2018: https://www.open.edu/openlearn/nature-environment/energy-buildings/content-section-0?active-tab=description-tab. Author constructed.)

and it will be seen that they are like the equivalent Canadian profiles apart from the energy used for lighting in commercial buildings.[9]

### 1.3.2 Increasing Demand for Cooling

The data presented so far in this chapter have been for developed and wealthy countries, but how do these compare with developing countries' Building energy profiles? In a recent publication, the IEA has stated that, *Worldwide, energy consumption for cooling is growing faster than for any other end use in buildings*, this being the case then, to characterize some of the differences between developing and developed countries Building energy profiles, the cooling aspects are now considered (IEA BERC 2019). A 2016 IEA compilation of the use of air-conditioning and cooling capacity for nine countries and two regions (the European Union and the Middle East) *is shown in Figure 1.8*, and it can be seen that the People's Republic of China (China) has more installed air-conditioner units than the United States, and that in 2016, the annual sales of such units in China were 220% more than in the United States and over 13 times more than those in the world's second most populous country, India (IEA Paris 2018). In terms of installed capacity, it is China's residential buildings that experienced the most growth and on extremely hot days can account

| | Installed stock | | | | | | Annual sales | | | | | |
|---|---|---|---|---|---|---|---|---|---|---|---|---|
| | Million units | | | GW output capacity | | | Million units | | | GW output capacity | | |
| | Res | Com | Total | Res | Com | Total | Res | Com | Total | Res | Com | Total |
| United States | 241 | 132 | 374 | 2 295 | 2 430 | 4 726 | 16 | 8 | 24 | 314 | 129 | 443 |
| European Union | 43 | 53 | 97 | 192 | 654 | 847 | 9 | 3 | 12 | 34 | 41 | 75 |
| Japan | 116 | 33 | 148 | 407 | 352 | 759 | 9 | 2 | 11 | 47 | 14 | 61 |
| Korea | 30 | 29 | 59 | 129 | 220 | 348 | 2 | 2 | 4 | 19 | 15 | 34 |
| Mexico | 7 | 9 | 16 | 40 | 65 | 105 | 1 | 1 | 2 | 5 | 6 | 10 |
| China | 432 | 138 | 569 | 2 092 | 807 | 2 899 | 41 | 12 | 53 | 305 | 81 | 386 |
| India | 14 | 13 | 27 | 77 | 72 | 149 | 3 | 2 | 4 | 14 | 12 | 25 |
| Indonesia | 7 | 5 | 12 | 32 | 27 | 59 | 1 | 1 | 2 | 5 | 4 | 9 |
| Brazil | 14 | 14 | 27 | 59 | 68 | 127 | 1 | 0.3 | 1 | 5 | 1.4 | 6 |
| South Africa | 1 | 1 | 3 | 6 | 15 | 22 | 0.1 | 0.1 | 0.3 | 0.9 | 1.1 | 2.1 |
| Middle East | 30 | 18 | 47 | 147 | 153 | 299 | 4 | 2 | 6 | 29 | 16 | 45 |
| World | 1 093 | 529 | 1 622 | 6 181 | 5 491 | 11 673 | 94 | 40 | 135 | 848 | 359 | 1 207 |

**FIGURE 1.8** The use of air-conditioning units and cooling capacities, compiled by 2016 IEA survey. (Birol, F. for IEA. 2018. *The Future of Cooling: Opportunities for Energy Efficient air-conditioning.* OECD/IEA Report, May, 99 p, https://www.iea.org/reports/the-future-of-cooling.)

[9] In the UK, the terms *domestic* and *service* are used in describing Building energy and are equivalent to the US residential and commercial terminology; the figure is a reworking of that published in *OpenLearn*® 2016.

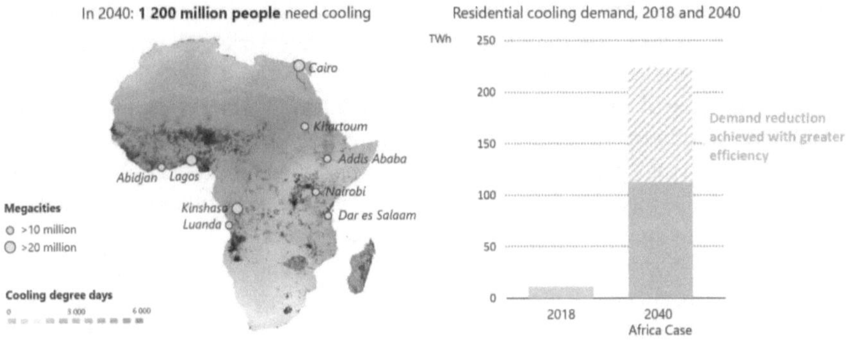


**FIGURE 1.9** Africa's cooling energy projections based on UN Agenda 2030. (IEA. Africa Energy Outlook 2019 presentation (November), https://www.iea.org/reports/africa-energy-outlook-2019.)

for as much as 50% of peak electricity demand (IEA BERC 2019). However, by 2040, the projected population of the continent of Africa could reach 2 billion, of which 60% (IEA Africa 2019) will need cooling under the *Africa Union*'s ambitious Agenda 2063, the master plan to make Africa a global powerhouse (African Union 2013). If the plan came to fruition, then Africa's energy demand for building cooling, *summarized as shown in Figure 1.9*, would far exceed that of the current needs in the United States and the European Union combined (IEA Africa 2019).

### 1.3.3 Cooking

Other than in the opening paragraphs of this chapter, the Building energy consumed for *cooking* has not been mentioned. Of all the factors that together constitute Building energy demands, cooking is one which is difficult to precisely quantify on a global basis. According to the 2013 IEA report *Transition to Sustainable Buildings*, in the 36 developed countries as represented by the Organization for Economic Co-operation and Development (OECD) cooking represents 5%–10% of residential energy use (OECD/IEA 2013). However, in such countries, many food products are pre-cooked, wholly or partially, and then re-heated if necessary, in residential households, but the bulk of the energy use will likely take place outside the household. *Take-out*, *carry-out, cooked food delivery*, and *eating-out* are also popular in the OCED countries. The UK data depicted in Figure 1.7 shows that 3% of domestic (residential) energy is used for cooking, but that 10% is used in the services (commercial) sector for *catering*. The amount used for cooking in Canada is not specifically identified, but the energy for *appliances* is, which will also include ovens, stoves, and toasters as well as other electrical and electronic devices. In 2016, just less than 14% of total Building energy was used on all appliances (NRCan 2019 Energy Use 16th Edition). These are noteworthy amounts but are far less than that used in non-OCED countries[10] where

[10] Generally, non-OCED countries are less wealthy in terms of GDP/PPP per capita and personal income than the OCED countries.

cooking consumes almost a third of residential Building energy, and in some regions as much as 80% (OECD/IEA 2013), and in Sub-Saharan Africa, the cooking energy portion could be higher than 90% (IEA Africa 2019). At least 19 African countries have less than 5% of their households using what the World Health Organization (WHO) term *clean cooking* fuels (WHO 2016, Burning Opportunity).

The problem is not restricted to African nations, since south-east Asia and the western Pacific also have regions where clean cooking fuels are either not available or not used. In most cases, the fuels used are solid fuels, such as the somewhat controversially defined *carbon-neutral* biomass (mainly firewood), coal, and animal dung (Daley 2018). In residential settings, when burned, all these energy sources are health hazards leading to millions of premature deaths globally (WHO 2016, Ambient Air Pollution). Despite its many social and economic advances over the past decade, India is the exemplar of non-clean residential cooking energy use and its effects, with almost 800 million people using such sources, causing 1.3 million premature deaths annually from household air pollution (HAP) (WHO 2016, Burning Opportunity). Country-wide, 49% of households use firewood, as the primary fuel for cooking, with rural communities accounting for 62.5% of this amount. The second most used fuel source is LPG/natural gas, at 28.5%, of which 65% in consumed in urban areas and whose portion of cooking energy use has been increasing since the start of the 21st century (USEIA 2016). A snapshot, based on the country's 2011 census, of the primary energy sources *used for cooking in India is shown in Figure 1.10.*[11]

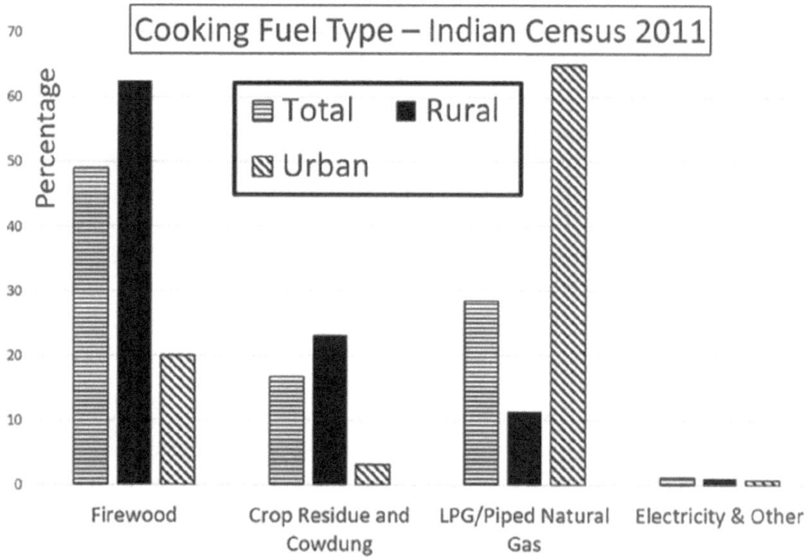


**FIGURE 1.10** Sources of energy for cooking in India. (Author produced from data in USEIA. 2016. Country Analysis Brief: India, updated to 14 June 2016, 28pp, https://www.eia.gov/beta/international/analysis_includes/countries_long/India/india.pdf. Author constructed.)

[11] An author chart based on data published in USEIA 2016.

### 1.3.3.1 Cooking with Polluting Fuels

The WHO estimate globally that 43% of the population use *polluting* fuels for cooking, even in Europe and the Americas (North, Central, and South), 12% and 16% of households, respectively, do not cook with *clean* fuels (WHO 2016, Burning Opportunity). Cooking, as an intentional human activity, started with wood burning fires at least 500, and maybe, more millennia ago and today at least a million people are still using it, and upward of a combined 3.3 million using polluting fuels for cooking, representing 20% of Building energy use, and a considerable amount of harmful emissions. In trying to address the issues associated with burning wood for cooking, an obvious obstacle is that biomass is stated to be a renewable energy source by intergovernmental agencies, the European Union, and several countries and regions (Reader 2020, 1–45). Nonetheless, until eventually subsumed by the global natural carbon cycle or sequestrated by new growth, the $CO_2$ produced in the biomass combustion processes will add to the level of global atmospheric GHG inventories.

Biomass use protagonists argue their *carbon-neutral* case based on its replacement of conventional fossil fuels and sequestration, so that atmospheric $CO_2$ content will be reduced after a period called the *carbon payback time* or *time to carbon parity* (Manomet Centre 2010). The payback time for biomass can vary between 1 year and 1,000 years (Mitchell et al. 2012), for example, burning wood from Norwegian boreal forests can result in a payback delay of 150–230 years (Holtsmark 2010). Apart from the *renewable* and *payback* time issues, there is also an economic dimension to the use of biomass in developing countries, for instance, it has been calculated that by 2030 the charcoal industry alone in the desperately poor Sub-Saharan Africa will employ 12 million people and have an economic value of US$12 billion (Sander, Hysenik, Haider 2011, 9). To attempt to convince regional populations to forego employment and income will not be easy. Likewise, the anthropogenic climate change urgings, embodied in the many IPCC documents and agreements, to reduce global $CO_2$ emissions may be as equally unconvincing to *countries reliant on biomass, since as shown in Figure 1.11,* these countries are low $CO_2$ emitters.[12] So, while almost half the global population use polluting fuels for cooking, this proportion is close to zero in high-income countries. Maybe this explains why energy for cooking has not attracted the same amount of focused attention as other elements of global energy consumption, including Building energy. This situation has started to change over the last decade or so as such organizations as the WHO and the World Bank (Ekouevi and Tuntivate 2012) have taken far more interest in the challenges presented in meeting clean household cooking and heating energy requirements globally with a view to addressing specific UN SDGs (formerly Millennium Development Goals) (United Nations 2015). One objective will be to develop cleaner methods of using biomass and biofuels rather than solely concentrating on the provision of grid or other forms of accessible electricity to all the world's population. This approach is an implicit acknowledgment that economics are always at the core of energy transitions (Smil 2010). This is evident in the energy mixes of

[12] https://en.wikipedia.org/wiki/List_of_countries_by_carbon_dioxide_emissions_per_capita (accessed December 7, 2019).

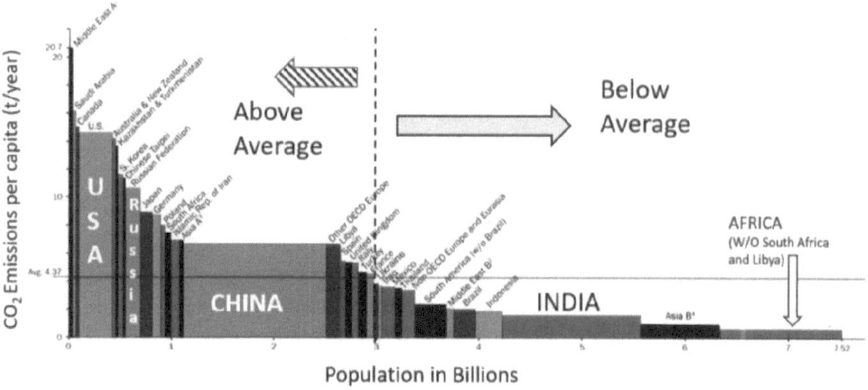


**FIGURE 1.11** Country and regions annual $CO_2$ emissions per capita versus population. (https://en.wikipedia.org/wiki/List_of_countries_by_carbon_dioxide_emissions_per_capita, accessed 7 December 2019. Slight modifications by Author.)

different regions of the world, for example, although biomass combustion is part of all such mixes there is a significant disparity, as previously mentioned, in its use for cooking between rich and poor regions *as illustrated in Figure 1.12* (adapted from, WHO 2016 Burning Opportunity). Regardless of whether biomass and biogas are specified and accepted as carbon-neutral in energy mixes and transitions, the air pollution caused by the burning of these biofuels is not neutral in terms of the damaging human health effects. In recent WHO reports, it has been stated that, *Measured in years of healthy life lost attributable to it, HAP is the single most environmental health risk factor worldwide* and is responsible for as many as 4.3 million premature deaths each year of which 60% are women and children (WHO 2016 Burning Opportunity; WHO 2016 Ambient Air Pollution).

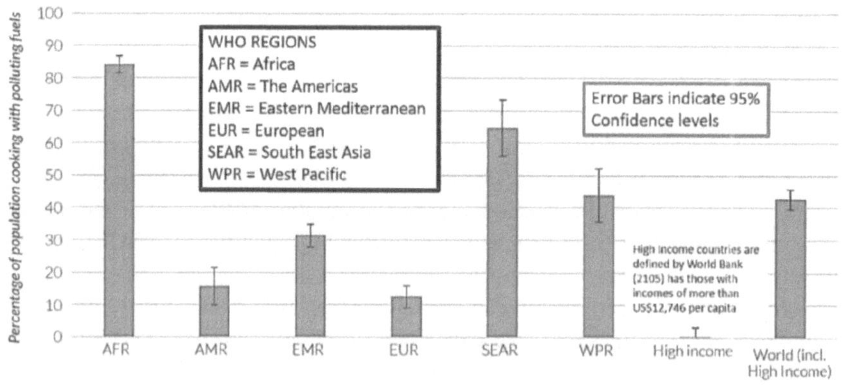


**FIGURE 1.12** Percentage of population cooking with polluting fuels in WHO regions. (Author modified from WHO. 2016. *Burning opportunity: Clean household energy for health, sustainable development, and wellbeing of women and children.* 113 pp, ISBN 9789241565233.)

### 1.3.4 Lighting

Building energy is also used for lighting and usually represents between 3% and 10% of the total energy use, but as can be seen in Figure 1.7, the percentage can be as high as 21%. The introduction of light-emitting diode (LED) bulbs has made a significant impact on Building energy usage for lighting as they can be up to 75% more efficient than the traditional incandescent[13] and fluorescent lamps, they have longer lifetimes, and are far cheaper to operate but more expensive to buy. The main energy source is grid electricity, but there are still approximately 1 billion people without access to electricity, and most of these will use kerosene or paraffin lamps for lighting, which also contribute to HAP (Ritchie 2019). Even households with access to electricity may still retain their kerosene lamps because of the unreliability of the local electrical supply system or advantageous government subsidies (Lam et al. 2016). Where electricity is available for lighting, it is also likely that households may use electricity to power a myriad of different appliances from washing machines to televisions, computers, and communications.

### 1.3.5 Elevators

In commercial buildings including high-rise residential units, power will also be required for *Elevators,* and depending upon the capacity, speed of movement, and the height range, elevator motors vary in size from 5 to 100 HP (3.7 to 75 kW), (Imperial Electric 2019). The energy consumption will be determined by how many hours per day the elevator is in use. According to Nina Rastogi (2009), the annual consumption for a typical hydraulic elevator in a three-floor multistory office building is 3.8 MWh or approximately as much as an average US household uses in four months. For a 30-story high-rise building, the elevator energy consumption could be as high as 43 MWh per year.

Elevators are now obligatory in many jurisdictions, driven to a large extent by legal accessibility requirements for users with mobility issues. The regulations usually apply to new building construction and buildings undergoing extensive renovations (Canadian Consulting Engineer 2015). In most instances, the regulations are quite complex, especially where *exemptions* are applicable. The main point is that as global urbanization and population increases the power for elevators will become a *not-to-be ignored* factor in determining Building energy requirements. There could also be some contributions to residential Building energy use in high-income countries as their populations continue to age and seniors, wishing to remain in their residences, compensate for losses of agility and mobility by installing personal elevators (Canada Mortgage and Housing Corporation 2018). However, it is not just existing home owners seeking elevator options, there appears to be a growing trend to consider fitting elevators in new homes, especially in urban areas, where building lots are becoming smaller, and to compensate, homes with an increasing number of floors are being constructed, termed *building-up rather than building-out* (Kinbar 2019). The energy needs for single-family residential elevators, especially in high-income countries, while not of current significance may be in the future.

[13] No longer manufactured in many countries.

### 1.3.6 Energy Source Transition

It has been seen that in low-income and developing countries, solid fuels are used to provide some large populations with Building energy, but is this the case for OECD countries and other higher income countries? Take Canada, for example, the main sources of Building energy in 2016 were natural gas (45.9%) and electricity (41%) with 78.9% of electricity production being by hydro and nuclear power, and only 9.4% from coal (NRCan 2019 Energy Handbook 16th Edition). In the United States in 2018, the energy source for 63.6% of electrical generation was fossil fuels, mainly natural gas (35.2%) and coal (27.5%), with 53% of the remainder being provided by nuclear power (USEIA 2019 What is US electricity). Natural gas and electricity sources also dominate both the residential (43%, 42% respectively) and commercial (38%, 52%) building sectors in the United States (USEIA 2019 U.S. energy facts).

As electricity is likely to play a dominant, if not the dominant, role in Building energy consumption in the future, it is a sobering challenge for those seeking to achieve a zero-carbon world, with a view to mitigating anthropogenic climate change, that fossil fuels were the energy sources for almost 65% of global electrical generation *as recently as 2017, as depicted in Figure 1.13* (adapted from, IEA 2019 Key World Energy Statistics). If this situation continues, then the United Nations Environmental Programme's (UNEP) fervent calls for many rich countries to achieve a 100% carbon-free electricity supply and implement building standards requiring new buildings from 2030 to be fully electrified will likely go unheeded as they are palpably unrealistic, at least politically, if not technically (United Nations Environmental Programme 2019, 11).

Energy transition away from fossil fuels is inevitable as such raw resources become, not so much unattainable, as more expensive than other energy sources. However, a word of caution, *proven or proved reserves* and actual *resources* are two different measures of how much natural gas and oil remain in the planet's crust.

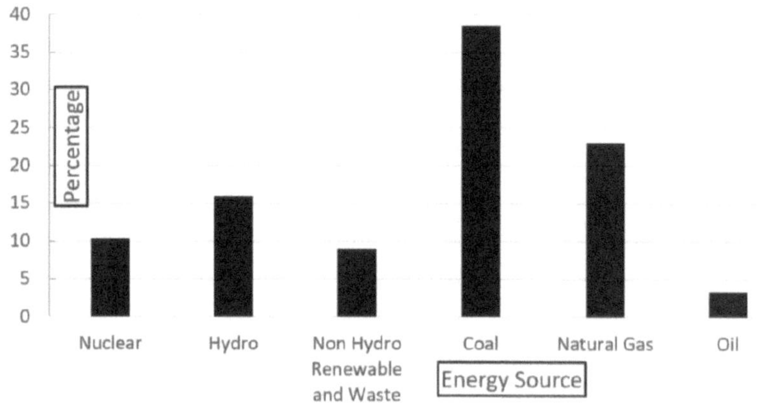


**FIGURE 1.13** Global electricity generation by energy source. (IEA. Key World Energy Statistics, 81pp, 26 September 2019, https://webstore.iea.org/key-world-energy-statistics-2019. Author constructed.)

*Proven reserves* are a measure of how much of a natural commodity can be extracted in an economically profitable manner. Nevertheless, further care must be taken with the published annual data on reserves and resources. For example, the USEIA stated that, in 2017, the major oil and natural gas producers reported the largest increases in their proven reserves since 2013 (USEIA 2018 Oil Company additions). Moreover, the last comprehensive assessment of coal resources in the United States was conducted 35 years ago, and the USEIA now uses three measures to estimate how much coal is left in the United States, (a) Demonstrated Reserves, (b) Estimated Recoverable Reserves, and (c) Recoverable Reserves at producing mines. Using the recoverable reserves data alone there is enough coal to last about *332 years* from January 1, 2019, and the United States only has 22% of the world's proven reserves (USEIA 2019 Coal Explained). Although there is a significant degree of uncertainty in estimating resources, it would appear that if these resources were to become proven reserves, then at the present rate of consumption coal could last for over 2 millennia, natural gas for a millennia, and oil into the 21st century (Reader 2019).

Any forecasts about the *end* of fossil fuels in the next three decades, or indeed by the end of the 21st century, based on their availability, are flawed. Still, most certainly if the global population increases as forecast, along with global efforts to reduce poverty and create a cleaner environment, then a transition, in whole or in part, to non-fossil energy forms will be necessary. To enable other forms, such as solar, wind, and renewable hydropower to eventually play significant roles in replacing fossil-fueled electrical power generation, they must become universally affordable and globally accessible. A possible pathway to this scenario for buildings is to use existing energy streams more efficiently and reduce the real, rather than subsidized, cost of all energy forms. Buildings must also be constructed to better retain thermal energy and be more resistant to extreme outside temperatures, high and low, i.e., *better insulated.*

## 1.4 ENERGY EFFICIENCY

As in other major energy use sectors, there will be a demand for increasing amounts of Building energy in the future as the population increases. Of the many laudable goals of the UN's *Agenda 2030* involving building construction and operations is the elimination of global poverty and the creation of *a world where human habitats are safe, resilient, and sustainable and where there is universal access to affordable, reliable, and sustainable energy* (United Nations 2015 Transforming Our World, 4). It is now almost five years since the signatories of the *Paris Agreement* confirmed Agenda 2030 and many countries submitted Nationally Determined Contributions (NDCs) to the UN as part of the pathways to achieving the SDGs embedded in the Agenda and the Agreement (Dzebo et al. 2019). These NDCs are due to be updated in 2020 and recent IEA analyses have indicated that the hope for achievements in the targets for decarbonization, GHG emission reductions, improved energy intensity, and better energy efficiency have not been fully realized, especially with regard to the building sector (Global Alliance for Buildings and Construction 2019; IEA 2019 World Energy Outlook). To overcome these deficiencies, the IEA have called for a doubling of efforts on energy efficiency and a switch to renewable energy sources in order to decarbonize the building sector (Global Alliance 2019).

Almost three-quarters of the countries who committed to the *Paris Agreement* have *mentioned buildings in their NDCs, yet few have specified the actions they will use to reduce emissions* (Global Alliance 2019). There is now just a decade left to take the necessary actions to achieve these commendable objectives, which is likely to prove a capricious timeline.

Could internationally enforceable regulations, at least among the *Paris Agreement* countries, be imposed along the lines of those introduced in 1975 for various classes of the US on-road motor vehicles regulating how far vehicles must travel on a gallon of fuel? The US Corporate Average Fuel Economy (CAFE) rules have led to significant improvements in energy efficiency in the transportation sub-sectors covered by the regulations (US Department of Transportation 2019). However, as part of the work being undertaken by a recently formed IEA Global Commission,[14] a public survey was conducted involving respondents in over 80 countries, and in response to the question of why potential energy efficiency improvements were not being fulfilled, the most important factor was identified as *How to raise the profile of energy efficiency to ensure it is higher on government agendas* (IEA 2019 Global Commission). The Commission intends to develop a series of *recommendations* by summer 2020 so that improvements in energy efficiency can be *rapidly accelerated* by government actions. It is the acceleration that is called for since improvements in Building energy efficiency have been made both in the residential and non-residential sectors according to the *IEA, as shown in Figure 1.14* (Global Alliance 2019).

Building energy efficiency is a key element in achieving a sustainable energy future and the bulk of the technology that could strongly influence this efficiency is already available, from better insulation to smart sensors controller for electrical

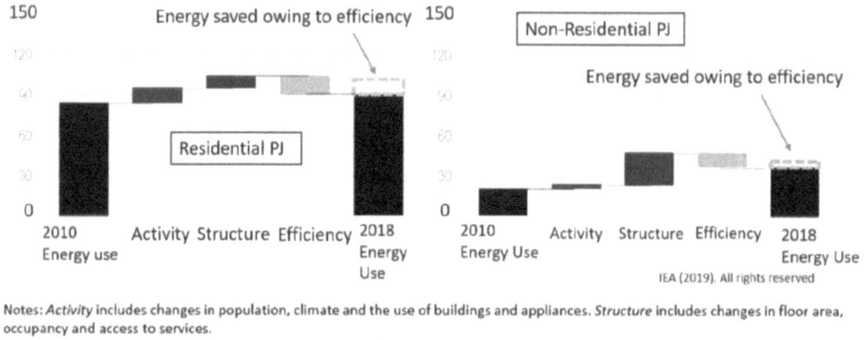


**FIGURE 1.14** Factors influencing Building energy use. Global alliance for buildings and construction. (IEA and UNEP. 2019. *2019 global status report for buildings and construction: Towards a zero-emission, efficient and resilient buildings and construction sector,* 41 pp, ISBN 978-92-807-3768-4.)

[14] The IEA formed the *Global Commission for Urgent Action on Energy Efficiency* on June 24, 2019, for the purpose of examining how progress on energy efficiency can be rapidly accelerated through new and stronger policy action by governments across the globe *https://www.iea.org/programmes/global-commission-for-urgent-action-on-energy-efficiency.*

appliances and indoor environmental conditions. The energy requirements for heating, cooling, cooking, lighting, etc., can be significantly reduced using smart devices by building occupants while retaining individual and collective comfort. For new buildings, these reductions can be more readily achieved than for the existing building stocks. Unfortunately, although in some instances buildings, mainly commercial, can be repurposed, the retrofitting of existing buildings will be required on a large scale, as, for example, over 75% of the present US building stock will still be in use in 2030 (Campbell and Calhoun 2016), although the average lifetime of residential building is only some 50–60 years, about half of that in Europe (Hertwich et al. 2019). Following the reconstruction of mainland Japan in the aftermath of the destruction of the 2nd World War, building lifetimes have been in the region of 30–40 years, still higher than in China with lifetimes of just 25 years (Hertwich et al. 2019). Globally, many of the younger buildings can be found in developing and low-income countries and the increases in future floor area of buildings in India, China, and Africa will far outpace the US and Europe for the next four decades (Dulac 2017). Energy efficiency in these new developments could have major impacts on the transition to sustainable buildings and low-carbon energy economies, but only one-third of countries have mandatory Building energy codes (IEA 2019 Tracking Buildings). So, whether old or new, significant financial investments will need to be made to ensure that all buildings are either zero energy or at least low carbon. Where the funds will come from for less wealthy regions remains to be seen.

Associated with increasing the efficiency of the energy usage in the *internals* of buildings and buildings spaces, there is a need to address the types of primary energy sources used to provide the Building energy if GHG emissions are to be reduced. While the favored source for future building construction and retrofitting is electricity, this is not a primary *fuel* source and it is the fuel source which delineates the generation of the unwanted emissions. Nevertheless, improving the energy efficiency of a building could reduce overall harmful emissions even if the fuel source of its supplied energy generates large quantities of GHGs such as $CO_2$. To wholly eliminate such emissions in a timely manner, say by 2050, will be very difficult, but perhaps not impossible, for some countries, but an almost insurmountable challenge for others. *This disparity can be illustrated, Figure 1.15, by comparing* the fuel sources for the global production (IEA 2019 Key World Energy Statistics) of electricity with those of Canada (NRCan 2019 Electricity Facts). It can be seen that globally 64.8% of electricity is generated by carbonaceous fuels, while for Canada, it is only 19%. However, in both cases, nuclear energy use is significant and if the societal fears of this form of carbon-free energy were ever to be allayed, countries with the wherewithal to build nuclear power stations could probably reach their *Paris Agreement* emission targets in a timelier manner than seems possible with renewable and sustainable energy sources and convertors (Shellenberger 2020).

### 1.4.1 The Role of Building Materials

A building structure itself, usually referred to as the *building envelope*, also plays a meaningful role in Building energy use and not only during construction activities. The envelopes have historically been defined by the occupants. Humans have

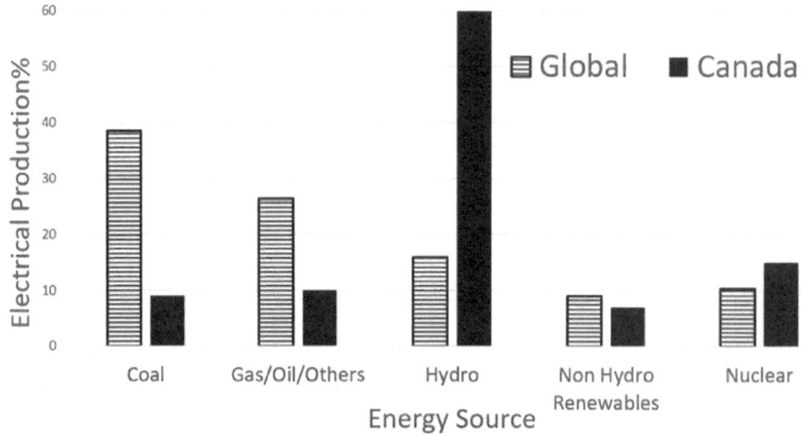


**FIGURE 1.15** Comparing electrical production in Canada and globally. (IEA. *Key World Energy Statistics*, 81 pp, 26 September, https://webstore.iea.org/key-world-energy-statistics-2019 and NRCan. 2019. *Electricity Facts*. https://www.nrcan.gc.ca/science-data/data-analysis/energy-data-analysis/energy-facts/electricity-facts/20068, modified 9 August. Author constructed from data sources.)

always needed shelter to protect them from the vagaries of the prevailing weather and the threat of other animals. For hunter-gatherers, the shelter would be temporary, but the agricultural revolution enabled more permanent shelters to be constructed. The materials used in the construction usually reflected what was available locally. However, in the long history of building regulations, the primary focus has been *health* and *safety* and, to a lesser extent, occupant welfare (Meacham 2016). This is not too surprising since costly building destruction, especially in urban areas by *fire* and *earthquakes,* has led to subsequent changes in building regulations, particularly regarding construction materials. For example, wood is obviously prone to fire hazards, but it is resilient to earthquakes, whereas stone and especially brick materials are resistant to fires but very susceptible to earthquakes. Subsequently, building codes in certain global areas required a combination of the two materials (Muir-Wood 2016). For a variety of reasons, concrete, first used as a building material over 8 millennia ago in the middle east, and made extensive use of during the Roman eras, fell into disuse for several centuries after the fall of the Eastern Roman Empire (Pepin 2017).

Timber, brick, stone, and mud remained popular for residential building construction even after the re-emergence of concrete. Subsequently, iron and steel became common building materials with the advent of high-rise and tall buildings in the early 20th century. These materials provide structural strength for such buildings, and with the increasing global urbanization, it was inevitable that more metallic materials would be used, and this trend will continue in the future. According to a recent article, there are already over 4 million high-rise buildings worldwide with a further 3.5 million proposed for the near future (Wray 2018). But does the choice of building construction materials make that much difference to GHG emissions?

The answer appears to be an unequivocal *yes* (United Nations Environmental Programme 2019 Emissions Gap Report).

In 2015, the GHG emissions from material production were almost a quarter of total global GHG emissions, with 40% associated with construction (Hertwich 2019 The Carbon Footprint). However, the present data for construction include both buildings and infrastructure, but precise attribution between the two sectors is not yet possible. However, it has been stated that, *Buildings often contain more energy-intensive materials such as concrete, steel, and glass than technically required. There is a documented tendency for overdesign in larger steel-frame structures of around 20%–30%* (United Nations Environmental Programme 2019 Emission Gap Report). To address all these issues, the UNEP, Global Alliance, and Hertwich have suggested that greater use should be made of lightweight materials such as timber, bamboo, and other plant fibers and that building materials from partly, or wholly, demolished buildings be recycled. What is clear is that Building energy is a measure of not only how energy is used inside buildings but also how this energy is generated, how much is necessary to construct the building, and how much is required to manufacture the constructional materials. All these elements need to be accounted for in the efforts to improve Building energy efficiency and reduce GHG emissions. However, can these improvements come at a cost affordable to all, and can they help ensure that all regions enjoy energy security?

## 1.5 ENERGY POVERTY AND SECURITY

Energy use has been a major focus of climate change and fuel security studies and regulations, especially regarding harmful atmospheric emissions, but the same level of attention has not been paid to *energy poverty* (González-Eguino 2015). The deficiency may be due to the lack of universally accepted definitions of exactly what constitutes *energy poverty*. In many instances, the term is used to describe those populations that have a lack of access to modern energy services, but this description depends on the interpretations of what are *services* and what is *modern*. In terms of buildings, does the description apply equally to the USEIA's definition of residential and commercial? In Canada, perhaps a more unambiguous definition is used for households, whereby, if 10% of household income is spent on utilities such as electricity, natural gas, and heating oil, it is considered to be experiencing energy poverty (Canada Energy Regulator (formerly, National Energy Board) 2019). In 2015, the average household expenditure was close to 3% of income. It may be expected in a wealthy country such as Canada, both in terms of income and energy resources, that few households would encounter energy poverty, but this is not the case. On average, 8% of households spend more than 10% of their income on energy *and as shown in Figure 1.16, the fuel poverty* rates vary between 6% and 13% across Canada (adapted from, Canada Energy Regulator 2019). If an energy-rich country like Canada has such levels of energy poverty, then what is the global situation?

There is no universally accepted definition of energy poverty, making it problematic to precisely assess national and global energy poverty rates. However, a proxy measure used by such organizations as the IPCC, IEA, the World Bank, and WHO is the lack of access to electricity, which is cited as the main cause for energy poverty. Using such a

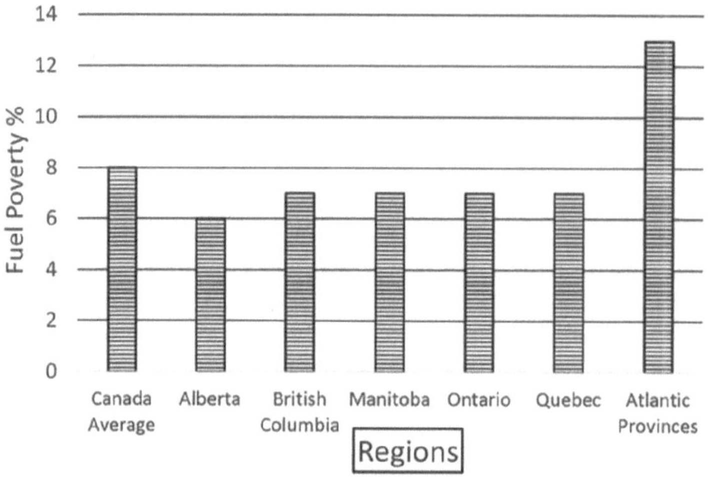


**FIGURE 1.16** Household fuel poverty in Canadian provinces in 2015. (Canada Energy Regular. 2019. Market snapshot: Fuel poverty across Canada – Lower energy efficiency in lower income households, https://www.cer-rec.gc.ca/nrg/ntgrtd/mrkt/snpsht/2017/08-05flpvrt-eng.html?=undefined&wbdisable=true, modified 28 August. Author constructed.)

measure, it can be estimated that about 13%–20% of the global population experiences energy poverty (Jessel, Sawyer and Hernández 2019; IEA 2017). But is it possible not to live with energy poverty when generally impoverished? The World Bank (2018), while acknowledging that fewer people now live in levels of extreme poverty, estimates that 3.4 billion people, 44% of the global population, *still struggle to meet basic needs* a figure echoed in a recent article in the Stanford Social Innovation Review about *the true extent of global energy poverty* (Herrling and Moss 2019).

Grid electrification for everyone may be a noble goal, but its consummate prioritizing could be as harmful as it is helpful, especially for those 2.7 billion who rely on burning biomass, especially for cooking (Foell et al. 2011). Maybe it would be more effective to seek improvements in household biomass burning stoves along the lines proposed by the US EPA and the WHO (USEPA 2019; WHO 2016 Burning Opportunity). After all, as previously stated, the IEA defines energy security *as the uninterrupted availability of energy sources at an affordable price,* which implies that a key issue is resource availability, but does this apply equally to individual countries and regions or globally shared resources? (IEA 2019 Energy Security). In some geographic areas where significant amounts of direct energy resources or fuels are imported, a cessation, or reduction, of these supplies for whatever reasons can quickly lead to regionalized energy poverty. On the other hand, if the need for *imported* Building energy can be reduced per capita, either by energy efficiencies or the development of local renewable resources, then energy security can be improved and the levels of energy poverty shrunk. Examples of how these situations can be addressed are given in a paper by Diana Urge-Vorsatz et al. (2013).

If concentrated efforts on all facets of Building energy efficiencies can enable more people to afford access to clean household energy, and simultaneously reduce

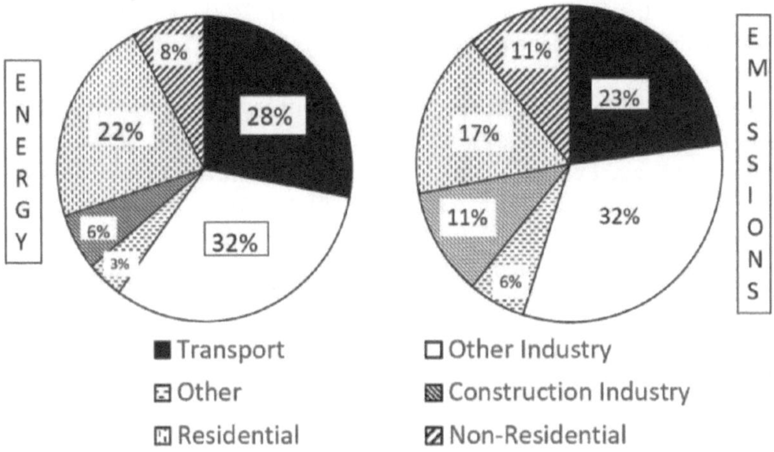


**FIGURE 1.17** Energy use and associated emissions by sector. Global alliance for buildings and construction. (IEA and UNEP. 2019. *2019 global status report for buildings and construction: Towards a zero-emission, efficient and resilient buildings and construction sector*, 41pp, ISBN 978-92-807-3768-4. (Author constructed from published data.)

global GHG emissions, then achieving global electrification using both grid and non-distributed systems may become far more palatable to all. Yet, despite all the complex inter-relationships between Building energy problems and possible solutions, it is important to remember that the proportion of the global energy usage used in the building and construction sector and the associated emissions is *considerable, as clearly illustrated in Figure 1.17* and is worthy of attention (Global Alliance 2019).

## 1.6 CONCLUDING REMARKS—BUILDING ENERGY IN A CHANGING WORLD

The significant role played by the operation and construction of buildings in global energy consumption, as previously cited in Section 1.1 (UNEP 2018), means that Building energy *practices, policies, and prospects* must be viewed in the context of the overall UN plans and vision to achieve emission reductions, progressively reverse the negative effects of anthropogenic climate change, and simultaneously fulfil all 17 SDGs, particularly the two that impact buildings, SDG 7 and SDG 11. The plans are elucidated in the United Nations' Agenda 2030 (UNGA 2015). Inherent in this agenda is an apparent commitment for the UN to regulate the actions of all UN parties, for the common good and to guarantee success. If this is an accurate interpretation, the *Transforming Our World* approach (UNGA 2015) would be analogous to *Global Governance* (Benedict 2001, UN 2014), which may not be universally acceptable, raising for some the unwelcome specter of *Global Socialism* (Newman 2016). Regardless of how possible political machinations could affect Agenda 2030, the main challenges will likely be more of a practical nature, because of future global demographics, country-specific population dynamics, disparities in regional wealth, and wide variations in inherent energy resources and subsequent power generation.

For example, the forecast population of Canada in 2100 could be 30 times less than that of the projected most populous country, India.[15] Furthermore, while the total population of North America is set to increase by about a third by the end of the century, the population of the African continent will be eight times larger than that of North America. By 2100, the two largest economies, based on GDP-PPP, will be India and China, and by then, these two countries are forecast to have passed their population peaks, China in 2030 and India in 2070 (Dr. Stats 2019), with the Western Europe population set peak by 2040 (Chen 2019). It also seems that in terms of individual wealth, measured in GDP-PPP per capita, neither China nor India will be in the top 60 countries (Top 105 Countries 2019).

Nevertheless, to attempt to project populations and wealth as far ahead as eight decades, based on defined social-political-economic scenarios, is fraught with difficulties and uncertainties which become almost overwhelming when projections for energy production and consumption are undertaken. The main reasons being that the interdependency of social changes, political changes, and uncertain government action policies, technical advances, the energy requirements of our biosphere, as well the efficacy of natural and external (solar) energy exploitation, is so complex that, as Smil (2017) stated in his book *Energy and Civilization*, the complexities *make specific scenarios of distant futures mere speculation*. Hence the IEA-WEM models, previously cited, rather than providing forecasts, define sets of scenarios that can be used to evaluate different futures and the actions that could bring them about (IEA 2019). As with many other agencies and international organizations, these days it appears that their studies of energy use and associated energy transition models tend to report the findings only out to 2040 or 2050, which is perhaps prudent.

However, some United Nations publications, such at the recent 2019 UNEP Emissions Gap Report (UNEP 2019), still predict climate parameter values out to 2100, such as a *66% chance* that global average temperature rises could reach between 3.0°C and 3.2°C by the end of century, even if countries fully implement their *Paris Agreement* unconditional NDCs and others effectively implement their conditional NDCs. Still, it must be remembered, that such predictions emanate from climate models. It is pointed out in the *Emissions Gap* report that 78% of all atmospheric GHG emissions come from the wealthy members of G20.[16] With the ever-present possibility of lawmakers changing political and policy complexions every four to eight years in the 18 democracies of G20, a single unifying scenario for addressing global energy transitions and emission reductions to the end of the 21st century, despite the persistent cajoling of organizations such as the UN, is unlikely to emerge.

For the moment, notwithstanding the growing public acceptance of the human effects on climate change, voters appear to be more concerned about a nation's economy, personal taxation levels, and, in many cases, access to universal health-care services (e.g., Newport 2018). What does this mean for Building energy consumption demands and associated emissions if the majority of G20 building occupants

[15] As population forecasts vary from one agency to another, although all use the UN publications, for consistency a single source has been used, namely, https://www.populationpyramid.net/.

[16] An international forum heads-of-state and heads-of-national governments plus senior ministers and governors of national banks of 19 countries plus the EU, albeit that the countries also include four EU members.

are more concerned with the individual amounts they have to pay for their Building energy services, rather than the projected global costs of addressing anthropogenic climate change? Can occupant behavior be regulated and enforced by political governance? As Smil (2010, 2017) has repeatedly emphasized, our chances of success in *creating a new energy system* and *one that would combine reduced demand with a difficult, but eventually rewarding, quest for a civilization powered by renewable energy flows* is dependent on *our willingness to moderate our energy expectations* combined with *a commitment to change.* If Smil's conclusions prove to be perspicacious, then buildings will need to become all-electric or zero-carbon, more energy efficient, and energy affordable.

## REFERENCES

African Union. 2013. Agenda 2063: The Africa that we want: Overview. https://au.int/en/agenda2063/overview.

Arnfield, A.J. 2019. Köppen climate classification. https://www.britannica.com/science/Koppen-climate classification, published April 11, 2019 (accessed December 1, 2019).

Barford, V. 2013. 10 ways the UK is ill-prepared for a heatwave. *BBC News Magazine.* July 18, 2013. https://www.bbc.com/news/magazine-23341698 (accessed November 14, 2018).

Belda, M., E. Holtanová, T. Halenka, and J. Kalvová. 2014. Climate classification revisited: From Köppen to Trewartha. *Climate Research* 59: 1–13 (published February 4, 2014).

Benedict, K. 2001. Global Governance. *International Encyclopedia of the Social & Behavioral Sciences*, 6232–6237. https://www.sciencedirect.com/science/article/pii/B0080430767044995.

Berry, C., G. Lawson, and M. Woodward. 2018. Highlights from the 2015 RECS: Energy consumption, expenditures, and end-use modeling. Webinar July 31, 2018. https://www.eia.gov/consumption/residential/webinar_slides/highlights_from_the_2015_RECS.pdf.

Campbell, I., and K. Calhoun. 2016. Old buildings are U.S. Cities' biggest sustainability challenge. *Harvard Business Review.* https://hbr.org/2016/01/old-buildings-are-u-s-cities-biggest-sustainability-challenge (accessed December 13, 2019).

Canada Energy Regulator. 2019. Market snapshot: Fuel poverty across Canada – lower energy efficiency in lower income households. Modified August 28, 2019. https://www.cer-rec.gc.ca/nrg/ntgrtd/mrkt/snpsht/2017/08-05flpvrt-eng.html?=undefined&wbdisable=true.

Canada Green Building Council®. 2017. Zero carbon building standard, (May): 32. https://www.cagbc.org/cagbcdocs/zerocarbon/CaGBC_Zero_Carbon_Building_Standard_EN.pdf.

Canada Mortgage and Housing Corporation. 2018. Accessible housing by design – Lift and elevators. Document 65542, March 31 2018. https://www.cmhc-schl.gc.ca/en/data-and-research/publications-and-reports/accessible-housing-by-design-lifts-and-residential-elevators.

Canadian Consulting Engineer. 2015. Ontario building code now has rigorous accessibility rules. *CCE Magazine.* January 13, 2015. https://www.canadianconsultingengineer.com/buildings/ontario-building-code-now-rigorous-accessibility-rules/1003399452/.

Chen, J. 2019. Heating degree day, updated April 14, 2019. https://www.investopedia.com/terms/h/heatingdegreeday.asp.

Chen, J. 2018. Weather derivative. https://www.investopedia.com/terms/w/weatherderivative.asp (updated May 12, 2019).

Daley, J. 2018. The EPA declared that burning wood is carbon neutral: It's actually a lot more complicated. *Smithsonian*.com: SMARTNEWS. April 24, 2018. https://www.smithsonianmag.com/smart-news/epa-declares-burning-wood-carbon-neutral-180968880/.

Deru, M., K. Field, D. Studer, K. Benne, B. Griffith, P. Torcellini, B. Liu, M. Halverson, D. Winiarski, M. Rosenberg, M. Yazdanian, J. Huang, and D. Crawley. 2011. *U.S. Department of Energy commercial reference building models of the national building stock*. 1–118. Available at: https://digitalscholarship.unlv.edu/renew_pubs/44.

Dr. Stats. 2019. *Top 10 Countries in 2100 by GDP PPP (1981–2100)*. Posted July 14, 2019. https://www.youtube.com/watch?v=21KYVQw09f8.

Dulac, J. 2017. Energy technology perspectives: Transitions to sustainable buildings. IEA report. https://unfccc.int/sites/default/files/01_unfccc_unep_ws_iea_john_dulac.pdf.

Dzebo, A., H. Janetschek, C. Brandi, and G. Iacobuta. 2019. *Connections between the Paris Agreement and the 2030 Agenda: The case for policy coherence*. Stockholm Environment Institute. https://www.sei.org/wp-content/uploads/2019/08/connections-between-the-paris-agreement-and-the-2030-agenda.pdf.

Ekouevi, K., and V. Tuntivate. 2012. *Household Energy Access for Cooking and Heating: Lessons Learned and the Way Forward*. World Bank Studies. Washington, D.C. https://openknowledge.worldbank.org/handle/10986/9372.

Enbridge. 2019. What is energy poverty? *Energy Matters*, https://www.enbridge.com/energy-matters/energy-school/what-is-energy-poverty (accessed December 16, 2019).

Foell, W., S. Pachauri, D. Spreng, and H. Zerriffi. 2011. Household cooking fuel and technologies in developing economies. *Energy Policy* 39: 7487–7496.

Ganti, A. 2019. Weather future. https://www.investopedia.com/terms/w/weatherfuture.asp (updated September 18, 2019).

Global Alliance for Buildings and Construction, IEA and UNEP. 2019. *2019 global status report for buildings and construction: Towards a zero-emission, efficient and resilient buildings and construction sector*. ISBN 978-92-807-3768-4.

González-Eguino, M. 2015. Energy poverty: An overview. *Renewable and Sustainable Energy Reviews* 47: 377–385.

Green Building Canada. 2019. What is green building? https://greenbuildingcanada.ca/green-building-guide/what-is-green-building (accessed November 18, 2019).

Herrling, S., and T. Moss. 2019. Scaling power for global prosperity. *Stanford Social Innovation Review: Informing and Inspiring Leaders of Social Change* Spring 2019: 59–60. https://ssir.org/articles/entry/scaling_power_for_global_prosperity.

Hertwich, E.G. 2019. The carbon footprint of material production rises to 23% of global greenhouse emissions. *Centre for Open Science*. https://econpapers.repec.org/.

Hertwich, E.G., S. Ali, L. Ciacci, T. Fishman, N. Heeren, E. Masanet, F.N. Asghari, E. Olivetti, S. Pauliuk, Q. Tu, and P. Wolfram. 2019. Material efficiency strategies to reducing greenhouse gas emissions associated with buildings, vehicles, and electronics- a review. *Environmental Research Letters* 14(4): 043004.

Holtsmark, B. 2010. Use of wood fuels from boreal forests will create a biofuel carbon debt with a long payback time. *Statistics Norway*. Research Department, Discussions Paper 637.

IEA Africa. 2019. Africa energy outlook 2019: Africa's energy future matters for the world. November 2019. https://www.iea.org/reports/africa-energy-outlook-2019.

IEA BERC. 2019. The future of cooling in China: Delivering on action plans for sustainable air conditioning. A joint report of the *IEA* and *Tsinghua University Building Energy Research Center (BERC)*. June 2019. https://www.iea.org/reports/the-future-of-cooling-in-china.

IEA Paris. 2018. The future of cooling: Opportunities for energy efficient air-conditioning. *OECD/IEA report*. May 2018. https://www.iea.org/reports/the-future-of-cooling.

IEA report. 2019. Energy security in ASEAN+6. https://webstore.iea.org/energy-security-in-asean6.

IEA. 2017. Energy access outlook: World energy outlook special report OCED/IEA. https://webstore.iea.org/download/direct/274.

IEA. 2019. Energy security. https://www.iea.org/topics/energy-security (accessed December 15, 2019).
IEA. 2019. Global commission for urgent action on energy efficiency: Results of public input survey. https://iea.blob.core.windows.net/assets/5a41e612-1b44-425f-97a6-7834b682bbc3/IEA-Global-Commission-Survey-Results.pdf (accessed December 1, 2019).
IEA. 2019. Key world energy statistics. September 26, 2019. https://webstore.iea.org/key-world-energy-statistics–2019.
IEA. 2019. Perspectives for the clean energy transition. IEA, Paris, www.iea.org/publications/reports/PerspectivesfortheCleanEnergyTransition/.
IEA. 2019. Tracking buildings. https://www.iea.org/reports/tracking-buildings.
IEA. *What is Energy Security?* https://www.iea.org/topics/energysecurity/whatisenergysecurity/ (accessed November 13, 2019).
IEA-WEM. 2019. World energy model documentation, 2019 Version. https://www.iea.org/weo/weomodel/docs/.
IEA. World energy outlook reports. https://www.iea.org/topics/world-energy-outlook.
IEA. 2019. *World Energy Outlook 2019.* PDF ISBN 978–92-64–97300-8. https://webstore.iea.org/world-energy-outlook–2019.
Imperial Electric. 2019. Elevator motors brochure. http://www.imperialelectric.com/products/IEBrochure_Motors_121117(LR-SINGLES).pdf.
International Code Council. Driving growth and affordability through innovation and safety. https://www.iccsafe.org/wp-content/uploads/11-05391_ICC-Tradeshow-Brochure_Final1_lores.pdf (accessed December 1, 2019).
International Code Council. International Energy Conservation Code® Resource Page. https://www.iccsafe.org/advocacy/international-energy-conservation-code-resource-page/ (accessed December 2, 2019).
Jessel, S., S. Sawyer, and D. Hernández, 2019. Energy, poverty and health in climate change: A comprehensive review of an emerging literature. *Frontiers in Public Health*, 7: 1–16, Article 357. https://www.frontiersin.org/articles/10.3389/fpubh.2019.00357/full.
Kinbar, S. 2019. Going up: Elevators in new homes are on the rise. *NewHomeSource*. https://www.newhomesource.com/guide/articles/going-up-elevators-in-new-homes-are-on-the-rise (accessed December 9, 2019).
Kuyper, J., H. Schroeder, and B. Linnér. 2018. The evolution of the UNFCC. *Annual Review of Environment and Resources* 43: 343–368. doi:10.1146/annurev-environ-102017-030119.
Lafrance, R., and L. Schembri. 2002. Purchasing-power parity: Definition, measurement and interpretation. *Bank of Canada Review* (Autumn): 27–33. Available on-line https://www.bankofcanada.ca/wp content/uploads/2010/06/lafrance_e.pdf.
Lam, N.L., S. Pachauri, P. Purohit, Y. Nagai, M.N. Bates, C. Cameron, and K.R. Smith. 2016. Kerosene subsidies for household lighting in India: what are the impacts? *Environmental Research Letters*, 11(4): 044014.
Lester, P. 2015. The history of air-conditioning. *USDOE*. 20 July 2015. https://www.energy.gov/articles/history-air-conditioning.
Lorch, R. 2017. Climate policies for buildings. *Building Research & Information* 45(5): 475–477.
Manomet Center for Conservation Sciences. 2010. Biomass sustainability and carbon policy study. June 2010. NCI -2010-3. https://www.mass.gov/files/documents/2016/08/qx/manomet-biomass-report-full-hirez.pdf.
Meacham, B.J. 2016. Sustainability and resiliency objectives in performance building regulations. *Building Research & Information* 44(5–6): 474–489.
Mitchell, S.R., M.E. Harmon, and K.E.B. O'Connell. 2012. Carbon debt and carbon sequestration parity in forest bioenergy production. *GCB Bioenergy*. doi:10.1111/j.1757-1707.2012.01173.x.
Muir-Wood, R. 2016. *The Cure for Catastrophe: How We Can Stop Manufacturing Natural Disasters*. New York: Basic Books. ISBN 9780465060948.

NRCan. Home page. https://www.nrcan.gc.ca/home.

NRCan. National energy use database. http://oee.nrcan.gc.ca/corporate/statistics/neud/dpa/data_e/databases.cfm (accessed November 24, 2019).

NRCan. 2014. Survey of household energy use 2011 (SHEU-2011), M144-120/2-2011-E-PDF online, http://oee.nrcan.gc.ca/publications/statistics/sheu/2011/pdf/sheu2011.pdf (accessed November 28, 2019).

NRCan. 2019. Electricity facts. Modified August 9, 2019. https://www.nrcan.gc.ca/science-data/data-analysis/energy-data-analysis/energy-facts/electricity-facts/20068.

NRCan. 2019. Energy use data handbook. http://oee.nrcan.gc.ca/corporate/statistics/neud/dpa/data_e/downloads/handbook/pdf/2016/HB2016e.pdf (published 2019).

NRCan. 2019. Energy use data handbook: 1990–2016, 16th Edition. http://oee.nrcan.gc.ca/corporate/statistics/neud/dpa/menus/trends/handbook/tables.cfm (accessed November 28, 2019).

Newport, F. 2018. Top issues for voters: Healthcare, economy, immigration, *GALLUP™*. https://news.gallup.com/poll/244367/top-issues-voters-healthcare-economy-immigration.aspx.

Newman, A. 2016. UN Agenda 2030: A recipe for global socialism. *New American*. https://www.thenewamerican.com/tech/environment/item/22267-un-agenda-2030-a-recipe-for-global-socialism.

OECD/IEA. 2013. Transition to sustainable buildings: Strategies and opportunities to 2050. https://www.oecd-ilibrary.org/energy/transition-to-sustainable-buildings_9789264202955-en, also available as a pdf file (accessed December 7, 2019).

OpenLearn®. 2016. Energy in buildings: Course content, section 1. https://www.open.edu/openlearn/nature-environment/energy-buildings/content-section-1 (accessed November 14, 2018).

Passive House Institute US. 2019. https://www.phius.org/home-page, (accessed November 23, 2019).

Peel, M.C., B.L. Finlayson, and T.A. McMahon. 2007. Updated world map of the Köppen-Geiger climate classification. *Hydrology and Earth System Sciences* 11: 1633–1644.

Perkins-Kirkpatrick, S.E., and P.B. Gibson, 2017. Changes in regional heatwave characteristics as a function of increasing global temperature. https://www.nature.com/articles/s41598-017-12520-2.pdf.

Penistone, A. 2019. 2018 UK greenhouse gas emissions, provisional figures. *UK Department for Business, Energy and Industrial Strategy*. March 28, 2019. https://assets.publishing.service.gov.uk/government/uploads/system/uploads/attachment_data.

Pepin, R. 2017. Evolution of concrete. *The Concrete Hub*. July, 28, 2017. https://www.giatecscientific.com/education/the-history-of-concrete/.

Peterson, A. 2016. https://commons.wikimedia.org/wiki/File:Canada_K%C3%B6ppen.svg (accessed December 4, 2019).

Rastogi, N.S. 2009. Energy and elevators. *The Green Lantern*, https://slate.com/technology/2009/04/when-people-take-the-elevator-does-earth-get-the-shaft.html.

Reader, G.T. 2019. Energy and sustainability: Policy, politics and practice. In *The Energy Mix for Sustaining our Future*, ed. A. Vasel and D.S.-K. Ting, 1–17. Switzerland: Springer International Publishing. ISBN 978-3-030-00104-9.

Reader, G.T. 2020. Energy, renewable alone? In *Sustaining Resources for Tomorrow*, ed. J.A. Stagner and D.S.-K. Ting, 1–45. Switzerland: Springer International Publishing. ISBN 978-3-030-27675-1.

Ritchie, H. 2019. Number of people in the world without electricity falls below one billion. January 18, 2019. https://ourworldindata.org/number-of-people-in-the-world-without-electricity-access-falls-below-one-billion (accessed December 9, 2019).

Ritchie, H., and M. Roser. 2019. Urbanization. *Published online at OurWorldInData.org*. Retrieved from: https://ourworldindata.org/urbanization (accessed November 14, 2019).

Sander, K., B. Hysenik, and W. Haider. 2011. Wood-based biomass energy development for Sub-Saharan Africa: Issues and approaches. *Energy Sector Management Assistance Program (ESMAP)*, Washington D.C., *World Bank report*, http://documents.worldbank.org/curated/en/843941468009629566/Wood-based-biomass-energy-development-for-Sub-Saharan-Africa-issues-and-approaches.

Scott, G. 2019. Cooling degree day (CDD). https://www.investopedia.com/terms/c/cold-degreeday.asp (updated May 2, 2019).

Shellenberger, M.D. 2020. Testimony to the house committee (US Congress) on Science, Space and Technology, 15 January 2020. https://science.house.gov/imo/media/doc/Shellenberger%20Testimony.pdf.

Smil, V. 2010. *Energy Transitions: History, Requirements, Prospects*. Santa Barbara, CA: Praegar Publishers. ISBN 978-0-313-38177-5.

Smil, V. 2017. *Energy and Civilization*. Cambridge and London: The MIT Press. ISBN 978-0262035774.

The Passive House Institute. 2019. https://passiv.de/en/02_informations/02_passive-house-requirements/02_passive-house-requirements.htm (accessed November 23, 2019).

The UK Passive House Organization. 2019. http://passivhaustrust.org.uk/what_is_passivhaus.php (accessed November 23, 2019).

Thomas, N. 2019. UK power from green energy set to surpass fossil fuels in 2019. *Financial Times*. June 20, 2019. https://www.ft.com/content/0ae5464e-9348-11e9-aea1-2b1d33ac3271 (accessed December 4, 2019).

*TOP 105 Countries GDP per capita Ranking Projection (2010~2100)*. Posted September 2, 2019. https://www.youtube.com/watch?v=RLre3OYKjCk.

Torcellini, P., S. Pless, M. Deru, and D. Crawley. 2006. Zero energy buildings: A critical look at the definition. 2006 ACEEE Summer Study of Energy Efficiency in Buildings, Pacific Grove, CA, USA. *NREL Conference paper NREL/CP-550–39833*: 12.

Underwood, C.P., and F.W.H. Yik. 2004. *Modelling Methods for Energy in Buildings*. Oxford, UK: Blackwell Science Ltd. ISBN 0-632-05936-2.

United Kingdom Government. Energy efficiency in buildings. https://www.gov.uk/housing-local-and-community/energy-efficiency-in-buildings (accessed December 11, 2019).

United Nations. 2014. *Global Governance and Global Rules for Development in the Post-2015 era*. United Nations publication Sales No. E.14.II.A.1. ISBN 978-92-1-104689-2.

United Nations. 2015. *Sustainable Development Goals*. https://www.un.org/sustainabledevelopment/sustainable-development-goals (accessed November 2, 2019).

United Nations Climate Change. *The Paris Agreement*. https://unfccc.int/process-and-meetings/the-paris-agreement/the-paris-agreement (accessed November 21, 2019).

United Nations Environment Programme. 2018. UNEP-IEA global status report: Towards a zero-emission, efficient and resilient buildings and construction sector. ISBN 978-92-807-3729-5

United Nations Environment Programme. 2019. Emissions gap report 2019. Table ES.2, 11. UNEP, Nairobi. http://www.unenvironmental.org/emissionsgap.

United Nations General Assembly. 2015. Transforming our world: The 2030 agenda for sustainable development. *A/RES/70/1,* October 21, 2015, https://sustainabledevelopment.un.org/post2015/transformingourworld.

United Nations, Department of Economic and Social Affairs, Population Division. 2019. World population prospects 2019: Data Booklet (ST/ESA/SER.A/424).

United Nations, Department of Economic and Social Affairs, Population Division. 2019. The World Urbanization Prospects: The 2018 revision (ST/ESA/SER.A/420). New York: United Nations.

United States Environmental Protection Agency. 2016. Green building website, last updated February 20, 2016 (now archived).

United States Environment Protection Agency. 2019. 2019 on-line resource, https://www.epa.gov/heat-islands (accessed November 14, 2019).

Urge-Vorsatz, D., K. Petrichenko, M. Staniec, and J. Eom. 2013. Energy use in buildings in a long-term perspective. *Current Opinion in Environmental Sustainability* 5(2): 141–151.

US Department of Transportation. 2019. Corporate average fuel economy. Last updated September 19, 2019. https://www.nhtsa.gov/laws-regulations/corporate-average-fuel-economy (accessed December 12, 2019).

USDOE. 2010. Building technologies program, Vol 7.1: Guide to Determining Climate Regions by County, https://www1.eere.energy.gov/buildings/publications/pdfs/building_america/ba_climateguide_7_1.pdf (accessed November 27, 2019).

USDOE. 2019. Climate zones. https://www.energy.gov/eere/buildings/climate-zones (accessed November 27, 2019).

USEIA. 2016. Country analysis brief: India. Updated to June 14, 2016: 25. https://www.eia.gov/beta/international/analysis_includes/countries_long/India/india.pdf.

USEIA. 2017. What's new in how we use energy at home: Results from EIA's 2015 residential energy consumption survey (RECS). Revised May 2018. https://www.eia.gov/consumption/residential/reports/2015/overview/pdf/whatsnew_home_energy_use.pdf.

USEIA. 2018. Oil company additions to proved reserves in 2017 were the highest since 2013. *Today in Energy*. June 22, 2018. https://www.eia.gov/todayinenergy/detail.php?id=36532.

USEIA. 2019. Coal explained: How much coal is left? November 12, 2019. https://www.eia.gov/energyexplained/coal/how-much-coal-is-left.php.

USEIA. 2019. How much energy is consumed in U.S. residential and commercial buildings? FAQ and Glossary. https://www.eia.gov/tools/faqs/faq.php?id=86&t=1 (accessed May 14, 2019).

USEIA. 2019. U.S. energy facts explained. Updated August 28, 2019. https://www.eia.gov/energyexplained/us-energy-facts/.

USEIA. 2019. What is US electricity generation by energy source? Updated October 25, 2019. https://www.eia.gov/tools/faqs/faq.php?id=427&t=3.

USEPA. 2019. Choosing the right wood-burning stove. https://www.epa.gov/burnwise/choosing-right-wood-burning-stove (accessed December 15, 2019).

Wharton, J. 2018. Air con could become the norm in UK homes with more heatwaves expected. *Metro News*. July 27, 2018. https://metro.co.uk/2018/07/27/air-con-become-norm-uk-homes-heatwaves-expected-7766855/ (accessed November 14, 2018).

WHO. 2016. Ambient air pollution: A global assessment of exposure and burden of disease. ISBN: 9789241511353; available on-line https://www.who.int/phe/publications/air-pollution-global-assessment/en/.

WHO. 2016. Burning opportunity: Clean household energy for health, sustainable development, and wellbeing of women and children. ISBN 9789241565233; available on-line https://www.who.int/airpollution/publications/burning-opportunities/en/.

World Bank. 2018. Nearly half the world lives on less than $5.50 a day. *World Bank Press Release*, 2019/044/DEC-GPV. https://www.worldbank.org/en/news/press-release/2018/10/17/nearly-half-the-world-lives-on-less-than-550-a-day.

World Bank. 2019. Sustainable energy for all (SE4ALL) database: Access to Electricity, https://data.worldbank.org/indicator/eg.elc.accs.zs.

Wray, S. 2018. 'Mini Cities': The rise of tall buildings. *SmartCitiesWorld*, April 23, 2018. https://www.smartcitiesworld.net/opinions/opinions/mini-cities-the-rise-of-tall-buildings.

# 2 Green Design Effectiveness for a Mini Automotive-Repair Facility

*Figen Balo and Hasan Polat*
Firat University

## CONTENTS

## 2.1 INTRODUCTION

Buildings are generally designed as climate modifiers to do thermal comfort and accommodate occupants in the live area promoted up as necessary through mechanical systems for heating and air-conditioning. When properly managed and built, important energy savings could be achieved in structures. Conservation as a least energy cost policy should be encouraged in the energy's future. Energy will have saved for each energy unit generated by a given innovation measure, eliminating and reducing the yearly working expenses linked with the energy's unit generation. Therefore, building architects can help solve the energy problem when decisions about the choice and integration of building materials are taken early in development.

The thermal insulation materials are a significant conduce and a clear logical and practical first stage to succeed in energy performance, particularly in dominated envelope structures in areas with tough climate terms [1]. In vacuum, air-conditioning can have the resources' large share used in constructed buildings. For instance, in the mean American residential building, area cooling and heating accounts for 50%–70% of its power consumption. In rest of the world, with extreme climate terms and lower energy efficiency infrastructure, this proportion could be smaller. The amount of energy a building needs to heat/cool relies on how well the building envelope is treated, particularly in dominated envelope buildings such as residential buildings [2].

The building envelope's thermal efficiency is defined through the materials' thermal characteristics used in its build, identified also besides its ability to emit or absorb solar heat, the corresponding component's total U value with the inclusion of insulation. Insulation material installation within the structural element under transient heat flow can impact its performance. Through insulating material near the heat flow entry point, the highest efficiency can have succeeded. For climate zones where heating in the winter prevails and outdoors where cooling in the summer prevails, it includes insulation. However, insulation to the between wall cavities or inside is common for practicality. The outer enclosing wall's massing is a consideration that is primarily on the building material's thickness and shape used and its ability to restrict heat transfer by the building construction after a certain time. It is another significant factor in evaluating the building's thermal performance, and thus, the energy needs to supply the occupied space with thermal comfort. Selecting the right form and shape of insulation materials depends upon the form of activity as well as the thermal, physical, and other features of the chosen material. Most of the thermal insulation materials undergo heat flows through modes' a mixture (i.e., radiation, convection, and conduction) resulting from variation surface emission of materials thickness characteristics, the define a pure conductive mode is not valid, the term obvious implied in the condition insulating material thermal conductivity [3–5]. Reports by manufacturers and published thermal conductivity values are usually evaluated for comparative thermal performance evaluation under laboratory standard temperature and humidity conditions.

Turkey today generates around 15 million tons of fly ash from 15 coal-fired energy facilities each year. The key waste management alternatives today being utilized for fly ash are ash lagoon's disposal and landfill. With important negatron environmental effects, these two methods are associated [6]. About 5% Turkey fly ash is now used as a substitute feedstock in cement production, partially replacement for cement concrete, for stabilizing and grouting mine operations, and as a filling product in structural and general civil engineering implementations [7–9]. The clay raw material is a substance that occurs naturally, consisting mainly of fine-grained minerals. It is a cheap natural mineral utilized for years as a plastic and rubber filler. With clay complexes, it can chemically change to produce organic polymers and monomers compliant [9–11]. Structural and chemical materials' adjustment during firing typically increases brick mechanical durability and strength [12,13].

The unsaturated renewable plant oils such as Crambe, castor oil, soybean, linseed, and corn oil compose one primary group of sustainable natural sources [14]. The unsaturated and saturated fatty acids are the major structure of these vegetable-based oils. These oils can polymerize to create an elastomeric system and promise alternative petrochemical resin product options [15]. Applications that include cosmetics, coating, painting, varnishes, and more lastly composites and polymer are emerging directly using these fatty acids or their derivatives from some chemical changes.

Corn oil has attracted raised attention in terms of industrial-aimed applicability among sustainable vegetable-based oils due to their inherent sustainability, good biodegradability, low price, high mechanical properties, shape memory effect, and universal availability [16]. Corn oil is a vegetable-based oil that is widely available in many countries of the world. Because of the large unsaturation ratio in corn oil with carbon–carbon double mounts, corn oil has the capacity to copolymerize it with various monomers. It has three diverse side chains of fatty acids made up of linolenic acid, oleic acid, and linoleic acid [17]. Corn oil can activate transformation polyols through a modified epoxy progression, thanks to hydroxylation [18]. The epoxidized maize oil determined in this type can be used to generate composite polymer materials with mechanical strength, thermal stability, and good coating properties [19]. For plastic production-insulation content, Balo and Sua researched renewable insulation materials' source on updated maize oil for green construction strategy and conducted a multi-criteria study of the choice of modified vegetable-based oils [9,16]. In the recent studies, several papers on different vegetable-based oils, fly ash, perlite, and clay had been published. Balo et al. developed fly ash, clay, and vegetable-based building insulation materials (sunflower, linseed, olive, soybean, palm, castor, canola, and tall) and studied these composite materials' thermal, mechanical, and technological properties. Increasing the fly ash and vegetable-based oil proportions resulted in poor thermal conductivity. As a result, it stated that abrasion strength, weight, and compressive strength diminished. Rising the clay proportion, on the other hand, induces reverse impacts on the above [20, 21]. Queralt et al. mixed for the manufacture of ceramic composite materials proper for use as bricks, paving, tiling, and stoneware [22]. Ceylan and Ebeoglugil used clay and perlite to produce lightweight construction materials with well mechanical and thermal properties in compliance with TSEN771-3 [23] and TS699 [24,25] requirements and investigated these materials' properties [26]. Through free cationic and radical copolymerization of soy and maize oils plus various cheap fillers, Gallacher and Larock created bioplastic composites [27]. Through polymerizing soy and maize oils, Sheares and Larock developed a remarkable variety of exciting new rubber, adhesive, elastomeric, and plastic materials. Sustainable maize oil provides natural, environmentally friendly, and cost-effective materials. In addition, these composite materials offered physical, thermal, and mechanical properties that aren't currently available in traditional plastics based on fuel [28].

This study purposed to compare the renewable material-based (epoxidized corn oil, perlite, clay, and fly ash) heterogeneous and homogenous external wall components with the diverse traditional wall types using the obtained energy cost and efficiencies with the help of Autodesk Revit Architecture Simulation Program. In this aim, the materials based on renewable raw materials produced by Balo et al. applied to build external walls as insulation or building materials. A mini automotive-repair facility projected as the applied building of both traditional materials and renewable-based materials. The expanded polystyrene and gas concrete were preferred as traditional insulation and building materials due to their wide applicability.

## 2.2 MATERIAL PRODUCTION AND PROPERTIES

One of the biggest challenges with the novel building material's introduction questioned sustainability, which applies to not only the economic figures but also the ecological effects. Though there is significant work on the application of different organic materials, much remains to do from the viewpoint of emerging green hybrid material that will offer both economic viability and ecological protection while complying with technological norms.

### 2.2.1 Material Production

To product, the build/insulation materials, fly ash, epoxidized corn oil, clay, and perlite are used as raw materials. Fly ash which is found from Afsin Power Facility in Kahramanmaras can be classified as Class C (according to ASTM C 618) on account of this material's chemical ingredient. The density and thermal conductivity coefficient are 1.5 g/cm$^3$ and 0.511 W/mK, respectively. The clay (density and thermal conductivity coefficient are 2.24 g/cm$^3$ and 0.93 W/mK, respectively) and perlite were found in Elazig (village of Sarıyakup) and the company of Izper in Izmir of Turkey, respectively. The raw materials' chemical ingredients in this analysis and the perlite's physical features are presented in Tables 2.1 and 2.2, respectively. The epoxidized corn oil is taken from the company of Konsan in Adana of Turkey. In Figure 2.1, the epoxidized corn oil's chemical content is shown. In Table 2.3, the general and chemical features of the epoxidized corn oil used in this analysis are given.

As primary raw materials applied in this chapter, 60% fly ash, 30% clay, and 10% perlite are used to produce the sample. Fifty percent of epoxidized corn oil is added into the mass as the binder in the sample. In the laboratory mixer, this is made by first dry mixing the fly ash, clay, and perlite. Later, the mixes involving epoxidized corn oil are ready in lab mixer with counter-current for a total of 7 minutes. For each mix, a prism-shaped ($6 \times 15 \times 2$ cm) sample and three cube-shaped (10 cm$^3$) samples are made. The samples' full compaction is provided through vibration after molding.

**TABLE 2.1**
**The Chemical Contents' Percent Range of Obtained Fly Ash, Perlite, and Clay**

| | Perlite | Clay | Fly Ash |
|---|---|---|---|
| $Al_2O_3$ | 13.4 | 20.259 | 4.025 |
| $SiO_2$ | 76.5 | 43.645 | 21.33 |
| CaO | 1.3 | 10.150 | 36.480 |
| $Na_2O$ | 3.0 | - | - |
| $Fe_2O_3$ | - | 12.954 | 2.606 |
| $K_2O$ | 4.1 | - | - |
| MgO | 0.30 | 1.534 | 1.032 |
| $TiO_2$ | 0.20 | - | - |
| FeO | 1.2 | - | - |
| Ignition loss | - | 9.650 | 29.798 |
| Unknown | - | 1.808 | 4.729 |

**TABLE 2.2**
**Perlite's Physical Characteristics**

| Color | White |
|---|---|
| Free moisture | % max 0.50 |
| Refractive index | 1.50 |
| Specific gravity | 2.2–2.4 |
| Mesh size | 4–8 mesh and finer |
| Fusion temperature | 1,260°C–1,343°C |
| pH of water slurry | 0.50–8.0 |
| Specific heat | 837 j/kgK |
| Bulk density (loose) | 50–400 kg/m³ |
| Thermal conductivity at 24°C | 0.04–0.06 W/mK |
| Softening point | 870°C–1,093°C |
| Solubility | • Soluble in hot conc. alkali and in HF<br>• Moderately soluble in <10% in IN NaOH<br>• Slightly soluble in <3% in mineral acid (IN)<br>• Very slightly soluble (<1%) in water or weak acids |

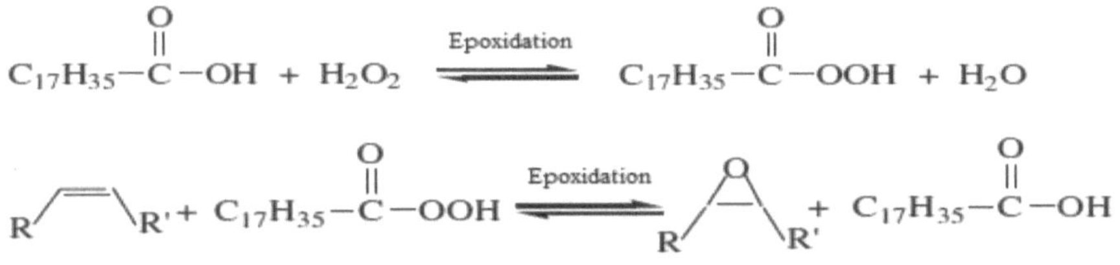


**FIGURE 2.1** The epoxidized corn oil's chemical structure.

**TABLE 2.3**
**The Epoxidized Corn Oil's General Properties**

| Properties | ECO |
|---|---|
| Brilliance | < (Pt-Co): 399 |
| Appearance at normal temperature | Thin to yellow liquid |
| Iodine value | 130.2 mg $I_2$ per 100 g |
| Heat conductivity coefficient | 0,158 W/mK |
| Saponification number | 187–193 |
| Acid value | (KOH/g): 1.27 mg |
| Flow point | 6°C |
| Ignition point | 319°C |
| Density (25°C) | 0.986–1.001 g/cm$^3$ |
| Oxirane value | % 9.78 |
| Viscosity | 42 centipoise (at 40°) |
| Boiling point | 158°C |
| Melting point in water (25°C) | < % 0.029 |
| Refractive index (25°C'de) | 2.144 |
| Weight per equivalent | 168.2 g |
| Loss on heating | <% 0.67 |
| Resin | 135.1 g |
| Crosslinker | 10.0 g |
| Pigment | 79 g |
| Additives (catalyst, antioxidant, flow agent, and anti-foaming agents) | 10.7 g |

The sample is completed through a towel. The prism-shaped (6 × 15 × 2 cm) samples are molded to find the sample's thermal conductivity coefficient. To determine abrasion loss and compressive strength values, the standard 10 cm$^3$ cube samples are utilized. In the electric oven, the unfired samples are baked. First, they are baked at 100°C for 10 hours, and then this is baked at 220°C for 8 hours in an electric oven [16]. The schematic overview of the analysis is given in Figure 2.2.

### 2.2.2 Analysis of Physical and Mechanical Properties of Produced Material

A rapid thermal conductivity coefficient meter (Kyoto Electronics Manufacturing Co. Ltd., Shotherm QTM500 apparatus, Japan) with hot wire method (DIN 51046) is used to gauge the thermal conductivity coefficient value. The sample's value obtained is 0.224 W/mK. The sample's density determined is 1.382 g/cm. In the Construction-Control Labs, the strengthening checks are presented in Elazig of Turkey. In the TS 699 index, the density and compressive strength are evaluated by utilizing the test indexes obtained [24]. In the TS699 index, the results are applied by utilizing equations. Through utilizing compressive testing machine (BC100-Beskom in Turkey), the sample's compressive strength is tested. It's value obtained is 6.95 MPa.

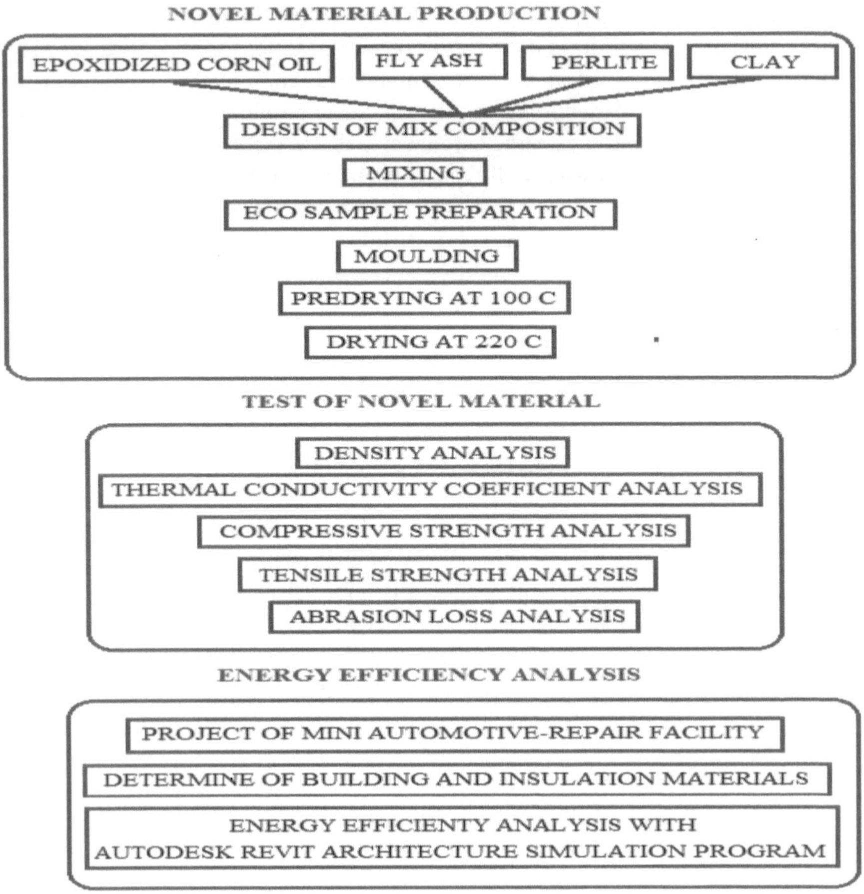


**FIGURE 2.2** The schematic overview of the analysis.

For this test machine, the pressure's maximal ratio applied is 200 tones. The result is provided through computer combined with the compressive-strength test apparatus. Through formula {2.1}, the sample's tensile strength is found as 0.909 MPa [7].

$$F_{\text{Tensile}} = 0.35\sqrt{F_{\text{compressive}}} \tag{2.1}$$

Through the sample's volume loss, the abrasion loss value is found. Through utilizing the Bohme abrasion loss test technique from superficies, the abrasion resistance is tested as per under TS699 (according to the Turkish standards). The sample's loss volume owing to the superficies-abrasion is computed. Through utilizing formula {2.2}, the abrasion loss is found as 4.54% [23–25].

$$\%\text{Abrasion loss} = \left[\left(\text{First mass} - \text{Last mass}\right)/\text{First mass}\right]\cdot 100 \tag{2.2}$$

In this study, the materials produced by Balo [16] to use as renewable building or insulation material in buildings and analyze their thermal-physical performance are applied to different designed walls of mini automotive-repair facility. To reduce heat conduction and energy expenditure in the walls, new building-insulation materials with low thermal conductivity coefficients are modeled in the Revit program based on internal, external, and sandwich wall types. Together with the Expanded Polystyren Foam (EPS)–gas concrete duo, one of the most broadly used building-insulation materials, the changes in the energy performance and cost of a mini automotive-repair facility with walls designed in different combinations are investigated [16].

## 2.3 METHODOLOGY

Revit is a creative tool used to create knowledge modeling, built by Autodesk. It is used in different disciplines: architects, MEPs (electrical, plumbing, and mechanical), engineers, contractors, designers, and structural engineers [29]. It has four primary properties listed below.

- *Design:* Construction of model parts, structures can have created and evaluated simulated by simulation results. It is important to update the models. The simulation automatically generates a report from the models created.
- *Cooperation:* To function on collectively distributed models, an online environment is created and access allows multiple project participants. This property supplies improved correlation to decrease rework and decrease clashes. This property supplies improved correlation to decrease rework and clashes cut.
- *Viewing:* For the models created, high-quality 3D visualization is given.
- *Interoperability:* Assistance is given to members of the expanded project team. It can export and import data from the group (such as MicroStation Design File (DGN), International Finance Corporation (IFC), DraWinG (DWG), and Drawing interchange [X] Format (DXF)) is widely used formats.

Revit analysis also offers opportunities for energy and thermal modeling of entire buildings through energy effective's all stages of construction plans. This is achieved by Autodesk Green Building Studio's Energy Analysis unit, which is a scalable cloud-sourced service that uses the simulation engine DOE2. For a series of further studies' popular form like gbXML, the power modeling generated for theoretical or complex architectonic plans can be extruded by various implications (e.g., to EnergyPlus, DOE2) [30]. Training is the mobile component, and the web-based part of the Ecotect Research package is Green Building Studio. Analysis and Green Building Studio can share the same gbXML file or other BIM resources imported from Revit folder. The detailed analytical capabilities provided by these two methods are summarized below [29]. In "Autodesk Green Building Studio Analysis" multiple plan options, cost and water use predictions, carbon emissions predictions, Leed day-lighting potential, Energy Star scoring, building energy analysis, wind energy, natural ventilation, thermal performance, and photovoltaic collection can be rated. Buildings are multiplex systems and good Building Performance Analysis needs (and the methods it uses) to consider and account overtime for a number of factors' complicated interrelationship. Some of these considerations are materials, form, building use structures, building use, climate, and systems.

The summary below introduces to the anatomy of a dynamic energy simulation with a series of schematic diagrams that build on each other.

- *Building Geometry*: A building's form is one of the primary factors that influence energy consumption. This covers its field, layout, volume, orientation, shading, and glazed fields and captures in an energy simulation using an Energy Analysis Model (EAM) (Figure 2.3a).
- *Thermal Comfort and Spaces*: These spaces change heat with the outside ecology and with each of others (the dotted lines). An EAM has discrete spaces that represent the air within a building (the pink boxes). A series of thermal comfort points are placed within these spaces (the dots) (Figure 2.3b).
- *Heat Transfer and Surfaces*: These spaces change heat with the outside ecology and with each of others (the dotted lines). This thermal transfer happens across the model's surfaces, abstracted from the geometry and representing the walls, roof, floors, and windows of the building (blue and green lines) (Figure 2.3c).
- *Material Properties*: The behavior of the heat transfer across these surfaces dictates by the materials (concrete, glass, insulation, etc.). These affect the heating and cooling energy required to keep up comfort.

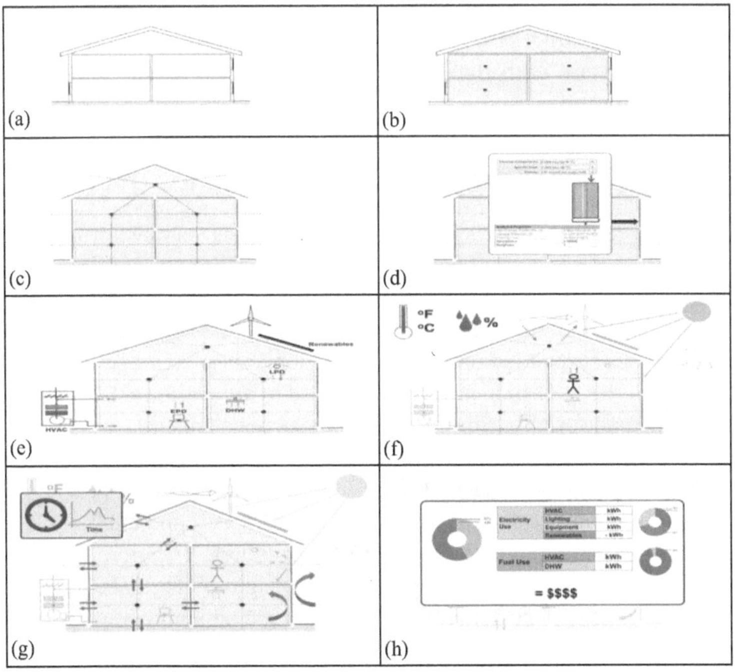


**FIGURE 2.3** Analysis working procedure.

- *Thermal properties* considered include density, specific heat capacity, and conductivity. The properties of each surface account for the joint thermal properties of the layered building constructions (Figure 2.3d).
- *Building Systems*: A building's systems consume and produce energy to meet the building's and occupants' needs. These include lighting, ventilation, heating, air-conditioning, equipment, renewables that produce power on area, and water heating (Figure 2.3e).
- *Exterior Conditions and Loads from Occupancy*: So far we've built-up a model that represents the building's spaces, surfaces, and systems. The other half of the modeling needed for energy simulation is the dynamic loads from people inside the building (based on building program assumptions) and the environmental conditions outside the building (based on climate data). How well will the building do under these loads? These factors include relative humidity, humidity, and heat from people, external air temperature, the spaces' target temperatures, infiltration assumptions, diffuse and direct solar radiation, wind direction, and speed (Figure 2.3f).
- *Heat Transfer Modeling*: Now that it is known both the building's properties and the loads it will be subjected to overtime, it can simulate heat transfer over time (usually an entire year, in hourly increments). This includes: radiation (yellow), convection (green), and conduction (red) (Figure 2.3g).
- *Cost Predictions and Energy Use*: Finally, based on all that it has modeled and simulated, it can extract energy use and cost predictions. These energy use values are broken down by energy source (electricity, fuel) and by end use (lighting, HVAC, etc.). This can translate to cost by multiplying predicted use with the predicted utility rates of each energy source (Figure 2.3h).

## 2.4 ANALYSIS OF MINI AUTOMOTIVE-REPAIR FACILITY

Mini automotive-repair facilities are known as "auto-king service" in Turkey. These are the services that are committed to making certain costs according to the payments made annually within the scope of motor vehicle insurance. In general, these small architectural buildings have workshop parts outside the staff rooms. Since motor vehicles are suspended and repaired in some cases, garage areas are built high. In addition, the workshop areas are designed to allow the vehicles to enter and exit the building comfortably. Since these repair sections are large and high areas, it is important to heat and cool the workers in the workshop so that they can work comfortably. These mini automotive-repair facilities are committed to providing insurance services to private customers. Even in Turkey's least populated city is present at least five such mini automotive-repair facility buildings. Therefore, such facilities are buildings that can be shown as an example of designs that can show the extent to which energy savings can be achieved by reducing the amount of energy consumed for heating large areas. The main research focus of this chapter is the extent to which the desired energy savings can be achieved with the right selected materials in these buildings and whether this saving rate can be achieved when the building is designed by supporting renewable materials. While designing this chapter, it is aimed to evaluate the performance of the building which is designed by using ecologically renewable products and the same building designed by using traditional building insulation materials. For this reason, the mini automotive-repair facility was chosen as a sample building model with high energy loss and too non-large building design.

### 2.4.1 The Projecting of Mini Automotive-Repair Facility

The mini automotive-repair facility for any damage that may occur on vehicles covered by automobile insurance (bodywork, paint, plastic bumper, interior upholstery, windshield, upholstery stains) during the one-year period of the service removes without a price. In this study, the mini automotive-repair facility designs to cover three repair units and service departments and other employee rooms.

First, the mini automotive-repair facility projects as seen in Figure 2.4. And then this facility animates in a virtual environment. The renders obtained from Revit analysis of the mini automotive-repair facility is shown in Figure 2.5.

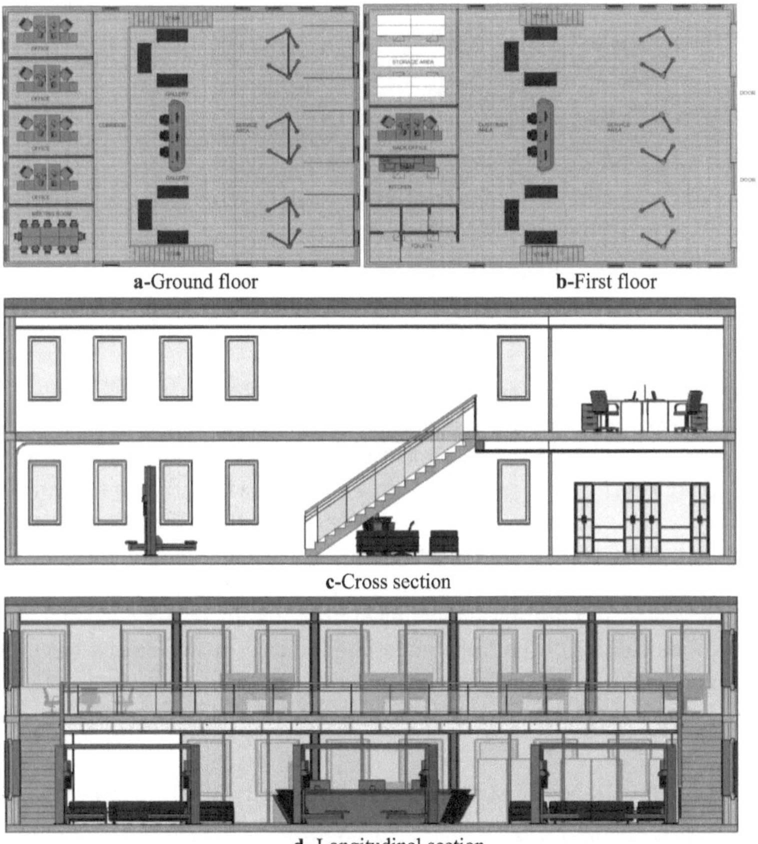


**FIGURE 2.4** The different views of the mini automotive-repair facility.

**FIGURE 2.5** The renders obtained from Revit Analysis of the mini automotive-repair facility.

### 2.4.2 The Wall Components Used in Analysis of Mini Automotive-Repair Facility

The analyzed building external wall types' heat transfer coefficients and the analyzed building external wall components' physical properties are given in Figures 2.6 and 2.7, respectively. After the settlement and the decisions are made on the plan level, the energy performance of the building analyzes with different building external wall types and elements. These analyze on the Autocad Revit platform in the building shell by selecting different types of walls and elements. In this study, three different external wall types (externally insulated, internally insulated, sandwich) and three different structure-insulation materials (EPS, ECO, gas concrete) analyze by stratification. In this chapter, the material produced by epoxidized corn oil, fly ash, perlite, and clay is called ECO. The different external wall stratification analyzed in the Revit simulation is display in Figure 2.8.

### 2.4.3 Ecological and Climate Factors Affecting the Revit Analysis

Malatya is situated in Turkey's eastern Anatolia region in the upper Firat river basin. For Malatya, the decimal longitude and latitude coordinates are 38.31667 and 38.35018, respectively. According to TSE 825, Turkey has four climatic zones and

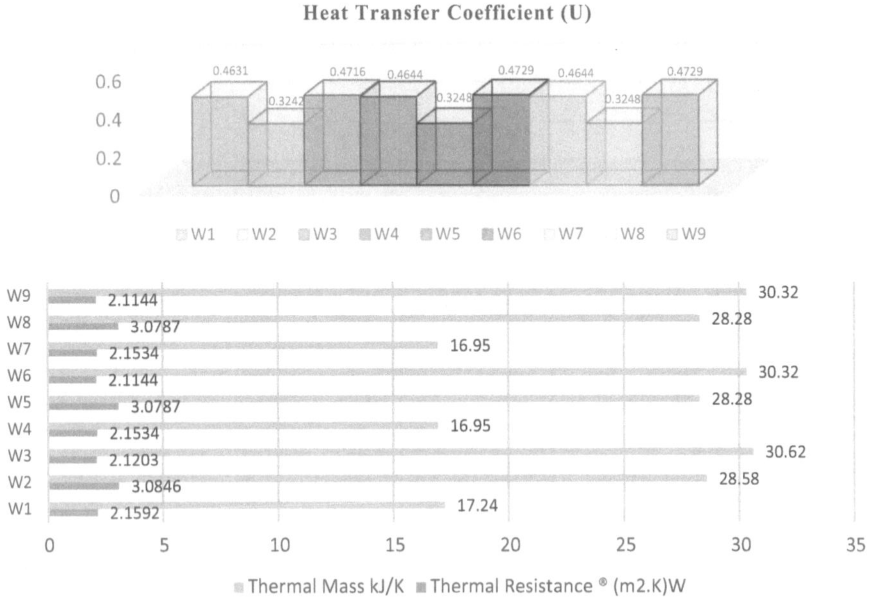


**FIGURE 2.6** The analyzed building external wall types' heat transfer coefficient [W/(m²K)], thermal mass (kJ/K), and thermal resistance (m²K)W.

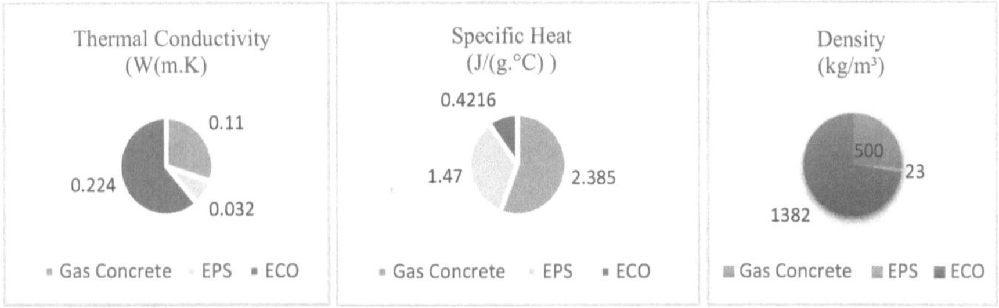


**FIGURE 2.7** The analyzed building external wall components' physical properties.

Malatya is in the third zone. The climatic zone (a) and place (b) on the Turkey map and of Malatya province are shown in Figure 2.9 [31].

Malatya is on the edge of a warm continental dry-summer clime [Köppen clime sorting Dsa], cold semi-arid weather [Köppen clime sorting BSk] with dry and hot summers and snowy and cold winters, and a cool Mediterranean clime [Köppen clime sorting Csa]. In Malatya, general terrestrial climate prevails in Central Europe, the climate is broadly in line with the weather. It's hot, muddy, and, over the course of a year, there are also several summer months. Malatya's mean yearly temperature is 18°C, with annual rain absolutely 248 mm. It is 4 for 240 days yearly UV index and dry with mean humidity of 52% [32,33]. The monthly average temperatures, minimum–maximum temperatures, rainfall, and humidity values in Malatya

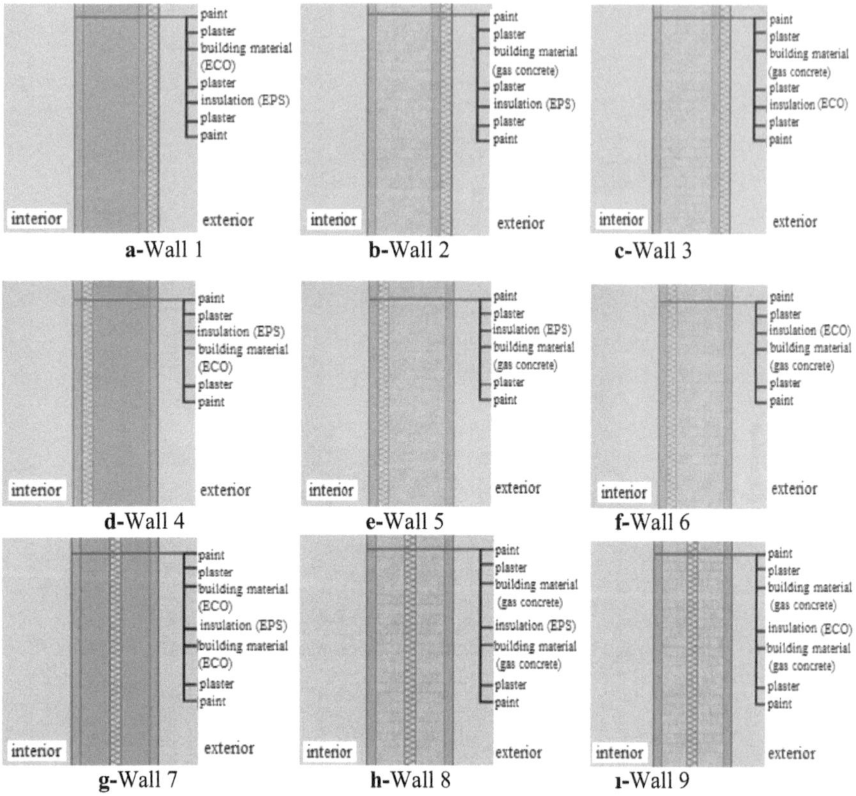


**FIGURE 2.8** Different external wall stratifications analyzed in the Revit simulation.

**FIGURE 2.9** The climatic zone and place on the Turkey map of Malatya province.

province are displayed in Figure 2.10 [32]. Using weather data analysis, a minimal humidity value is determined in August. For a year, the climatic values as months in Malatya province are shown in Figure 2.11 [33]. In Malatya province, cooling–heating degree days and annual design conditions used at analysis by Autodesk Revit Architecture Simulation Program are presented in Figure 2.12 [29]. As seen in next sections of this study, the higher months with heating load are January, December, and February in a high to low order, in regard to the cooling load, the months are August, July, and June. The cooling and heating loads analysis defines the cooling and heating needs of the structure modeling. The cold climes have a great deal more "heating degree days" than "cooling degree days." Maximizing insulation is the key to keeping it warm and using it. If the weather for one day is one means of 10 degrees warmer, or for ten days is a degree warmer, afterward the building will need ten days of cooling degree. If the weather for one day is 10 degrees below the comfortable least, the building will need ten days of heating degree [34].

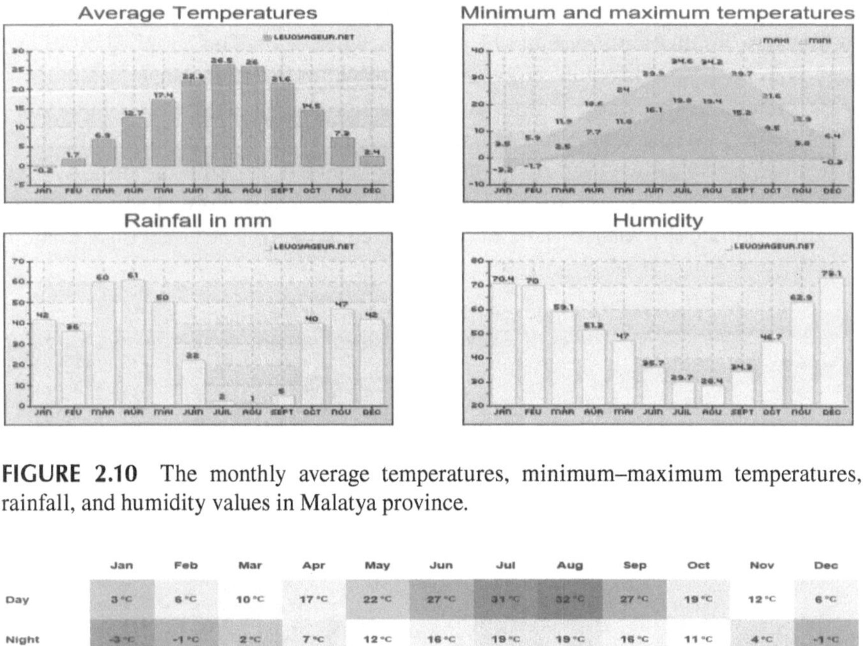


**FIGURE 2.10** The monthly average temperatures, minimum–maximum temperatures, rainfall, and humidity values in Malatya province.

| | Jan | Feb | Mar | Apr | May | Jun | Jul | Aug | Sep | Oct | Nov | Dec |
|---|---|---|---|---|---|---|---|---|---|---|---|---|
| Day | 3 °C | 6 °C | 10 °C | 17 °C | 22 °C | 27 °C | 31 °C | 32 °C | 27 °C | 19 °C | 12 °C | 6 °C |
| Night | -3 °C | -1 °C | 2 °C | 7 °C | 12 °C | 16 °C | 19 °C | 19 °C | 16 °C | 11 °C | 4 °C | -1 °C |
| Precip | 48 mm | 33 mm | 30 mm | 24 mm | 19 mm | 5 mm | 1 mm | 1 mm | 4 mm | 17 mm | 24 mm | 41 mm |
| Rain days | 13 | 11 | 13 | 13 | 11 | 3 | 2 | 0 | 3 | 7 | 7 | 12 |
| Snow days | 8 | 3 | 2 | 0 | 0 | 0 | 0 | 0 | 0 | 0 | 0 | 4 |
| Dry days | 10 | 14 | 16 | 17 | 20 | 27 | 29 | 31 | 27 | 24 | 23 | 15 |
| Sun hours per day | 6 | 8 | 9 | 11 | 12 | 12 | 12 | 12 | 10 | 8 | 7 | 6 |
| Wind force in Bft | 2 | 2 | 2 | 2 | 2 | 2 | 2 | 2 | 2 | 2 | 2 | 2 |
| UV-index | 1 | 2 | 3 | 5 | 5 | 6 | 7 | 7 | 5 | 4 | 2 | 2 |

**FIGURE 2.11** For a year, the climatic values as months in Malatya province.

| Cooling Degree Day | | Heating Degree Day | |
|---|---|---|---|
| Threshold | Value | Threshold | Value |
| 18.3 °C | 676 | 18.3 °C | 2764 |
| 21.1 °C | 378 | 15.6 °C | 2136 |
| 23.9 °C | 163 | 12.8 °C | 1578 |
| 26.7 °C | 38 | 10 °C | 1122 |

| Annual Design Conditions | | | | |
|---|---|---|---|---|
| Threshold | Cooling | | Heating | |
| | Dry Bulb(°C) | MCWB(°C) | Dry Bulb(°C) | MCWB(°C) |
| 0.1 % | 38.7 | 16.7 | -10.2 | -11.8 |
| 0.2 % | 37.8 | 16.8 | -9.2 | -10.6 |
| 0.4 % | 36.9 | 17.1 | -8.7 | -10.0 |
| 0.5 % | 36.4 | 17.6 | -8.4 | -9.7 |
| 1 % | 35.0 | 16.7 | -7.6 | -9.1 |
| 2 % | 33.3 | 16.3 | -6.1 | -7.8 |
| 2.5 % | 32.7 | 16.4 | -5.3 | -6.9 |
| 5 % | 30.5 | 15.6 | -3.2 | -5.1 |

**FIGURE 2.12** In Malatya province, cooling–heating degree days and annual design conditions are used at analysis by Autodesk Revit Architecture Simulation Program.

For Malatya province, the monthly design data (threshold of 2%), dry frequency distribution, and dry bulb cumulative distribution detected by Autodesk Revit Architecture Simulation Program are given in Figure 2.13. This graph shows how many hours per year dry-bulb temperatures fall within a certain range of temperatures. Sometimes this range is referred to as a bin of temperature. Dry-bulb temperature refers to a thermometer's actual temperature of the air. It is the temperature determined by an air-exposed thermometer but shielded from direct humidity and solar radiation. The temperature of the wet bulb is the temperature at which air can cool till it saturates by volatile water into it at a fixed pressure. The larger the variation

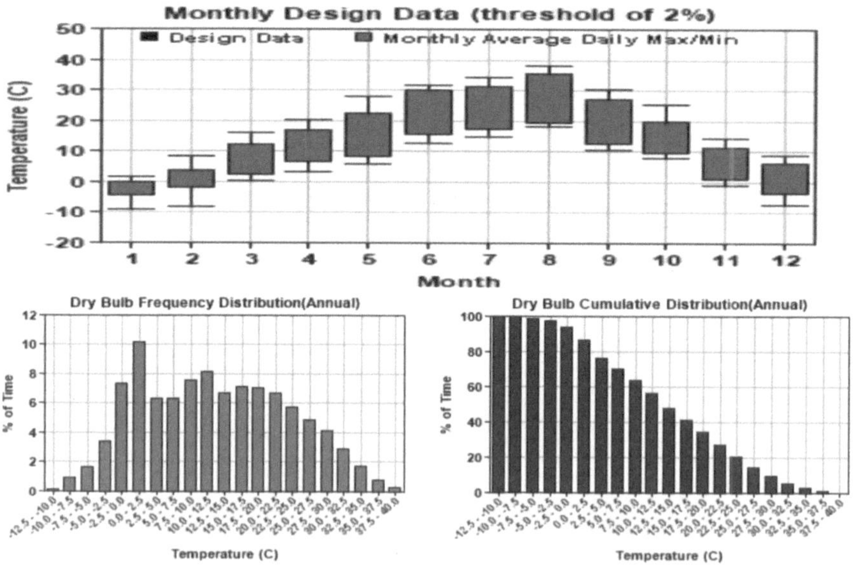


**FIGURE 2.13** For Malatya province, the monthly design data (threshold of 2%), dry frequency distribution, and dry-bulb cumulative distribution are detected by Autodesk Revit Architecture Simulation Program.

in the range of the temperature of the dry bulb and the wet bulb, the higher the relative moisture. The average daily range is the measure of the variation between maximal and minimal daily temperatures. In Malatya province, the most wet-bulb and dry-bulb temperatures are found in August and a minimum temperature is shown in January.

The wind roses detected by the same program is shown in Figure 2.14 [29]. This figure from REVIT sets the annual wind velocity for Malatya place as a function of time percentage in radiant scale. Such estimates also indicate the strength and intensity of prevailing winds in winter and summer. Revit analysis results show accordingly in the figure that the Malatya climate has blown from south-southeast directional wind in the winter period and exhibits the most prevailing wind from the north in the summer period.

Choosing the right weather data in energy analysis results in better prediction of energy. The simulation results' reliability using weather data (typical-meteorological-year) is investigated in the design stage. In addition to weather data, other realistic

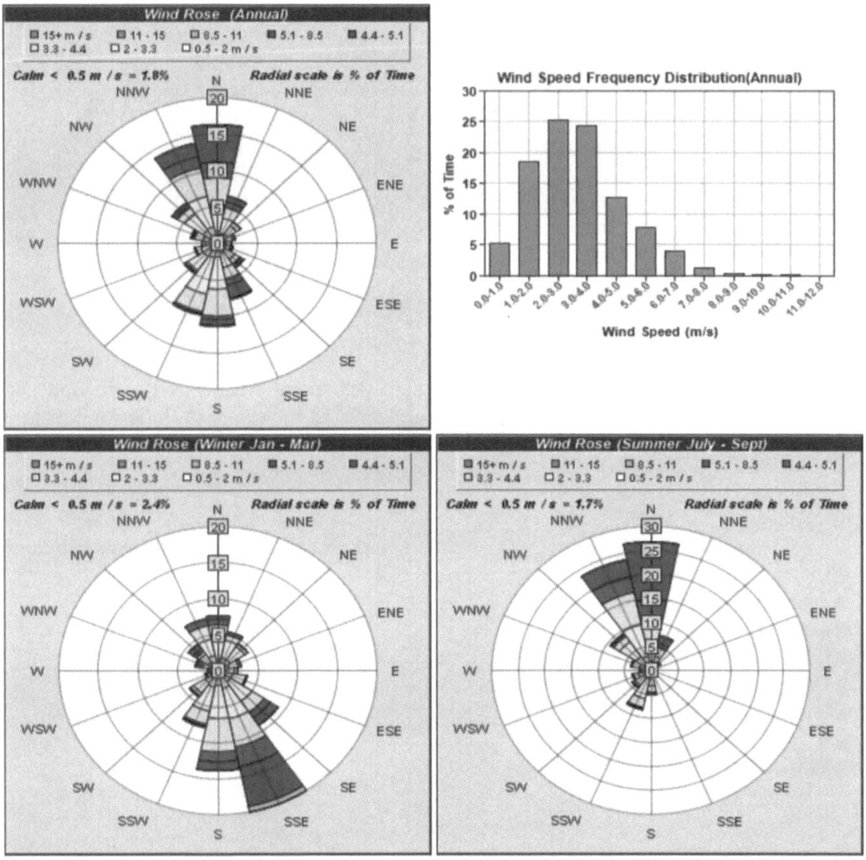


**FIGURE 2.14** For Malatya province, the wind roses are detected by Autodesk Revit Architecture Simulation Program.

development considerations need to be discussed. The temperature and angle of the device also influence the output of electricity. BIM-based measurement of energy is proper to use in this respect [29].

For Malatya province, the annual dew point frequency and annual relative humidity frequency distribution are indicated in Figure 2.15. A rise in relative humidity can be compensated for by a change in the temperature of the dew point [35].

The annual total sky cover frequency distribution and annual direct normal radiation frequency distribution are represented in Figure 2.16. The heavenly cloud-covered area is called the sky cover. To calculate it, it is necessary to find the percentage of the sky covered with clouds. For the purpose of this calculation, the regions of clouds that do not completely cover the sky are applied during the observation. For climatic measurements, the total area in the atmosphere is believed to be 10/10. Therefore, a 0–10 scale is used. According to this scale, a value of 0 is a sky without any cloud, because a value of 10 is a sky that is completely covered with clouds.

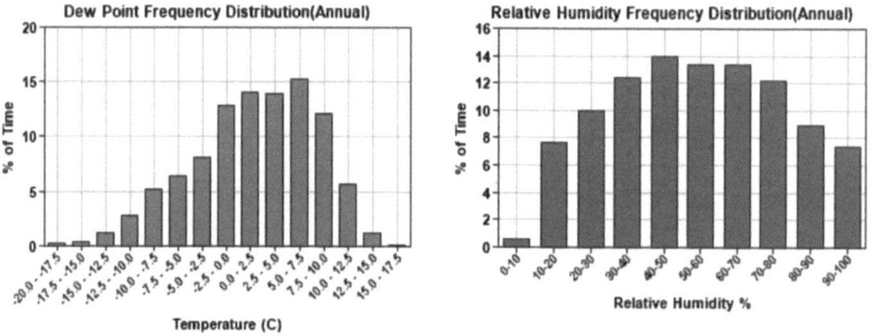


**FIGURE 2.15** For Malatya province, the annual dew point frequency distribution and annual relative humidity frequency distribution are detected by Autodesk Revit Architecture Simulation Program

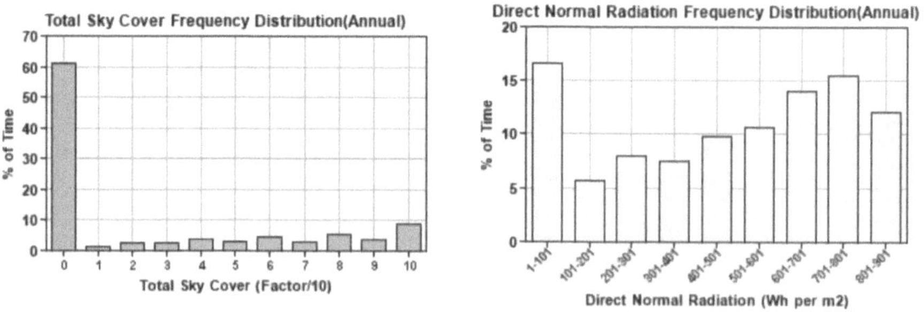


**FIGURE 2.16** For Malatya province, the annual total sky cover frequency distribution and annual direct normal radiation frequency distribution are detected by Autodesk Revit Architecture Simulation Program.

Sky cover of 1 is recorded when the sky observes slight cloud trails. It is, therefore, necessary to cover less than 1/10 of the sky with clouds. On the other end, when the sky is almost completely covered with clouds, a sky cover of 9 is recorded. The sky can only be seen partially through the clouds in this situation. An appropriate degree of the scale (2, 3, 4, 5, 6, 7, and 8) can be registered for the other numbers of clouds in the sky. On the other hand, sky cover is calculated for the night-time based on the cloud-covered star ratio. In the case of observers under the influence of natural events such as fog and snowstorm, sky cover is recorded as 10 because of the inability to estimate the covered part of the sky. When evaluating the energy savings of the building under investigation here, for an accurate assessment, the annual total sky encompasses the frequency distribution values considered in Revit.

The direct normal radiation is the amount of sun irradiation received per unit field by a surface that is always held regular (or perpendicular) to the rays coming directly from the direction of the sun at its current place in the sky [36]. The most important comments to the energy efficiency simulation must be provided in advance with reliable data on sun irradiation. Global horizontal radiation is the most beneficial sun irradiation data but also diffuse horizontal radiation or direct normal radiation. For example, when diffuse horizontal radiation or direct normal radiation is received in addition to global horizontal radiation, radiation is described on the surfaces of a solar cell or sun collector [37]. In this study, these three radiation values are used by the simulation program. The direct normal distribution chart of radiation frequency shows how much direct sunlight is available over a period of 12 months.

The annual diffuse and global horizontal radiation frequency distributions are displayed in Figure 2.17. The global horizontal radiation is the total amount of shortwave radiation emitted from above by a horizontal surface to the earth. Photovoltaic installations are particularly interested in this, involving both horizontal diffuse radiation and normal direct radiation. The diffuse horizontal radiation is sun irradiation, which does not come directly from the sun but is distributed by atoms and molecules in the atmosphere and comes from all ways equally. Essentially, it's the lighting that reaches the blue sky and the clouds [37].

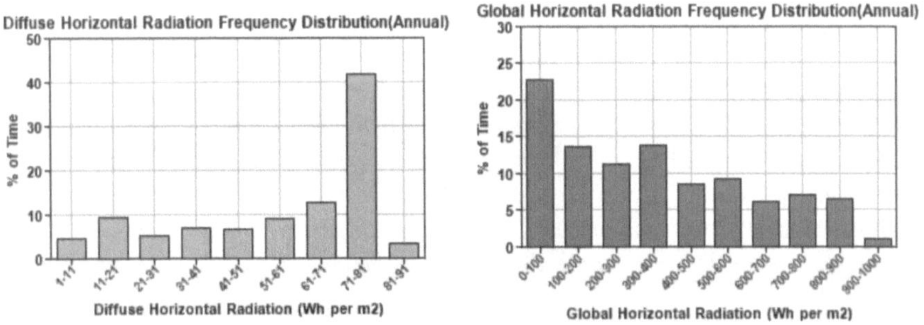


**FIGURE 2.17** For Malatya province, the annual diffuse and global horizontal radiation frequency distributions are detected by Autodesk Revit Architecture Simulation Program.

### 2.4.4 The Assessment of Energy Performances of Mini Automotive-Repair Facility Analyzed by Autodesk Revit Architecture Simulation Program

The modified energy profile causes a paradigm shift in terms of where the energy is being used and thus not only focuses on the energy in terms of heating since the main energy consumption is no longer coming from it. The emphasis on energy-efficient buildings should be preserved but adjusted to the way the energy is used [38].

Energy performance analysis is used to estimate how much energy the building consumes as a heating and cooling load. In this study, energy performance analysis is performed with building elements with Autodesk Revit Architecture Simulation Program. Building elements are created by BIM object modeling techniques. Further elaboration of the design is provided by conceptual modeling. In this modeling in Revit, the material qualities are selected in the rightist way to the designed project and the building to be applied and added with the qualities are specified in Section 2.4.2. In this way, the main limits of energy analysis have been formed within the scope of region properties and material properties. Revit conceptual analysis is performed through Autodesk's Green Building Studio cloud service. Monthly and annual energy analysis of the conceptual model is carried out in the energy analysis of mini automotive-repair facility building. Monthly year energy costs and European University Institute (EUI) values were then determined.

#### 2.4.4.1 Annual Energy Analysis of Mini Automotive-Repair Facility

Energy simulation can aid to evaluate energy flow out, in, and by a building model's volumes and spaces. This knowledge can aid designers to make educated, cost-effective decisions that fix performance and diminish the buildings' effect on the environment. Simulation of entire building energy tests' planned energy usage (electricity and fuel) depends on the climate, building geometry, building form, active systems (HVAC and lighting) of the building, and envelope properties. As a whole system, energy analysis takes into consideration the building's interdependencies.

Energy-effective structures area needs multiple domains' integration. The energy model utilizing Building Information Modeling (BIM) has the prospective to facilitate the modeling production process by taking advantage of smoothly the essential knowledge stored in a mechanical or architectural BIM. This BIM-based program is using a downloaded climate file to extract internal temperatures after constructing "thermal zones" in Revit, representing different building fields. Such hourly temperature data vary depending on building construction assembly, as well as any structure that would impact solar benefits, such as overhang. Such temperatures are better viewed as a long-term average, such as weeks or months, rather than an hourly one. These are also often used to measure the impact of a model or construction assembly over another, not so much as a number written in stone—these temperatures are extracted from a climate report using average temperatures over a span of 30 years.

The analysis is conducted by turning the overall form and structure of buildings into a "computational network" capable of effectively capturing all the primary heat transfer paths and processes throughout the building. It evaluates the results as to the likely energy performance of building models with existing external wall type, depending on what materials have been used in the building model creation. In analysis, the indoor and outdoor temperatures are used as 20°C and −12°C, respectively.

The annual energy performance of the mini automotive-repair facility is given in Figure 2.18. In the analysis of the external wall types, it was found that wall type 5 (116 908 kWh) gave the most effective results among the tested external wall types. The worst energy performance was obtained from wall type 7 (118 842 kWh). The highest consumption found to space heat (53 554 kWh) among energy-consuming elements of analyzed building.

### 2.4.4.2 Monthly Energy Analysis of Mini Automotive-Repair Facility

In Figure 2.19, the monthly results of total energy consumption are used as a basis for further comparison and show the highest energy consumption period for cooling loads being the three-month period for external wall types of mini automotive-repair building in December, January, and February. In January, the highest energy consumption value obtained as 32,492 kWh at wall type 7. This month, the highest consumption was determined as 11,356 kWh for space heat. The lowest consumption was determined as 322 kWh for external usage.

It indicates that the height of consumption of energy is in the month of August, while the lowest consumption is in December. This can be seen in Figure 2.19. Interestingly, the use of natural gas in the summer months is the lowest. This is due to the lack of need in these months for space heating, which is a major resource user. As expected due to the extreme cold weather and the strong need for space heating, the peak use for natural gas in January.

### 2.4.4.3 Annual Energy Cost of Mini Automotive-Repair Facility

Annualized information on energy consumption and costs will inform assessments of energy costs for construction and early design decisions. Construction energy cost calculation added to the cost of the life cycle offers the investors and owners a complete picture for making decisions. Using alternative systems or materials can increase the building's energy cost and produce a project's opposite result. Many measurements of building energy costs and early enforcement decisions could have

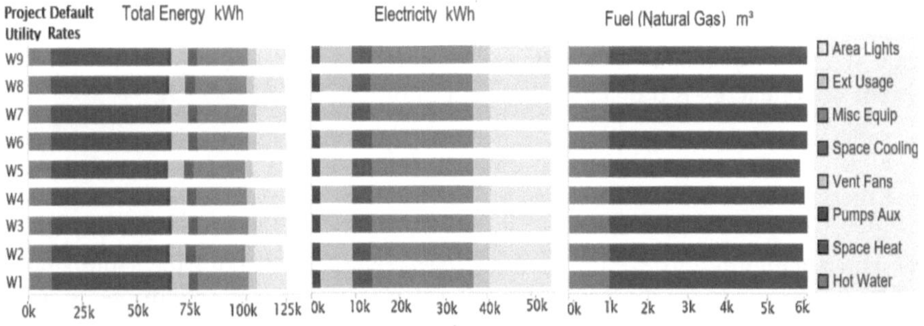


**FIGURE 2.18** Annual energy performance of the mini automotive-repair facility.

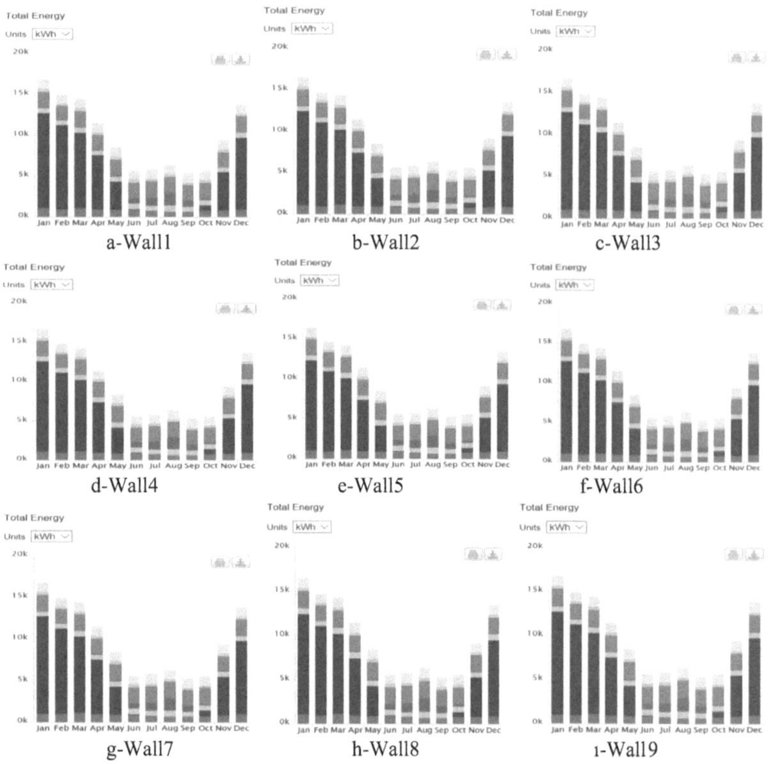


**FIGURE 2.19** Monthly energy performance of mini automotive-repair facility.

been made using annualized data on energy costs and usage. The costs are measured using average utility rates across the region, country, or nation-wide or the custom rates that may have added to the venture. It can challenge to compare costs because electricity is billed in kilowatt-hours (kWh) and natural gas in therm. Delivery and other charges make it even harder to assess cost differences accurately. Energy cost reports the current annual cost of energy per $ft^2$ or $m^2$. The input modification will vary this value instantly. Energy costs are forecasted utilizing territory-wide, nation-wide, or state-wide mean utility ratios.

The use of energy is transformed as below [29]:

1.0 kWh (kilowatt-hour) = 3.413 k Btu
1 therm = 100 k Btu

As shown in Figure 2.20, even if electric prices' range is considered, the natural gas's price is about three to four times lower than the electricity's price. In the analysis of the external wall types, the highest annual energy cost was obtained at wall type 4 (10,771 T). The lowest energy cost was defined from wall type 7 (10,631 T).

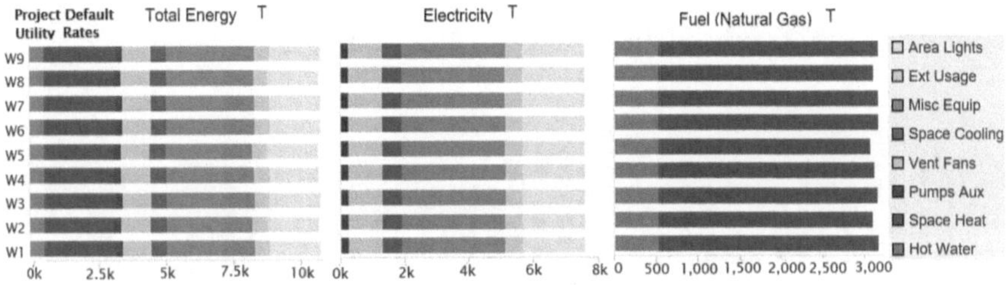


**FIGURE 2.20** The annual energy costs of mini automotive-repair facility.

### 2.4.4.4 Monthly Energy Cost of Mini Automotive-Repair Facility

A policy must be introduced at the design stage to mitigate a building's energy usage cost-effectively. The focus point should be on ventilation, air-conditioning (HVAC), and heating systems as well as the building envelope (roof, doors/windows, ground floor slab, and external walls). The building envelope includes elements such as the insulation type and insulation thicknesses as well as glazed areas' placement, types, and locations. Besides, if a smart design is made, it is possible to achieve a 50%–75% reduction in energy consumption relative to buildings built in 2000. As a result, the cost based on energy usage also will decrease [39].

For the analyzed external wall types, the monthly energy costs were obtained as a linear ratio with monthly energy needs. Among all analyzed external wall types, the highest electrical energy and natural gas consumption values were determined for June-July-August and December-January-February periods, respectively.

Examining the external wall types, this analysis was designed to find groups of influential limits within a development stage model and quantitatively assess how these groups could be influential on the projected use of energy. Conspicuously, Figures 2.20 and 2.21 display the change building energy costs with different external wall types that consist of high-performance envelope materials, thus revealing its important role in energy performance improvement of the right building shell.

In January, the highest total energy cost value was obtained as 2,292 T at wall type 7. This month, the highest cost was determined as 559 T for space heat. At wall type 7, the lowest energy cost was determined as 45 T for external usage among energy-consuming elements of analyzed building.

### 2.4.4.5 Annual EUI Values of Mini Automotive-Repair Facility

The energy intensity is literally energy requirement per unit field on the floor plan of the building, typically in square feet or square meters. It helps it to compare the energy requirement of diverse sizes of buildings, so it can see which applies better. Energy use intensity is an especially useful tool for setting targets and benchmarks for energy usage. This typically changes reasonably a bit depending on the climate, the construction plan, and the size of the building. Energy use intensity is the mean of the project's combined fuel and electricity per area (square feet in IP units or square meters in SI units) per year. The values of the energy intensity in Figure 2.22 above-mentioned only take into account the amount of fuel and

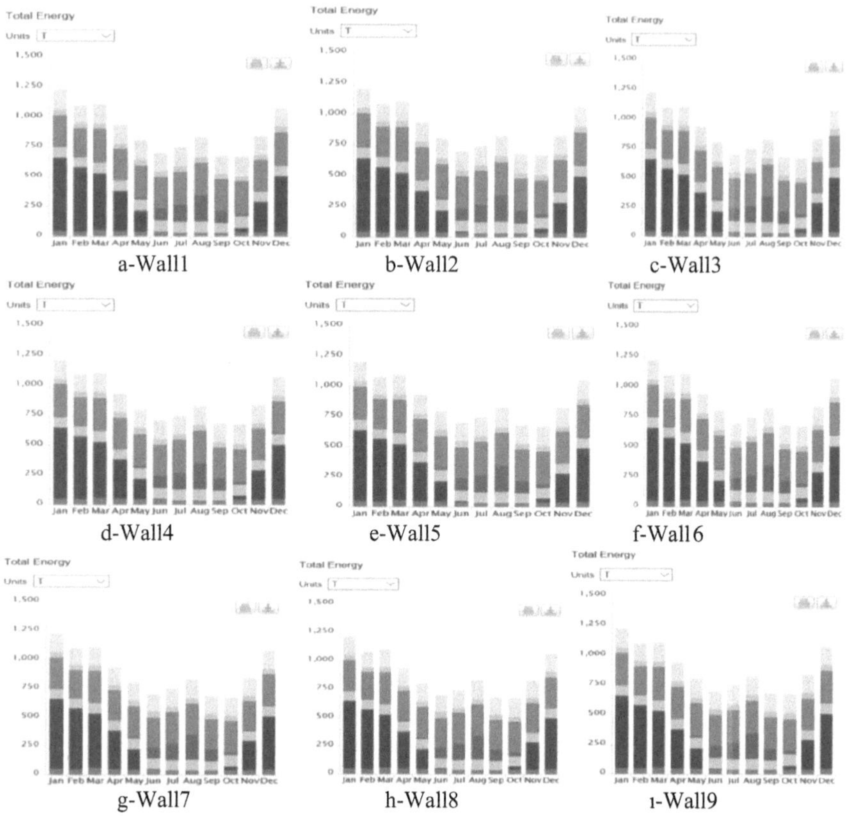


**FIGURE 2.21** The monthly energy costs of mini automotive-repair facility.

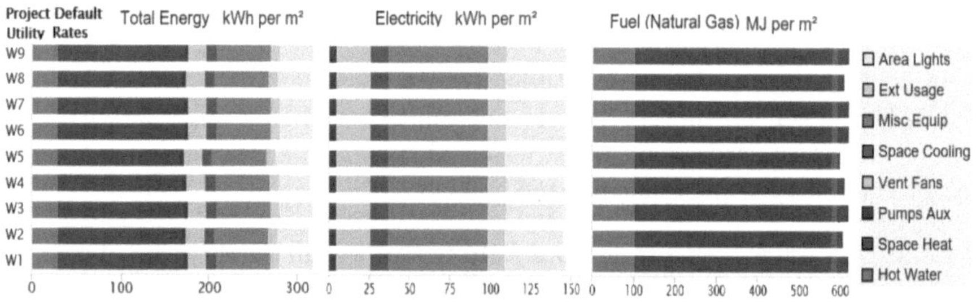


**FIGURE 2.22** The annual energy use intensity values of mini automotive-repair facility.

electricity ("location" or "secondary" energy) that is utilized on location. The fuel used to produce electricity or heat energy is not considered by them. Some energy use ıntensity tabulations and most energy use ıntensity building codes seek to capture the full effect of providing energy to a building by describing the word "source" or "main" energy that combines the fuel utilized to produce power on-location or at an energy station far away. For example, the 1 kWh site electricity from the solar panel on the roof of the building is equivalent to 1 kWh of "source" energy because the sun panel itself is the resource. The location energy is the most useful measurement when measuring energy utilized to supply visual or thermal comfort. But the source energy is the more accurate measurement when calculating total energy use to assess environmental effects.

Among the existing building external wall types, the lowest energy use intensities are determined as 312.87 kWh/m$^2$ at external wall type 5. In this wall type, the largest and least energy use intensity values are obtained as 61.88 (misc. equipment) and 4.7 (pump aux) kWh/m$^2$, respectively.

## 2.5 CONCLUSIONS

It is a known fact that the first investment costs in the construction and construction sector in our country are very important for employers and investors. This situation is considered an even more important criterion, especially for small- and medium-sized companies. In addition to the encouragement of legal frameworks for high-performance building design, the availability of expensive expertise and critical technologies is an important issue. Considering that the BIM models to be used are ready for all kinds of simulation and ready for rapid cost calculations and high-cost calculations after material modifications, it can be mentioned that they contain economic advantages especially in terms of construction, manufacturing, and supply chain management. At this point, sharing and adaptable BIM models become more accessible and constitute a solution approach to all these problems. This model is based on the approaches of developed and energy-efficient countries in the sectoral and public sectors and offers a technology-oriented solution by considering local problems and obstacles in the building sector. In our country, especially in the process of designing large and complex functional projects, the support of different disciplines and experts is needed. It is a fact that the sharing of knowledge of these different disciplines has been facilitated since the transfer of project processes to computer environments.

The sustainable green designs that are in line with the green concept that has gained importance in recent years, it has been researched to what extent both environment-friendly and energy-efficient building designs can have a positive impact on energy efficiency. For this purpose, the change in energy performance was analyzed comparatively if the building, which was designed and constructed in the urban sustainability concept, was supported with sustainable wall and traditional materials, which is one of the green design criteria for different building envelopes. These analyses are applied by using Autodesk Revit Architecture Simulation software.

A quantitative and qualitative research concerning green concept mini automotive-repair facility modeling's climate responsiveness is realized in this study. The obtained results are as follows:

The thermal quality performance's quantitative research relied on Autodesk Revit Architecture Simulation Program displays that building with wall type 5 (this external insulated wall designed EPS as insulation material and gas concrete as building material) is more effective than other wall types in cooling duration in addition to the building with wall type 5 one contributes to smaller heating load in winter duration.

- The qualitative research depends on comment/image display that building with wall type 5 is more efficient than existing buildings in cooling and heating duration.
- It was found that wall type 4 (this external insulated wall designed EPS as insulation material and gas concrete as building material) consumed more than 0.0096% and wall type 6 (this external insulated wall designed ECO as insulation material and gas concrete as building material) consumed 0.016% more energy compared to the wall type 5 with the best energy performance. The results show that ECO can be used as an alternative to traditional materials when used both as insulation and as a building material. The energy consumption rates obtained in internally insulated and sandwich wall structures are obtained in almost the same way with externally insulated wall types.

The implementation of a systematic BIM–Building Energy Model as suggested in this chapter could reduce the level of effort required in an iterative process for the conventional Building Energy Model and encourage preferable utilization of energy simulation to design high-performance buildings. Compiling and analyzing the key BIM–Building Energy Model methods over the years and the number of studies in each of the fields are already helping to fill the current gaps.

## REFERENCES

1. Al-Homoud M.S. Performance characteristics and practical applications of common building thermal insulation materials. *Building and Environment* 40 (2005) 353–366.
2. The US Department of Energy. Insulation fact sheet with addendum on moisture control, DOE/CE-0180, USA, 2002.
3. ASTM Standard C 168-97. Terminology relating to thermal insulating materials, 1997.
4. Peavy B.A. A heat transfer note on temperature dependent thermal conductivity. *Journal of thermal Insulation and Building Envelopes* 20 (1996) 76–90.
5. American Society of Heating, Refrigerating, and Air Conditioning Engineers (ASHRAE). *Handbook of Fundamentals*. ASHRAE, Atlanta, GA, 2001. (Chapter 23).
6. Carlson L., Adriano D.C. Environmental impacts of coal combustion residues. *Journal of Environmental Quality* 22 (1993) 227.
7. Akakın T. Betonun Mühendislik Özellikleri, 32–40 (2001). Available from: http://www.kalitekontvol.org/beton/betonun-muhendislik-ozellikikleri.htm.
8. Balo F. Feasibility study of "green" insulation materials including tall oil: Environmental, economical and thermal properties. *Energy and Buildings* 86 (2015) 161–175.
9. Sagbansua L., Balo F. Ecological impact & financial feasibility of energy recovery (EIFFER) model for natural insulation material optimization. *Energy and Buildings* 148 (2017) 1–14.
10. Khedari J., Suttisonk B., Pratinthong N., Hirunlabh J. New lightweight composite construction materials with low thermal conductivity. *Cement & Concrete Composites* 3 (2001) 65.

11. Balo F. Characterization of green building materials manufactured from canola oil and natural zeolite. *Journal of Material Cycles and Waste Management (JMCWM)* 17 (2015) 336–349.
12. Murad E., Wagner T. Clays and clay minerals: The firing process. *Hyperfine Interactions* 117 (1998) 337.
13. Balo F. Castor oil-based building materials reinforced with fly ash, clay, expanded perlite and pumice powder. *Ceramics-Silikáty* 55 (3) (2011) 280–293.
14. Pryde E.H. *Handbook of Soy Oil Processing and Utilization.* American Oil Chemists Society, Champaign, IL, 1990.
15. Nevin C., Moser B. Vinyl oil monomers. I. Vicinal methacryloxy-hydroxy soy oils. *Journal of Applied Polymer Science* 7 (1963) 1853.
16. Sagbansua L., Balo F. A novel simulation model for development of renewable materials with waste-natural substance in sustainable buildings. *Journal of Cleaner Production* 158 (2016) 245–260.
17. Lu Y., Larock R.C. Corn oil-based composites reinforced with continuous glass fibers: Fabrication and properties. *Journal of Applied Polymer Science* 102 (2006) 3345–3353.
18. Fajardo F., Dela Pe~na, J., Bayani, J., Magdaluyo, E. Synthesis and Characterization of Corn (Zea mays L.) Oil-based Polyurethane Fil. College of Engineering University of the Philippines Diliman, Quezon City, Philippines, 2016.
19. Li F., Hasjim J., Larock R.C. Synthesis, structure, and thermophysical and mechanical properties of new polymers prepared by the cationic copolymerization of corn oil, styrene, and divinylbenzene. *Journal of Applied Polymer Science* 90 (2003) 1830–1838.
20. Balo F. Investigation of the usability of flax oil-based materials as building or insulation material in buildings. 11th National Installation Engineering Congress and Teskon+Sodex Fair, pp. 752–762, İzmir, Turkey, 17–20 April 2013.
21. Balo F. Investigation of thermal conductivity and some of mechanical properties in castor oil-based building materials, *2nd Anatolian Energy Symposium*, pp. 752–762, Diyarbakır, Turkey, 2–4 May 2013.
22. Queralt I., Querol X., Lopez-Soler A., Plan F. Use of coal fly ash for ceramics: A case study for a large Spanish power station. *Fuel* 76 (1997) 787–791.
23. TS 699, Methods of Testing for Natural Building Stones. Turkish State Planning Organization, 1987.
24. TS 699, Tabii yapi tas¸ lari muayene ve deney Metotları, (Eight Five Years' development Program on Use of Industrial Materials Such as Pumice, Perlite, Vermiculite Phosphogypsum and Expanding Clays). Turkish State Planning Organization, Ankara, 2001.
25. TS EN 771-3, The Standart of the Brick Made of Using Natural or Artifical Light or dense Aggregates for Walls. Turkish State Planning Organization, Ankara, 2005.
26. Ceylan A., Ebeoglugil F. The investigation of the use of clay and expanded perlite in the production of lightweight and heat insulation enhanced construction bricks. *Dumlupinar University Science and Technology Journal* (2002) 33–38. The 10th anniversary Special Issue.
27. Larock R.C., Gallagher P.W. Development of Corn/Soy Plastic Composites. Department of Chemistry, Iowa State University, Ames, IA, 2012.
28. Larock R., Sheares, V. Center for Crops Utilization Research. Spring & Summer, 2002.
29. REVIT. http://www.autodesk.com/products/revit-family/overview, 2016.
30. Atam E. Current software barriers to advanced model-based control design for energy-efficient buildings. *Renewable and Sustainable Energy Reviews* 73 (2017) 1031–1040.
31. Retrieved from http://www.atermit.com/Sayfalar/teknik-bilgiler (Accessed: 8 Dec 2019).
32. Retrieved from https://www.besttimetovisit.co.za/turkey/malatya3743295/#Climate_Malatya (Accessed: 8 Dec 2019).

33. Retrieved from https://www.worlddata.info/asia/turkey/climate-malatya.php (Accessed: 8 Dec 2019).
34. Balo F. Energy and economic analyses of insulated exterior walls for four different city in Turkey. *Energy Education Science and Technology Part A: Energy Science and Research* 26:2 175–188.
35. Van Wijngaarden W.A., Vincent L.A. Trends in relative humidity in Canada from 1953–2003. Report prepared for Climate Research Branch, Meteorological Service of Canada, 2003.
36. https://www.vaisala.com.
37. Lee H.-J., Kim S.-Y., Yun C.-Y. Comparison of solar radiation models to estimate direct normal irradiance for Korea. *Energies* 10 (2017) 594.
38. "Passivhaus Institut.". http://passiv.de/en/02_informations/02_passive-houserequirements/02_passive-house-requirements.htm. (Accessed: 26 Oct 2018).
39. Clarke J.A. *Energy Simulation in Building Design*, 2nd edition. Oxford; Woburn, MA: Butterworth-Heinemann, 2001.

# 3 Green Hospitals and Sustainability

## *Case of Companion House of a Research Hospital*

*Figen Balo, Lutfu S. Sua, and Hasan Polat*
Firat University

## CONTENTS



## 3.1 INTRODUCTION

In recent years, businesses in all sectors have linked not only sustainability and profitability to product or service quality but also to the variables such as social responsibility, education, and environment too. They feel the obligation to implement social responsibility practices to obtain "social profit" along with the economic profit [1]. For environmental, economic, and social sustainability, it is inevitable to observe developments in health institutions that the society needs. The concept of sustainability is being used for maintaining the existence of a society or any system with continuity without overloading on its main resources [2]. Sustainable development can be achieved when environmental, economical, and social sustainability are achieved. The creation of a sustainable urban environment also creates a sustainable social life. A sustainable urban development encompasses all aspects of social justice, sustainable economies, and environmental sustainability [3].

Health buildings are important areas to ensure the continuity of the sustainable services that are required for social life, and they are built to increase the quality of life by maintaining the mental and physical health of the individuals. In hospitals, which are one of the health care facilities where health services are carried out, it is important to provide additional spaces for relatives who are taking care of patients, while maintaining the care and treatment of patients [4]. Companion houses with accommodation facilities in hospitals that seek solutions and develop new insights to provide quality health care services have become a necessity [5]. Especially in hospitals where people in rural areas apply for health problems, such buildings should be provided to the service of patients and their companions in the most appropriate way during the periods when they need to be in the hospital. In all scientific researches on this subject, it has been shown that physical and environmental conditions in hospitals and the companion areas attached to them have a significant effect on patient-companion satisfaction in the short term and on public health and social sustainability in the long term [6]. The internal and external spatial arrangement and physical conditions of hospitals are among the effective criteria in satisfaction assessments of individuals benefiting from the service. Inadequate room conditions, sloppy-planned hospitals, and lack of accommodation areas that are close to the hospital for companions have been subjects to complaints and dissatisfaction of companions and patients [7]. Therefore, a companion house with a quality and convenient accommodation is important for a hospital to reduce the negative psychological effects as well [8]. The quality of health care service delivery approach which gained importance recently in Turkey is supported by the "Health Transformation Program" enacted by the Ministry of Health in 2003 and service delivery parameters have been revised [9]. With this conversion program, a restriction across the health care system has started. The new regulations aim to provide health services carried out in the country more effectively and efficiently not only in terms of service but also in terms of patient and companion comfort [10]. For this purpose, a set of guiding principles such as sustainability, continuous quality improvement, people-centeredness, participation, reconciliation, volunteering, separation of powers, decentralization, and competition in service have been adopted [11,12].

Sustainability perspective has been used both in terms of urban and global point of views. Urban sustainability aims to design and put into service the spaces of hospital to meet the needs and demands of patient-companion relatives. Quality of urban sustainability in health care buildings can be defined as the ability of the services provided in accordance with the international standards of indicators to fulfill the expectations and needs of patients and patient relatives [13,14]. All hospitals that support the principle of providing sustainable and quality service in health with the necessary accreditation certificates in accordance with the international level will also have parameters that contribute to social and economic sustainability.

In this sense, urban sustainability requires the production of high-quality projects in which spatial planning for hospitals provides accommodation opportunities for patients as well as patient relatives who accompany them. Global sustainability, on the other hand, aims to build energy-efficient and environmentally friendly buildings, along with the rapid depletion of fossil resources in recent years.

The buildings account for 40% of the final fuel used worldwide [15]. The world has dedicated itself to reducing its emissions in the near future in an attempt to mitigate climate change [16]. Buildings will have to become considerably more energy efficient in order to achieve this goal [17]. This is backed by ecological policies that aim for nearly zero energy (or zero carbon) for all existing public sector buildings, hospitals, schools, and even all new buildings [17–19].

The design process often progresses from stand-alone and paper-based Computer-Assisted Design to Building Information Modeling (BIM) approaches. This means that all documents, records, and data about the project and resources must be stored in an electronic form, with some sharing of data [20]. The BIM is a well candidate as a basis for extracting data from various domains, such as energy or daylighting [21,22]. It also provides an opportunity to store information centrally, thereby reducing replication and duplication of data. Designers can iteratively develop their building as improvements to a building which could impact energy demand in real time. Work on the interoperability of BIM and Building Performance Simulation methods, however, shows several problems that remain unresolved [23]. This suggests that information is often transmitted using the Industry Foundation Classes and the gbXML schema. The advantage of the gbXML schema is that a lot of BIM and energy tools support it [24]. The Industry Foundation Classes system, however, has been described as the "non-proprietary, well-developed, and only free data model for existing architecture and buildings" [25]. It has a larger scope than gbXML and considered to be an important factor [26].

At the building design point, green building practitioners face a major challenge in fulfilling these standards, as evaluations are typically less reliable, unreliable, and time- and cost-consuming in terms of resources [27]. BIM's creative implementation has created incentives in recent years to support green building practices and is known as green BIM. A central concept of green BIM is "a model-sourced system to produce and manage organized and consistent information that promotes the achievement of defined sustainability targets" [28]. By this concept, green BIM encourages numerous quality evaluations and analyses such as carbon emission, acoustic analysis, waste management of building and demolition, lighting analysis, use of operating resources, and use of water. Because BIM includes multidisciplinary information for various analyses, sustainability metrics can be superimposed on a three-dimension modeling to assist credit assessment in Green Building Assessment Schemes [29].

In the early design stage of a project, storing and handling Green Building Assessment Schemes rating information within BIM resources can then help make practical design decisions using real project data. A large amount of academic research on green BIM has been conducted over the past two decades. Gao et al. submitted a BIM-based Building Energy Modeling analysis for the production of energy-efficient construction designs [30]. For example, between 2015 and 2018,

Pezeshki et al. published a comprehensive analysis of green BIM literature concentrating on the use of BIM software in Building Energy Modeling [31]. Sanhudo et al. proposed an analysis of BIM's energy retrofitting engineering capability [32]. Kamel et al. discussed the interoperability problems and approaches between the modeling systems of BIM and Building Energy Modeling [33]. From the perspective of building life cycle, Wong et al. presented a review of green BIM literature [34].

While significant contributions to mapping green BIM have been made by the abovementioned studies, none of them discussed practical studies on the use of BIM to determine Green Building Assessment Schemes requirements and gain sufficient credits. Lu et al. conducted a comprehensive review on the application of BIM to environmental assessment by the life cycle of the building project and various roles of BIM [35]. Nevertheless, their analysis concentrated less on combining BIM with the Green Building Assessment and failed to address the variety of requirements that could be met with BIM. In contrast, none of these studies considered commercial implementations for the evaluation of Green Building Assessment Schemes requirements and credits. Certain researches such as those carried out by Azhar et al. addressed the use of BIM in the Green Building Assessment Schemes criteria evaluation [36]. The report concluded that BIM did not achieve access to public transportation, rapidly reusable resources, or product recycling requirements, while BIM effectively tested these criteria [37–39].

In this study, Firat University Hospital building and an additional companion house with accommodation facilities were projected. The building was built considering the parameters that determine the framework of the Transformation Program of Ministry of Health. The building was projected in a way to contribute to the global energy, social life, and the developments in urban design within the framework of "urban and global sustainability." The project was designed by including the patients as well as the patients' relatives into the planning process, thus adding value to the social structure and continuous quality development. The qualifications of the patient rooms were determined according to the 2010 Guide for Health Buildings Minimum Design Standards of Turkey [13] and the design was carried out with these references. In the first stage of the study, the design criteria of the companion house with accommodation facilities which was designed in accordance with the criteria of the Ministry of Health were presented with all the details of the project. In the second phase of the study, the energy efficiency was evaluated comparatively if the building was constructed in accordance with the green hospital concept with green roof and green walls.

## 3.2 STRATEGIC POSITION OF FIRAT UNIVERSITY SCHOOL OF MEDICINE

Firat University School of Medicine Hospital has been serving since 1983 to the city where it is located as well as the surrounding cities. Firat University School of Medicine has trained more than 2,500 physicians to date. There are 79 professors, 37 associate professors, 70 faculty, 8 experts, and 314 research assistants working at the available 3 departments and 45 majors with 37 minors under these departments.

The University Hospital also provides tertiary services in its campus building. In the hospital, 55 units, including emergency and pediatric emergency units, provide outpatient services, 40 of which also offer inpatient treatment. The bed capacity of the medical center is 962. The number of people diagnosed and treated is around 300,000 annually.

Firat University Hospital receives intensive patient and treatment demand from the city as well as from neighboring provinces due to its central geographic location and service quality. Therefore, due to its position as a regional hospital, there are intensive patient applications.

However, after the hospitalization procedures, a significant portion of the patients' relatives spend the night in the hospital garden, corridors, waiting units, and emergency cafeteria due to their economic limitations and this situation sometimes continues for weeks.

The university has planned to build a companion house in order to solve this problem and has implemented it to contribute to the city. In order to cover the expenses and needs of the structure, commercial areas (pharmacy, restaurant, cafeteria) on the ground floor and a paid parking garage at the back of the building were planned. Thus, by using the rental income of the commercial areas and the closed parking fees for the companion house, it was planned to accommodate the relatives of patients who do not have financial means. It was also planned to provide significant contributions to the city both in terms of health services and sustainability. Within the framework of the planning process, it was aimed to reorganize the hospital companion house in order to create a start for the green hospital concept. This way, the project which was built to support urban sustainability, was re-projected by using green coatings on walls and roofs in order to determine the extent to which it can support global energy efficiency. The effects of the second design on energy efficiency were investigated comparatively. This study is intended to be the basis for future evaluation for the entire hospital. This way, it was aimed to raise awareness on the importance of this issue for other hospitals.

## 3.3 PROJECT ANALYSIS OF COMPANION HOUSE WITH ACCOMMODATION FACILITIES

The green hospital movement started years ago after the release of Energy and Environmental Design standards for building construction by the US Green Building Council. Although the initial cost of adopting green practices may be higher, such practices are in the best interest of the facility in the long run. Furthermore, there is a growing consensus among the health care professionals that pollutants produced by medical facilities must be diminished. In addition, green hospital development was related to better patient outcomes and retention of staff. A number of newly built and renovated hospital buildings have been receiving and pursuing certification for Leadership in Energy and Environmental Design (LEED) over the past few years. The green hospital is a hospital that takes the initiative to do one or more of the following things: utilizes efficient and sustainable designs, chooses an eco-friendly place, considers green in the course of construction, maintains the greening process, and utilizes green construction products and materials. A green hospital builds around a facility to recycle, reuse materials, reduce waste, and produce cleaner air [40].

In this study, the building with green concept is designed according to the criteria required by the green hospital concept. Then, the existing building and the building with green concepts are compared in view of energy performance.

### 3.3.1 Analysis in the Light of Project Criteria for Sustainability

The building is in the form of a single mass due to lack of available land. It consists of three basic functions: commercial workplaces (pharmacy, restaurant, and cafeteria), accommodation (for companions), and indoor parking. The building which was built just south of the existing hospital building is designed as 13,400 $m^2$ with a basement floor of 2,680 $m^2$, ground floor, and three upper floors.

The sections facing the main street (east direction) and the existing hospital building (north direction) are planned as a companion house on the upper floors and commercial areas on the ground floor. The sections facing the rear (west direction) are arranged as a closed and multi-story car park with access to the entrance and exit of the school of medicine at the bottom of the existing hospital building. The entrance and exit of the parking lot at the rear are given from the basement floor by taking advantage of the elevation difference. In the layout plan, three buildings in the back belong to the medical school. Located on the same level as the medical school, the indoor parking lot is reserved for the use of faculty members and staff. The layout plans for the companion house are provided in Figures 3.1–3.6.

The basement parking lot allocated for faculty members has a capacity of 68 vehicles. In the basement, there are also technical spaces for the companion house. Among these units, there is a laundry where the bed linen of companion house and companion clothes can be washed. In addition, warehouses, boiler room, electricity and transformer room, and water tank with hydrophore are arranged within this area. Vertical circulations from the technical spaces to the companion house and to the upper floors in the parking lot are provided through stairs and elevators. On the ground floor, there is a parking lot for 58 cars serving the hospital and a ramp for

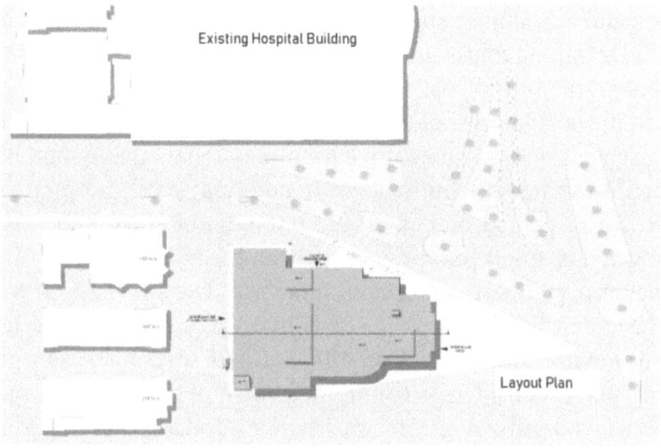


**FIGURE 3.1** Layout plan.

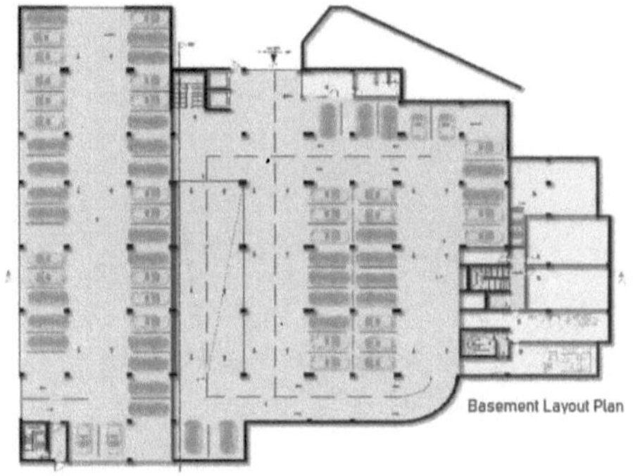


**FIGURE 3.2** Basement floor plan.

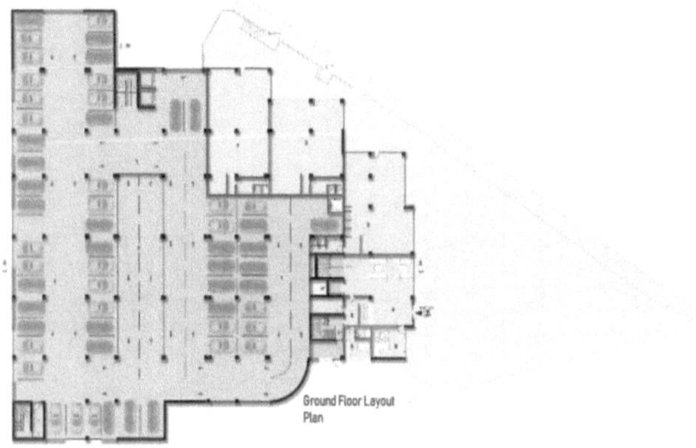


**FIGURE 3.3** Basement floor plan (together with close environment).

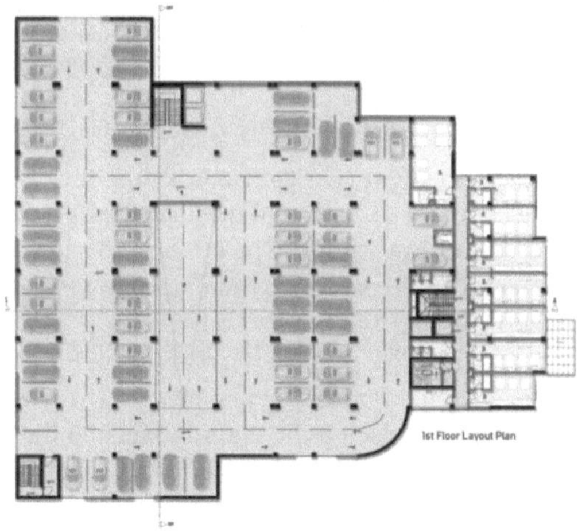


**FIGURE 3.4** 1. Floor plan.

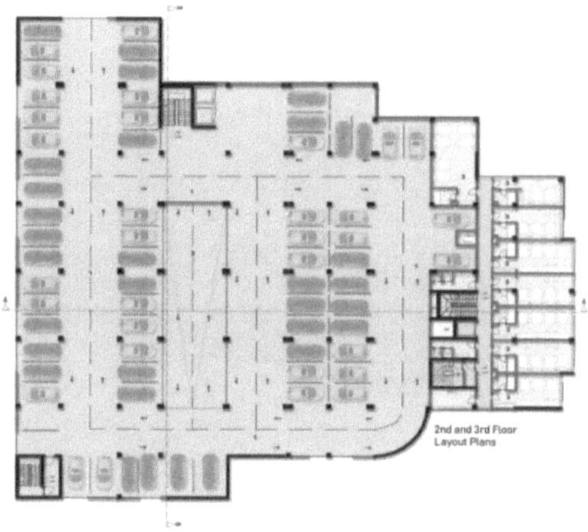


**FIGURE 3.5** Second and third floor plans.

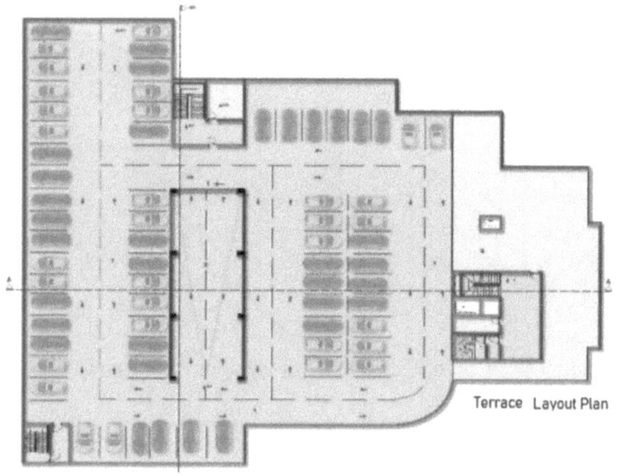


**FIGURE 3.6** Terrace floor plan.

exits and descent to the upper floors. On the front, facing the main facade and the hospital, there are three commercial areas and the entrance of the companion house. Commercially planned spaces are 156 $m^2$, 135 $m^2$, and 130 $m^2$. There is an 80 $m^2$ lobby, 15 $m^2$ reception area, 20 $m^2$ executive room, and 18 $m^2$ room with restroom-shower facility for the officer on duty.

On the first, second, and third floors, there is a parking lot for 195 vehicles with 65 vehicles on each floor and a ramp for landing and descent to the upper floors. There are eight bedrooms on each floor facing the main facade and the hospital. There is one bedroom for two people, two for three people, four for four people, and one for six people on each floor. Each floor has a capacity of 30 beds. All of the rooms have restroom shower, and there is a communal kitchen on the floors where companions can prepare food and drinks. There are also two communal restrooms on each floor.

The penthouse is planned as a terrace. The main reason for this is that the terrace area is also considered as a parking space. This terrace with a capacity of 65 vehicles is also the floor where the entire building ends. When designing the building facades, it is essential to provide integrity with the existing hospital building, which is functionally and organically related. The existing hospital building is heavily composed of glass surfaces. Solid surfaces of the building are clinker coatings. However, in order to emphasize the difference of design and construction periods, low-density andesite elements and horizontal chrome profiles connecting these elements were used when making the facade design. Figures 3.7–3.12 provide the views of the companion house from different perspectives.

In the facades where there is an indoor parking lot, partial parapet walls and metal railings were used to get out of the ordinary. This design has both resulted in an esthetic appearance with simple touches that will disrupt the ordinariness of the facade and enable partial view of the parking lot from outside.

Figures 3.13–3.17 show the perspectives of the companion house from different directions whereas Figure 3.18 provides the view of the completed structure.

Table 3.1 provides the information for the companion house project designed for the hospital.

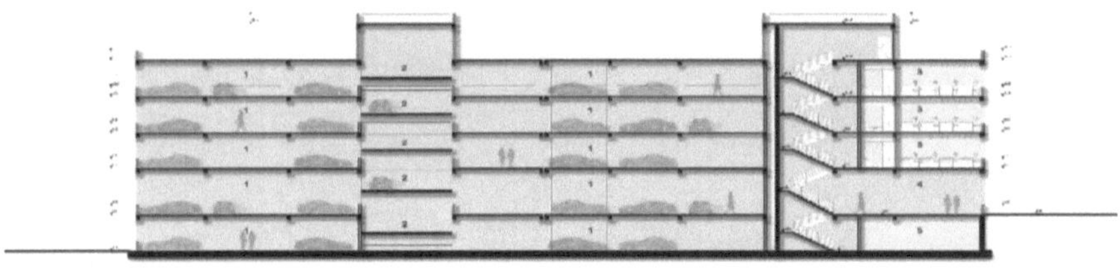

**FIGURE 3.7** A-A view.

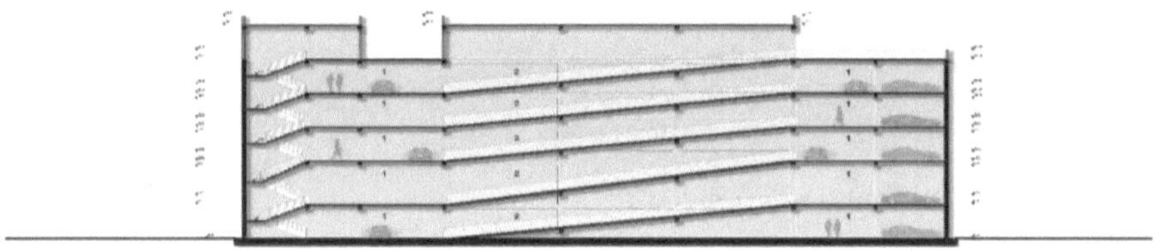

**FIGURE 3.8** B-B view.

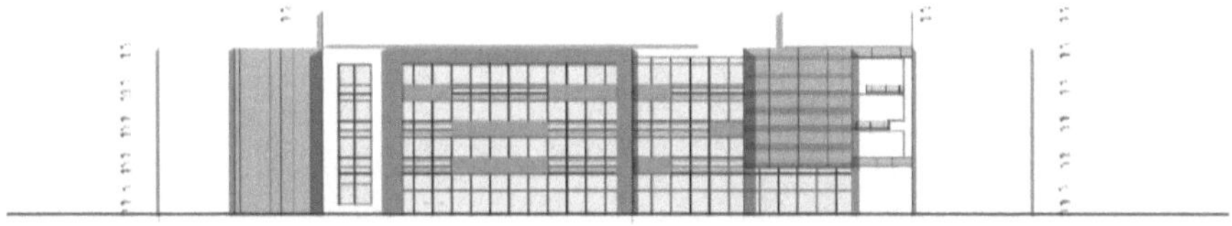

**FIGURE 3.9** Front facade (east view).

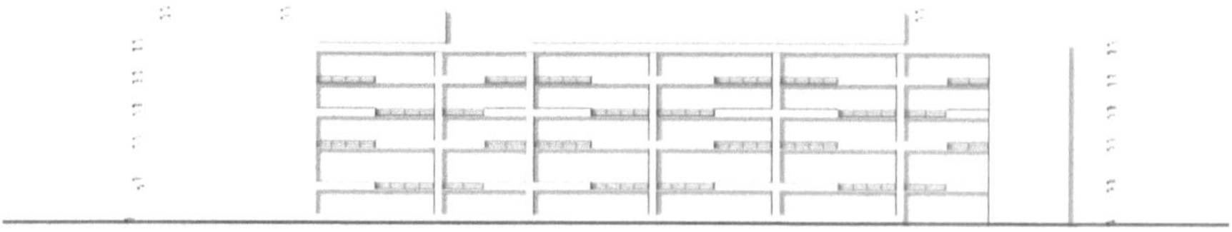

FIGURE 3.10 Rear facade (west view).

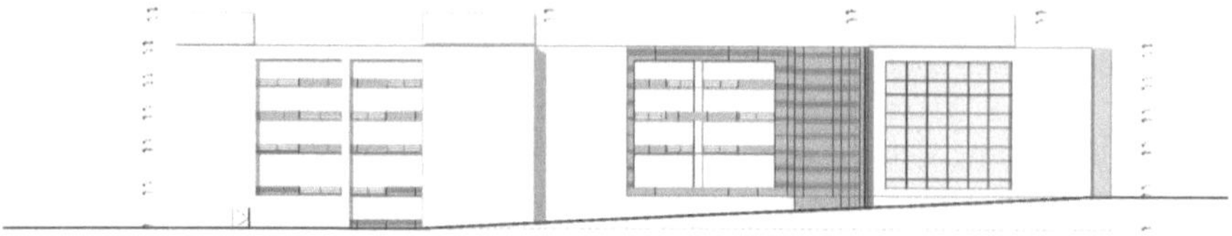

FIGURE 3.11 Left side facade (south view).

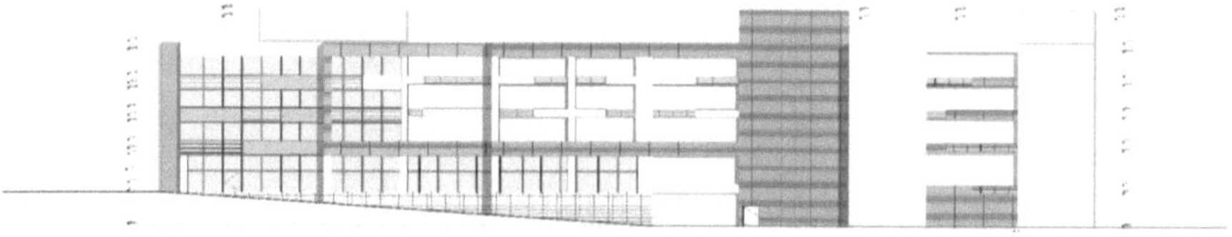

FIGURE 3.12 Right side facade (north view).

FIGURE 3.13 Perspective from southeast.

FIGURE 3.14 Perspective from southwest.

**FIGURE 3.15** Perspective from northeast.

**FIGURE 3.16** Perspective from east (main entrance and commercial areas).

**FIGURE 3.17** Perspective from northwest.

**FIGURE 3.18** View of completed structure.

**TABLE 3.1**
**Project Information**

| Architectural Design | Hasan POLAT, Dr. Architect |
|---|---|
| Application project | Hasan POLAT, Dr. Architect—Cetin IRMAK, Architect |
| Static project | Mirac GOKDOĞAN, Civil Engineer |
| Mechanic project | Cigdem BOZDAG, Mechanical Engineer |
| Electrical project | Erhan AZTEKIN, Electrical Engineer |
| Employer | Firat University Rectorate |
| Control | Firat Engineer Physical Plant and Technical Department |
| Contractor | Han Engineering |
| Contract date | June 7, 2017 |
| End date | October 25, 2018 |
| Contract amount | 8,638,000 TL |

### 3.3.2 Projecting in the Light of Green Hospital Criteria for Global Sustainability

Additional service projects such as providing free accommodation at a location close to the patients have started to be considered in the design, implementation, and operation processes of hospital projects. However, when considered globally, ecology, environment, and energy resources are also among the other important factors that need to be considered in building designs. In the construction sector, which is one of the sectors where energy is used most intensively, attention should be paid to this issue with the case studies about energy saving that can be provided by green concept and energy-efficient buildings. For the sectors that consume intense energy, these types of structures will be a valuable step for more conscious energy consumption. Moreover, it has been proven with many researches in recent years that increasing ecology, environment, natural life, and health concerns can be addressed by supporting structures designed with greener designs. Figure 3.19 presents the projected views of the existing hospital building in case of designing with green concept.

### 3.3.3 Energy Efficiency Analysis for the Existing Building and the Building Designed According to Green Hospital Concept

#### 3.3.3.1 Wall Components for the Existing Building and the Building Designed According to Green Hospital Concept

This section aims to provide information about the various components of a "green concept" used in the buildings studied in this research. Technical properties of the wall and glass materials are provided.

The wall section view of existing building is given in Figure 3.20. The analytical properties of the wall of existing building and technical properties of the wall components of the existing building are presented in Tables 3.2 and 3.3, respectively.

FIGURE 3.19 Views of the existing building in case of designing with green concept.

View of glass facade and green facade section is shown in Figure 3.21, while the green facade technical properties are displayed in Table 3.4.

The view of glass facade section of aluminum-framed entrances is shown in Figure 3.22. Glass material technical properties are given in Table 3.5.

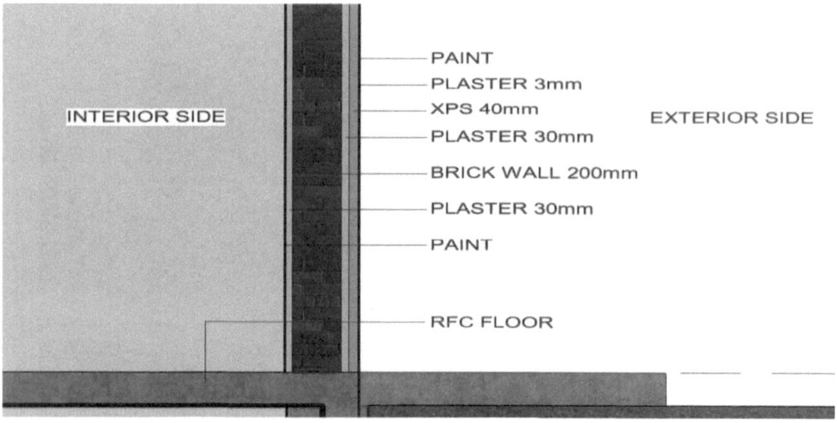


**FIGURE 3.20** The wall section view of the existing building.

**TABLE 3.2**
**The Analytical Properties of the Existing Building's Walls**

| **Analytical Properties** | |
|---|---|
| Heat transfer coefficient (U) | 0.5433 W/(m²K) |
| Thermal resistance (R ) | 1.8407 (m²K)/W |
| Thermal mass | 16.01 kJ/K |
| Absorptance | 0.7 |
| Roughness | 3 |

**TABLE 3.3**
**Technical Properties of the Wall Components of the Existing Building**

| | **Plaster** | **Brick** | **XPS** |
|---|---|---|---|
| Thermal conductivity [W/(mK)] | 0.510 | 0.330 | 0.0360 |
| Specific heat [J/g°C] | 0.960 | 0.84 | 0.96 |
| Density [kg/m³] | 1.120 | 600 | 100 |
| Emissivity | 0.9 | 0.95 | 0.9 |
| Permeability [ng/(Pa.s.m²)] | 0 | 182.4 | 0 |
| Porosity | 0.01 | 0.01 | 0.01 |

### 3.3.3.1 Autodesk Revit Architecture Simulation Software

Building design with reduced energy consumption and high-quality indoor environment at an acceptable cost is often achieved by predicting the buildings' energy efficiency. The energy efficiency of a building is often predicted by constructing a Building Energy Model for Energy Simulation results.

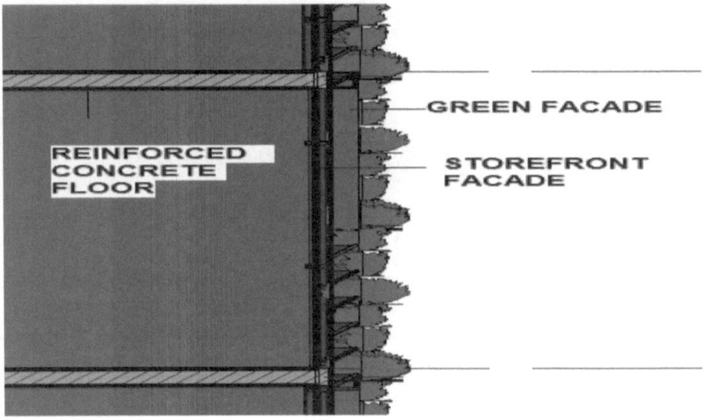


**FIGURE 3.21** View of glass facade and green facade section.

## TABLE 3.4
## Green Facade Technical Properties

| | Green Facade Technical Values | Plant on the Walls |
|---|---|---|
| Thermal conductivity [W/(mK)] | 0.331 |  |
| Specific heat [J/g°C] | 2.0920 | |
| Density [kg/m$^3$] | 920 | |
| Emissivity | 0.95 | |
| Permeability [ng/(Pa.s.m$^2$)] | 3.4200 | |
| Porosity | 0.01 | Hydrangea |

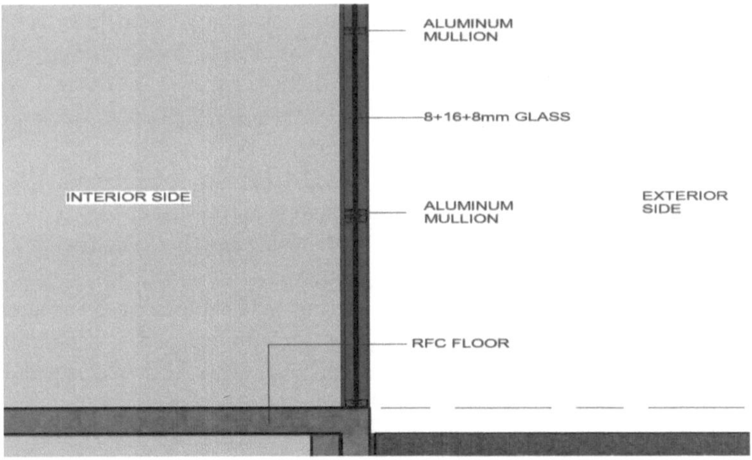


**FIGURE 3.22** View of glass facade section of aluminum-framed entrances.

**TABLE 3.5**
**Glass Technical Properties**

| Glass | C.O.G.[b] U-Factor | U-Factor[a] | Frame CRF[c] |
|---|---|---|---|
| 1″ IG | 0.47 | 0.58 BTU/hr·ft²·°F (3.29 W/m²·K) | 59 |

[a] U-factor and condensation resistance (CR) are based on a nominal size of 47.25″ (1,200 mm) × 59″ (1,500 mm) with 2L of glass using NFRC-100 and 500 - 2010.
[b] Intercept® Spacer.
[c] Based on AAMA 1503.1.

Using the advanced BIM tools, the analysis of the existing building and the building designed with green concept is conducted with the help of Autodesk Revit Architecture Simulation Software that can do energy analysis by considering all the details of the architectural project, all parameters related to the environment and ecology.

The energy model is an abstraction of a computational network of the overall form and structure of a building. This network collects all of the building's main heat transfer paths and processes.

The energy usage determined with Autodesk Revit Architecture Simulation Program provides guidance for improved building energy efficiency throughout the project, from concept to detail and layout to existing and new buildings service. The following strategies are used by it:

- An intuitive, range-based approach to the design of energy management
- Automatic energy model design, based on the architectural Revit model
- Innovative methods for parallel cloud computing
- Branch leading simulation engines (DOE 2.2 AND EnergyPlus) well developed and trusted.

Depending on the energy schema of building, the energy modeling has three primary components:

- Surfaces are the heat transfer paths to or from through room, including surfaces between indoor and outdoor spaces.
- Zones are groups of spaces that are used to create some commonality between these spaces. Zones may, for example, represent spaces with the same orientation, the same purpose, or the same Heating, Ventilating and Air Condititioning (HVAC) system operation.
- Spaces are distinct quantities of air suffering and heat loss. Such variations in heat are caused by internal processes such as occupancy, ventilation, appliances, and HVAC, as well as heat exchange with other spaces and the outside environment.

For example, as a computational network in the energy modeling, the following diagram in Figure 3.23 displays spaces (1), surfaces (2), and zones (3) [41].

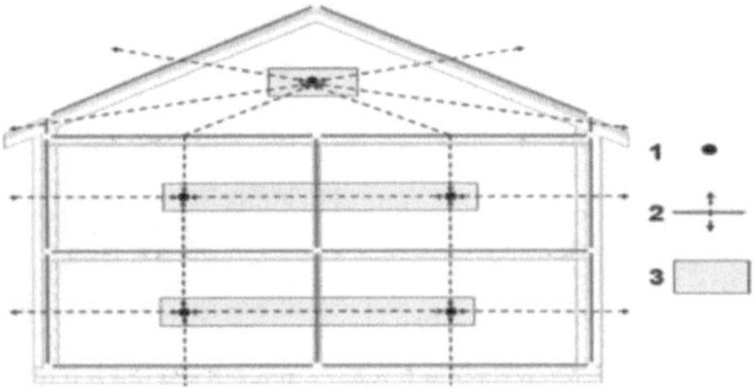


**FIGURE 3.23** The parameters in the energy modeling.

Bahar et al. displayed data exchange's summary between thermal simulation and energy modeling tools (including SketchUp, Revit and ArchiCad) and commonly utilized BIM tools [42]. Autodesk Revit Architecture Simulation Program's application fields are generally natural ventilation, carbon emissions, thermal analysis, water cost and usage, environmental design, yearly energy consumption, life cycle costing, daylighting, etc. Their building information transformations are simplified material information thermal region and building geometry, etc.

#### 3.3.3.2 Environmental and Climatic Factors Affecting the Analysis

Autodesk Revit Architecture Simulation Program presents passive building design climatic information and illustrates the principle of BIM. It uses the following climatic parameters to evaluate weather data: wet and dry bulb temperature, diffuse and direct solar, prevailing winds, and relative humidity. In Autodesk Revit Architecture Simulation Program, the location of the building will be listed on the Location Weather and Site dialog page. When using the Default City chart, the selected location's longitude and latitude coordinates are compared to an existing weather station database of World Meteorological Organization derived from the *2005 ASHRAE Fundamentals Handbook*. Together with its weather data, the nearest World Meteorological Organization weather station is selected and the values are displayed on the Climate section. For the comfort prediction, temperature and humidity are the most significant climatic parameters. For the city of Elazig, the dry frequency distribution and dry-bulb cumulative distribution in analysis are given in Figure 3.25. According to Figure 3.24, the maximum dry-bulb temperature is approximately 54.5°F.

Wind encourages natural ventilation and is a passive method for cooling in the summer. For Elazig, the effect of wind roses in analysis is given in Figure 3.25. This figure shows that the primary winds direction are west-northwest. The frequent wind speed is obtained to be between 3.9 and 5.8 in annual frequency distribution. Revit analysis results show accordingly in figure that Elazig climate blows from east-southeast directional wind in winter period and exhibits the most prevailing wind from west-northwest in summer period.

The effect of dew point frequency distribution and relative humidity frequency distribution (annual) in analysis is displayed in Figure 3.26. The selection of bio-climatic improvements is based on climate analysis of Elazig and sensibility study.

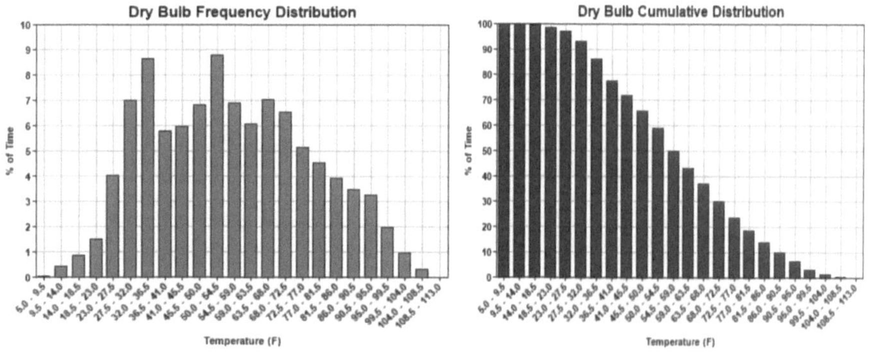


**FIGURE 3.24** The effect of dry frequency distribution and dry-bulb cumulative distribution in analysis.

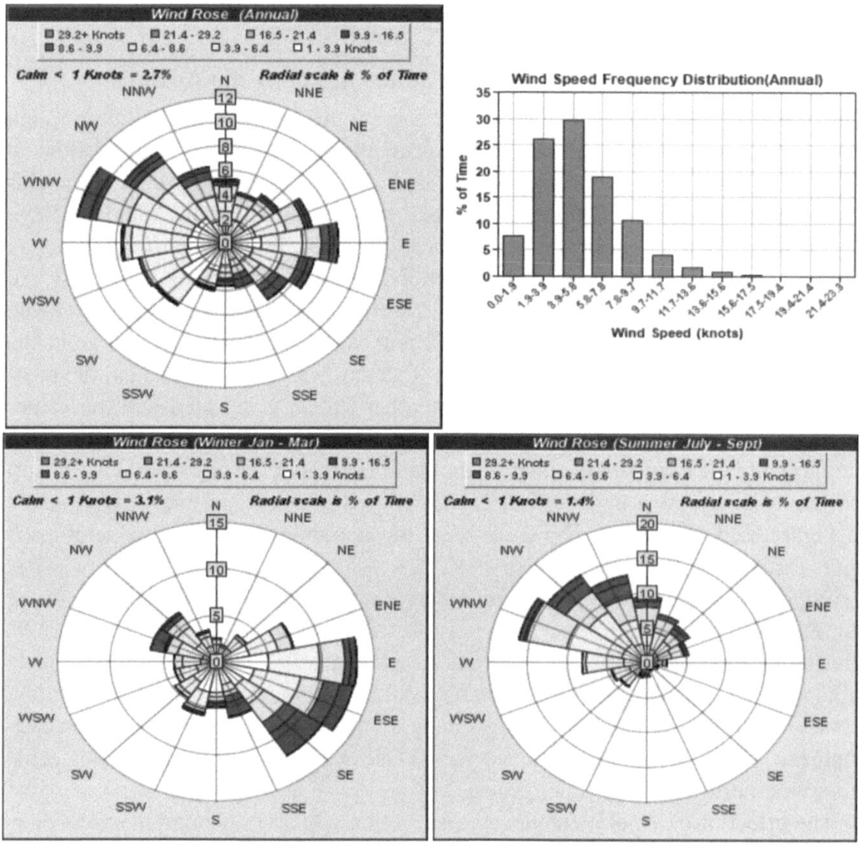


**FIGURE 3.25** The effect of wind roses in analysis.

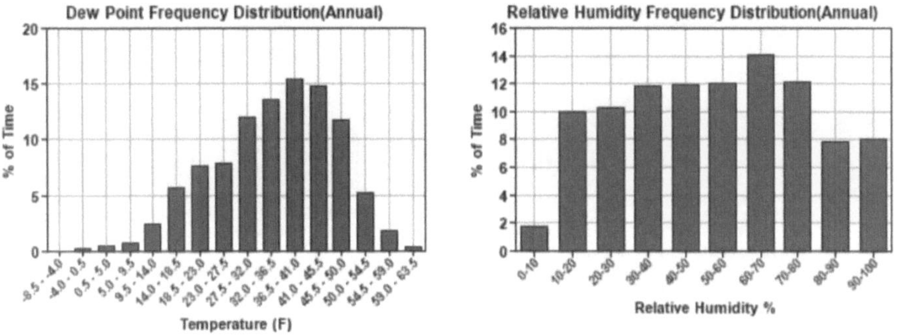


**FIGURE 3.26** The effect of dew point frequency distribution and relative humidity frequency distribution (annual) in analysis.

This sensibility is determined depending on the dew point frequency distribution and relative humidity frequency distribution. Thus, the achieved results show that the efficient bioclimatic design techniques of Elazig climate are limited to values in Figure 3.26.

The area covered by the clouds in the sky is called sky cover. The percentage of the sky covered with the clouds needs to be determined in order to calculate it. During the observation, the areas of the clouds that do not cover the sky completely are added for the purpose of this calculation. The total area in the sky is assumed to be 10/10 in climatology observations. Thus, a scale from 0 to 10 is used. According to this scale, a value of 0 represents a sky without no cloud, while a value of 10 represents a sky that is completely covered with clouds. Sky cover of 1 is recorded when slight trails of clouds are observed in the sky. Thus, less than 1/10 of the sky must be covered with clouds. On the other end, a sky cover of 9 is recorded when the sky is almost completely covered with clouds. In this case, the sky can only slightly be seen through the clouds. For the other amounts of clouds in the sky, an appropriate level of the scale (2, 3, 4, 5, 6, 7, and 8) can be recorded. For the night time, on the other hand, sky cover is determined based on the ratio of stars covered with the clouds. In the case of observer being under the influence of nature events such as fog and snow storm, sky cover is recorded as 10 due to the inability to estimate the covered portion of the sky. While analyzing the energy savings of the building under investigation here, annual total sky cover frequency distribution values that are considered within the framework of Revit is significantly important for an accurate assessment [43].

The effect total sky cover frequency distribution and direct normal radiation frequency distribution (annual) in analysis are presented in Figure 3.27. The direct normal radiation is the amount of sun irradiation obtained per unit field by a superficies that is always kept normal (or perpendicular) to the rays that come straight from the sun's direction at its existing position in the skies [44]. The reliable data on sun irradiation must be provided in advance as the most significant input to the energy efficiency simulation. The most beneficial sun irradiation data is global horizontal radiation but diffuse horizontal radiation or direct normal radiation is also significant.

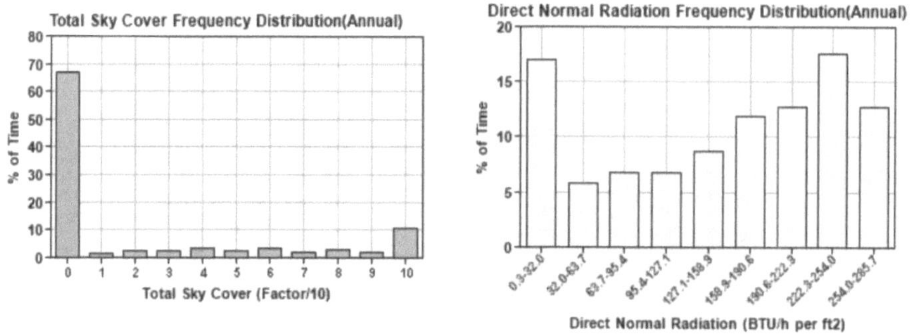


**FIGURE 3.27** The effect of total sky cover frequency distribution and direct normal radiation frequency distribution (annual) in analysis.

For instance, radiation on the superficies of a sun cell or sun collector is defined when either diffuse horizontal radiation or direct normal radiation is provided in addition to the global horizontal radiation [45]. In this study, the simulation program utilizes these three radiation values. The direct normal radiation frequency distribution chart shows how much direct sunlight is available over a 12-month period.

Choosing appropriate weather data in energy analysis results in better prediction of energy. The reliability of simulation results using weather data (typical-meteorological-year) was investigated in the design stage. In addition to the weather data, other realistic development considerations need to be discussed. The temperature and angle of the device also influence the output of electricity generation. BIM-based measurement of energy is appropriate to use in this respect [40]. The effect diffuse and global horizontal radiation frequency distributions in analysis are displayed in Figure 3.28. The total amount of shortwave radiation emitted from above by a horizontal superficies to the earth is the global horizontal radiation. This is of particular interest to photovoltaic installations and involves both horizontal diffuse radiation and normal direct radiation. The diffuse horizontal radiation is sun irradiation that does not come from the sun on a direct path but has been spread in

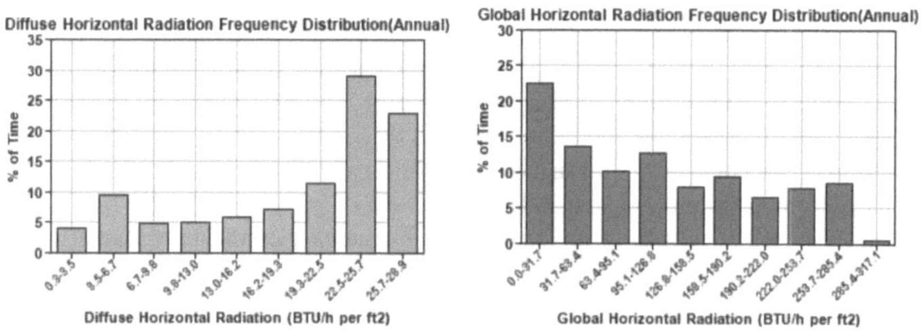


**FIGURE 3.28** The effect of diffuse and global horizontal radiation frequency distribution (annual) in analysis.

the atmosphere through particles and molecules and comes evenly from all aspects. In essence, it is the lighting that arrives on the blue sky and the clouds [45].

The threshold graphs show a well-studied, multi-directionally motivated class of graphs. It is necessary and vital to the well-being of its population to determine the threshold temperatures suitable for providing climatic and thermal comfort. The most common type of degree-day measurement depends upon indoor cooling and heating-related thresholds [46]. An effective and important predictor is cooling and heating degree days in various aspects. The index of the degree day as one of the most practical and simple indices is to determine the energy needed to provide comfort climate. The maximum mean deviation of the average temperature for the human comfort is known as the daily temperature of the threshold level. At temperatures greater than the minimum temperature, the atmosphere needs to be cooled while the air needs to be heated at temperatures lower than the minimum. These are known as the cooling and heating requirements [47]. The heating and cooling degree-day measurement can provide an accurate and clear picture of the building, the region, and the city's thermal needs. It also has a constructive role to play in maintaining thermal comfort and enhancing patterns of energy consumption [48]. The measurement and estimate of the mean degree-day requirement for cooling and heating as basic data for calculating the amount of energy needed to cool and heat the building in winter is a function of energy consumption planning [46]. Monthly design data (threshold of 2%) is provided in Figure 3.29.

#### 3.3.3.3 The Evaluation of Analyzed Green and Existing Building by Autodesk Revit Architecture Simulation Program

This research aims to develop the Firat University Research Hospital Companion House with accommodation facilities and to bring forward resolutions to decrease the urban landscape's energy consumption. It also aims to obtain a green building with low-energy usage, taking into account the environment and architecture

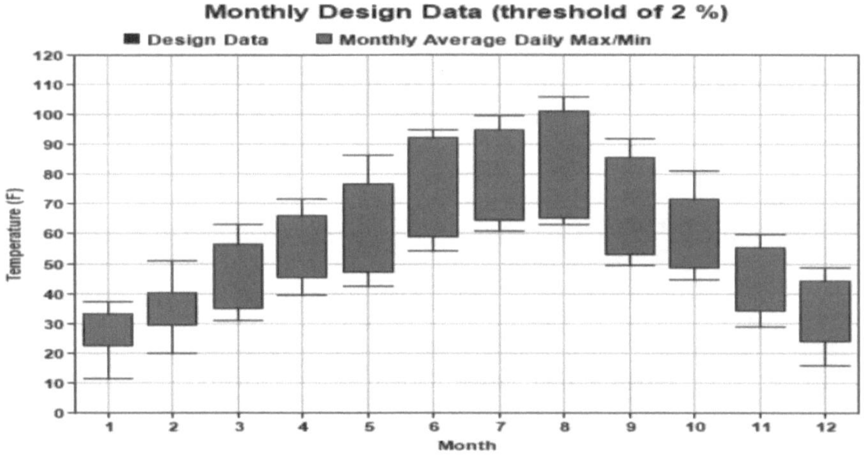


FIGURE 3.29 Monthly design data (threshold of 2%).

for the future generations. Thus, this chapter aims to decrease the cooling and the heating loads and to take advantage of responsive development approaches. Indeed, such recent strategies depend on the local climate's bioclimatic solutions. In fact, we do focal point to this chapter to obtain efficiency of existing and green building for cooling and heating load calculation by Autodesk Revit Architecture Simulation Program.

For existing building and building with green concept, the energy, carbon, cost summary and LEED, photovoltaic, wind energy, and natural ventilation potential are presented in Tables 3.6 and 3.7, respectively.

**TABLE 3.6**
**Energy, Carbon, and Cost Summary**

| | | Energy, Carbon, and Cost Summary | |
|---|---|---|---|
| | | **Existing Building** | **Building with Green Concept** |
| Annual energy cost | | 60,770 T | 55,082 T |
| Life cycle cost | | 827,684 T | 750,219 T |
| Annual $CO_2$ emissions | Electric | 0.0 mg | 0.0 mg |
| | Onsite fuel | 44.6 mg | 44.5 mg |
| | Large SUV equivalent | 4.5 SUVs/year | 4.5 SUVs/year |
| Annual energy | Energy use intensity (EUI) | 1,106 MJ/m$^2$/year | 1,030 MJ/m$^2$/year |
| | Electric | 349,189 kWh | 308,455 kWh |
| | Fuel | 893,785 MJ | 891,919 MJ |
| | Annual peak demand | 97.5 kW | 88.9 kW |
| Life cycle energy | Electric | 10,475,673 kW | 9,253,647 kW |
| | Fuel | 26,813,535 MJ | 26,757,555 MJ |

**TABLE 3.7**
**LEED, Photovoltaic, Wind Energy, and Natural Ventilation Potential**

| | | Existing Building | Building with Green Concept |
|---|---|---|---|
| LEED water efficiency | Indoor (TL/year) | 7,595 | 7,595 |
| | Outdoor (TL/year) | 621 | 621 |
| | Total (TL/year) | 8,217 | 8,217 |
| Wind potential | Annual electric generation (kWh) | 570 | 570 |
| Natural ventilation potential | Total hours of mechanical cooling required | 7,378 | 7,184 |
| | Possible natural ventilation hours (hours) | 5,191 | 4,129 |
| | Possible annual electric energy savings (kWh) | 104,805 | 59,358 |
| | Possible annual electric cost savings (TL) | 14,568 | 8,251 |
| | Net hours of mechanical cooling required | 2,187 | 3,055 |

For existing building and building with green concept, energy end use charts are given in Figure 3.30.

Many measurements of building energy costs and early enforcement decisions can be made using annual data on energy costs and consumption. The costs are measured using average utility rates across the region or country. The program uses a downloaded climate file to extract internal temperatures after constructing “thermal zones” in Revit, representing different building fields. Such hourly temperature data vary depending on the building construction assembly, as well as any structure that would impact solar benefits, such as overhang. Such temperatures are better viewed as a long-term average, such as weeks or months, rather than an hourly one. These are also often used to measure the impact of a model or construction assembly over another, not so much as a number written in stone. These temperatures are extracted from a climate report using average temperatures over a span of 30 years.

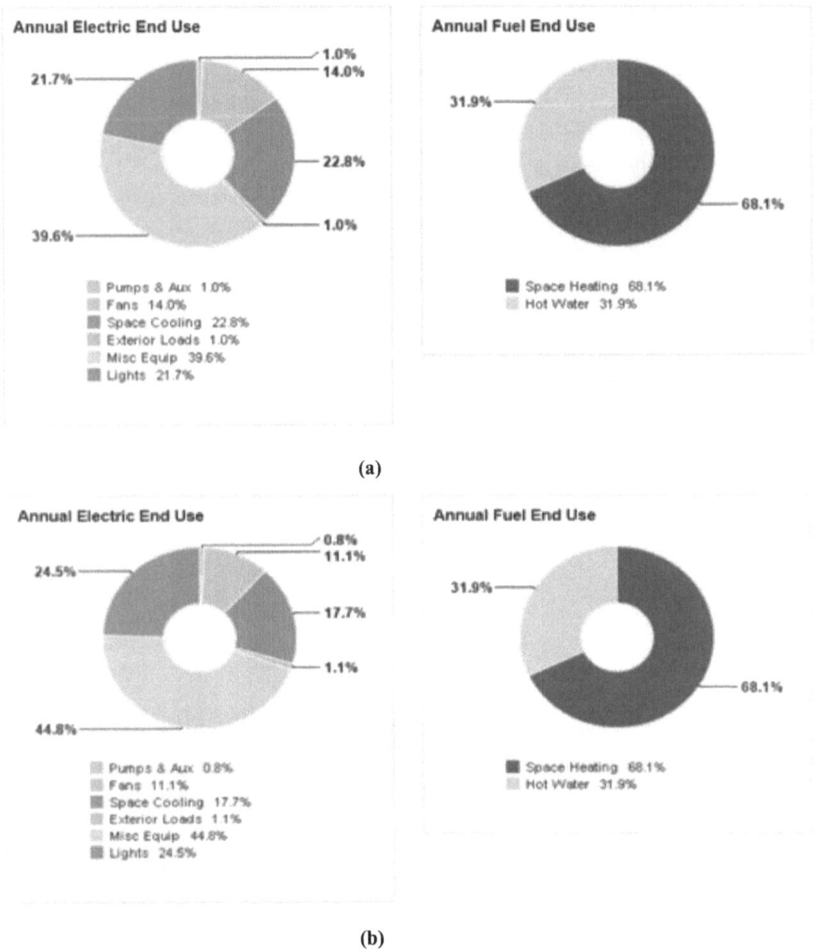


**FIGURE 3.30** Energy end use charts: (a) existing building and (b) building with green concept.

The analysis is conducted by turning the overall form and structure of buildings into a "computational network" capable of effectively capturing all the primary heat transfer paths and processes throughout the building. It evaluates the results as to the likely energy performance of building models with green versus existing concept, depending on what materials have been used in the building model's creation.

In Figure 3.31, the monthly results of total energy consumption were used as a basis for further comparison. It shows that the highest energy consumption period for cooling loads for the three-month period of existing buildings are December (61,375 kWh), January (65,382 kWh), and March (55,939 kWh). Total energy consumption value is obtained as 590,693 kWh. The highest consumption is determined as 169,155 kWh for space heat. The lowest consumption is determined as 48,883 kWh for ventilation fans. In Figure 3.32, the highest monthly results of total energy consumption on cooling loads were obtained for the three-month period of December (59,151 kWh), January (63,214 kWh), and February (53,880 kWh) for green concept building. Total energy consumption value was obtained as 549,177 kWh. The highest consumption was determined as 168,162 kWh for space heat. The lowest consumption was determined as 34,328 kWh for ventilation fans.

For existing building and building with green concept, annual and monthly energy requirements for all fuel and electricity amounts utilized are displayed in Figures 3.31 and 3.32, respectively. For existing building and building with green concept, annual and monthly energy costs are shown in Figures 3.33 and 3.34, respectively. Energy use

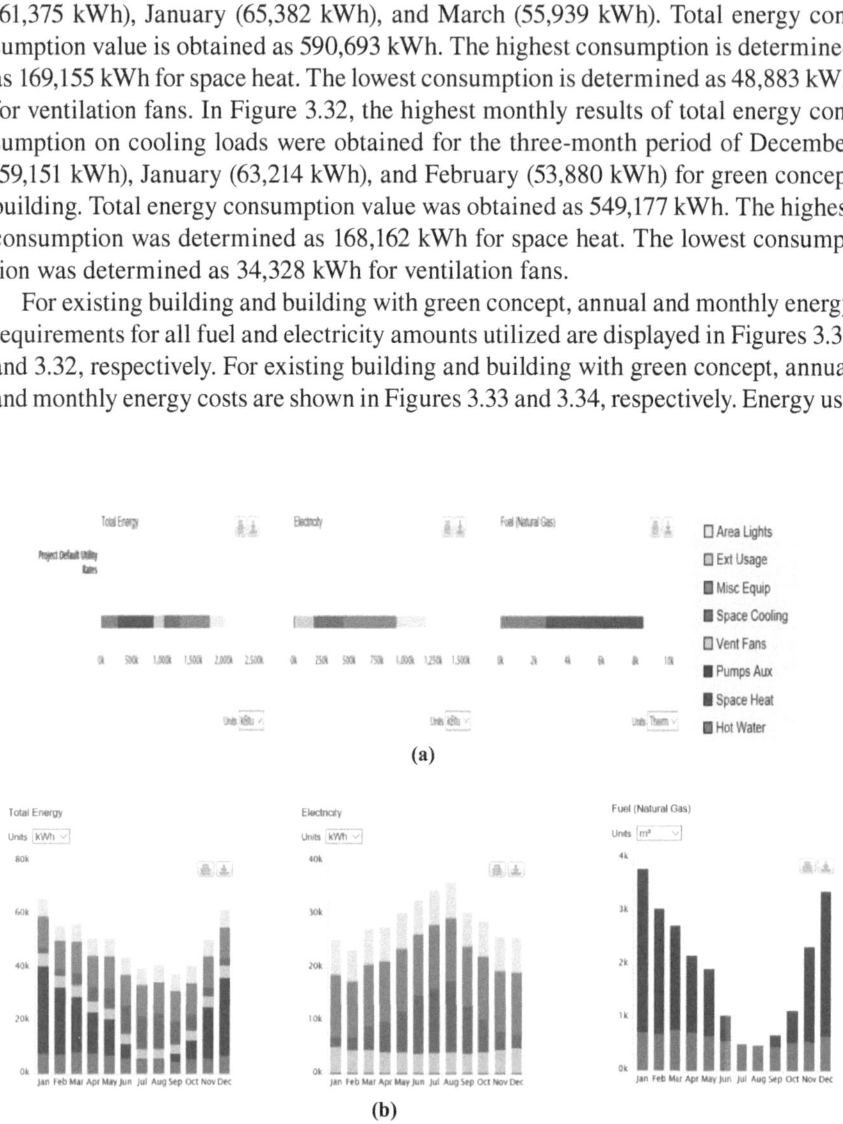


**FIGURE 3.31** The existing building: (a) annual energy requirement and (b) monthly energy requirement.

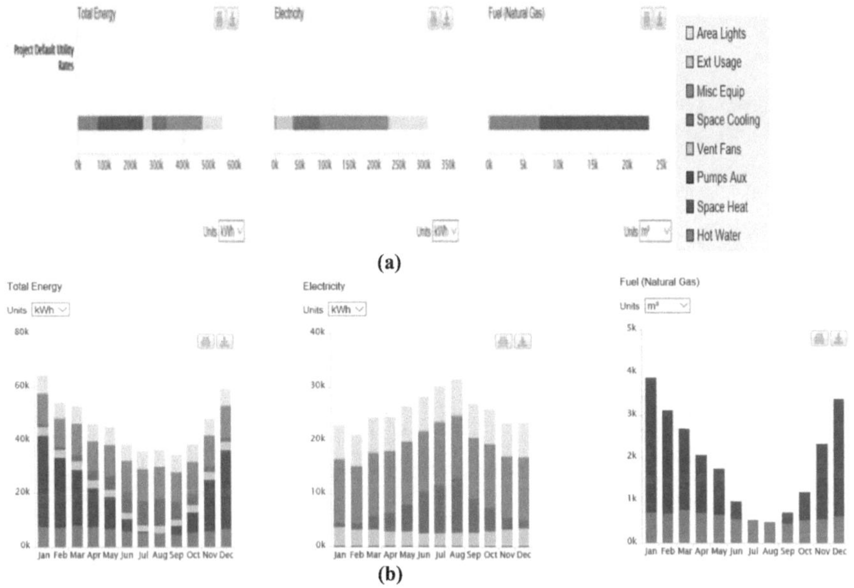


**FIGURE 3.32** The building with green concept: (a) annual energy requirement and (b) monthly energy requirement.

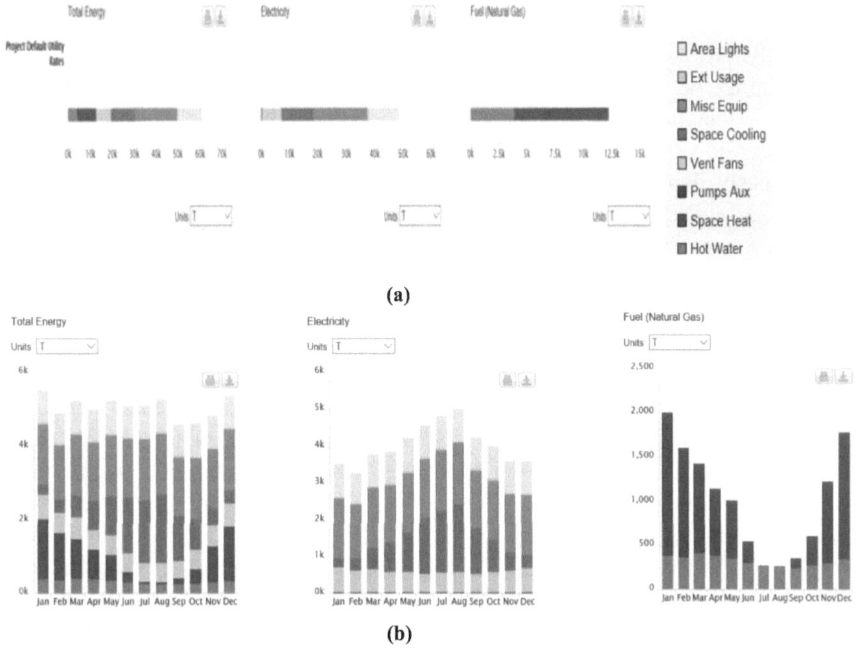


**FIGURE 3.33** The existing building: (a) annual energy cost and (b) monthly energy cost.

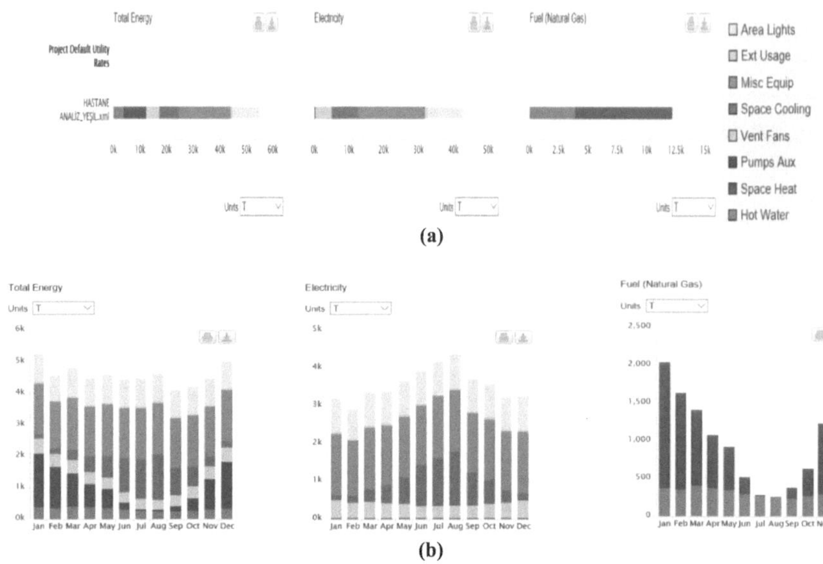


**FIGURE 3.34** The building with green concept: (a) annual energy cost and (b) monthly energy cost.

intensities are given in Figure 3.35. For both of the analyzed building types (existing and green concept buildings), the highest electrical energy and natural gas consumption values were June-July-August and December-January-February periods, respectively.

Examining the two building concepts, this analysis was designed to identify groups of influential parameters within a development stage model and quantitatively assess how these groups could be influential on the projected use of energy. Conspicuously, Figures 3.34 and 3.35 display the modified building with green concept that constitutes high-performance envelope materials, thus revealing its important role on energy performance improvement.

Many measurements of building energy costs and early compliance decisions can be made using annualized data on energy costs and usage. Costs are calculated using average utility rates across the city, region, or country.

For both of the analyzed building types (existing and green concept buildings), the monthly energy costs were obtained as a linear ratio with monthly energy requirement. For the existing building and the building with green concept, total energy costs were determined as 60,767.18 T and 55,079.82 T. The maximum total energy costs were determined in December and January. The maximum electricity energy cost was obtained in August. This is probably due to the fact that high-priced electricity is used more than low-priced natural gas to cool the building during the summer months.

Energy use intensity is the mean of the project's combined fuel and electricity per area (square feet in IP units or square meters in SI units) per year. For the existing building and the building with green concept, total energy use intensities were determined as 97.41 kBtu $ft^2$ (56.93 kBtu $ft^2$ for electricity and 40.48 kBtu $ft^2$ for natural gas) and

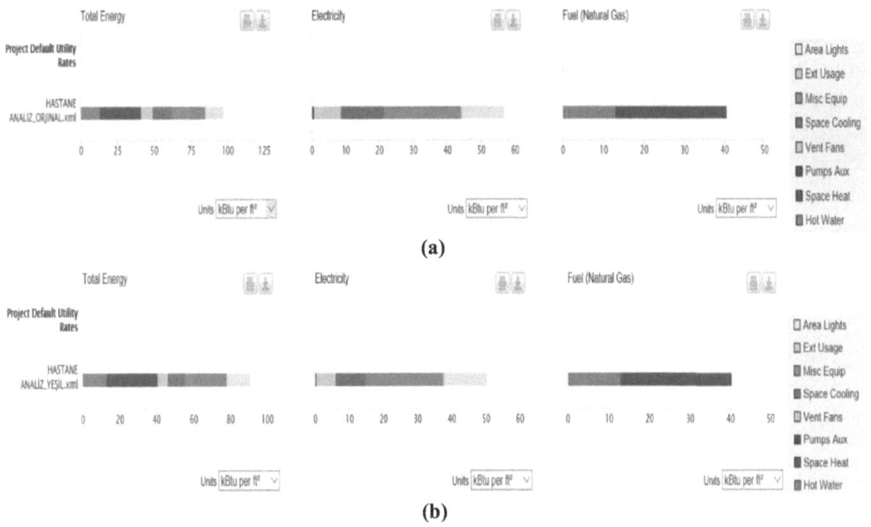


**FIGURE 3.35** Energy use intensity: (a) the existing building and (b) the building with green concept.

90.69 kBtu $ft^2$ (50.29 kBtu $ft^2$ for electricity and 40.40 kBtu $ft^2$ for natural gas), respectively. For the existing building and the building with green concept, the maximum and minimum energy use intensity values were obtained as 27.58 (space heat)-0.55 (pump aux) kBtu $ft^2$ and 27.5 (space heat)-0.4 (pump aux) kBtu $ft^2$, respectively.

## 3.4 CONCLUSIONS

The desire of the Ministry of Health to increase the service quality of hospitals within the scope of Health Transformation Program contributed to the consideration of different demands in projects. Thus, health services that contribute to spatial and social sustainability, which is an important issue in urban life, provide great advantages not only for that city but also to the neighboring provinces. In recent years, this type of hospital projects which are designed on the basis of quality, comfort, and efficiency is among the most common applications that include accommodation units for patients' relatives. Especially in developed countries, in order to increase the attractiveness of the hospital, five-star hotel comfort is planned for the hospitals. Luxury hotels are included in the planning to accommodate patient relatives. In some occasions, in developing countries, as a service delivery carried out by public and local government with the mission of being a social state, accommodation units for poor and disadvantaged patients' relatives are added to the projects.

In order to cover the cost of this service, some income-generating functions were added to this project so that the hospital could provide companion service with its own budget. As an example of hospital projects designed with this approach, the accommodation companion house project designed with the additional building function examined is an important example in terms of contributing to the spatial and social sustainability of the city and surrounding provinces and meeting the demand. At the same time, through the green hospital designs that are in line with the green concept that has gained importance in recent years, it has been researched to what extent both environment-friendly and energy-efficient building designs can have a positive impact on energy efficiency. For this purpose, the change in energy performance was analyzed comparatively if the building, which was designed and constructed in the urban sustainability concept of the health ministry, was supported with green wall and green roof materials, which is one of the green hospital criteria. These analyses are applied by using Autodesk Revit Architecture Simulation Software.

A quantitative and qualitative research concerning the climate responsiveness of the existing and green concept companion house with accommodation facilities is conducted in this study. The obtained results can be summarized as follows:

- The thermal quality performance analysis relying on Autodesk Revit Architecture Simulation Program displays that the building with green concept is more effective than the existing one in cooling duration. In addition, the building with green concept contributes to smaller heating load in winter duration.
- The qualitative research displays that building with green concept is more efficient than existing building in cooling and heating duration.

The implementation of a systematic BIM-Building Energy Model as suggested in this chapter could reduce the level of effort required in an iterative process for conventional Building Energy Model and encourage preferable utilization of Energy Simulation to design high-performance buildings. Compiling and analyzing the key BIM-Building Energy Model methods over the years and the amount of studies in each of the fields are already helping to fill the current gaps.

## REFERENCES

1. Yaş, Z. S. (2009). Toplam Kalite Yönetimi Anlayışının Sağlık Sektöründe Uygulanması: Trakya Üniversitesi Sağlık Araştırma ve Uygulama Merkezi İle Edirne Devlet Hastanesi'nin Karşılaştırılması, Trakya Üniversitesi Sosyal Bilimler Enstitüsü Yüksek Lisans Tezi, Edirne.
2. Yazar, K. H. (2006). Sürdürülebilir Kentsel Gelişme Çerçevesinde Orta Ölçekli Kentlere Dönük Kent Planlama Yöntem Önerisi, Doktora Tezi. Ankara.
3. Sağlık Bakanlığı, T. C. (2010). Türkiye Sağlık Yapıları Asgari Tasarım Standartları, 2010 Yılı Kılavuzu, T.C. Sağlık Bakanlığı Yayınları, Yayın No: 800, Ankara.
4. Karahan, A. (2008). Hastanelerde Kat Hizmetleri ve Memnuniyet Analizi, *Uşak Üniversitesi Sosyal Bilimler Dergisi*, 1(2), 155–171, Uşak.
5. Aslantekin, F., Göktaş, B., Uluşen, M. & Erdem, R. (2007). Sağlık Hizmetlerinde Kalite Deneyimi: Dr. Ekrem Hayri Üstündağ Kadın Hastalıkları ve Doğum Hastanesi Örneği, *Fırat Üniversitesi Sağlık Hizmetleri Dergisi*, 2(6), 55–71, Elazığ.
6. Büber, R. & Başer, H. (2012). Sağlık İşletmelerinde Müşteri Memnuniyeti: Vakıf Üniversitesi Hastanesinde Bir Uygulama, *Sosyal ve Beşeri Bilimler Dergisi*, 4(1), 265–274.

7. Çinal, A. (2009). Özel Hastanelerin Sınıflandırılması, Hasta Güvenliği ve Yeni Roller, T.C. Sağlık Bakanlığı Yayınları, Performans Yönetimi ve Kalite Geliştirme Daire Başkanlığı, Ankara.
8. Kıdak, B. L. & Aksaraylı, M. (2008). Yatan Hasta Memnuniyetinin Değerlendirilmesi ve İzlenmesi: Eğitim ve Araştırma Hastanesi Uygulaması, *Dokuz Eylül Üniversitesi Sosyal Bilimler Enstitüsü Dergisi*, 10(3), 87–122, İzmir.
9. SHGM (Sağlık Hizmetleri Genel Müdürlüğü), Sağlıkta Kalite ve Akreditasyon Daire Başkanlığı (2012). Sks Işığında Sağlıkta Kalite, Sağlık Bakanlığı Yayını, Ankara.
10. Tükel, B., Acuner, A., Önder, R. Ö. & Üzgül, A. (2004). Ankara Üniversitesi İbn–İ Sina Hastanesi'nde Yatan Hasta Memnuniyeti; Genel Cerrahi Anabilim Dalı Örneği, *Ankara Üniversitesi Tıp Fakültesi Dergisi*, 57(4), 205–214, Ankara.
11. Sağlık Bakanlığı, T. C. (2008). Türkiye Sağlıkta Dönüşüm Programı İlerleme Raporu, Sağlık Bakanlığı Yayınları, 20–21, Ankara.
12. Tatar, M. (2012). Sağlık Kurumları Yönetimi-I (Editör), T.C. Anadolu Üniversitesi Yayını, No: 2631, Eskişehir.
13. Sağlık Bakanlığı, T. C. (2010). Türkiye Sağlık Yapıları Asgari Tasarım Standartları, 2010 Yılı Kılavuzu, T.C. Sağlık Bakanlığı Yayınları, Yayın No: 800, Ankara.
14. Yazar, K. H. (2006). Sürdürülebilir Kentsel Gelişme Çerçevesinde Orta Ölçekli Kentlere Dönük Kent Planlama Yöntem Önerisi, Doktora Tezi. Ankara.
15. WBCSD (2013). Transforming the Market: Energy Efficiency in Buildings, World Business Council for Sustainable Development.
16. UK Parliament (2008). Climate Change Act 2008, The Stationary Office Limited.
17. McLeod, R. S., Hopfe, C. J. & Rezgui, Y. (2012). An investigation into recent proposals for a revised definition of zero carbon homes in the UK, *Energy Policy*, 46, 25–35.
18. King, P. & Clare, M. (2008). The Definition of Zero Carbon. Zero Carbon Task Group Report, UK Green Building Council.
19. The European Parliament and The Council of the European Union (2010). Directive 2010/ 31/EU of the European Parliament and of the Council of 19 May 2010 on the energy performance of buildings, *Official Journal of European Union*, L 153, 13–35.
20. Cabinet Office (2011). Government Construction Strategy.
21. Eastman, C. (1999). *Building Product Models: Computer Environments Supporting Design and Construction*. CRC Press, Boca Raton, FL.
22. Krygiel, E., & Nies, B. (2008). *Green BIM*. Wiley Publishing, Inc., Indianapolis, IN.
23. Osello, A., Cangialosi, G., Dalmasso, D., Di Paolo, A., Turco, M. L., Piumatti, P., & Vozzola, M. (2011), Architecture data and energy efficiency simulations: BIM and interoperability standards, *Proceedings of Building Simulation 2011: 12th Conference of International Building Performance Simulation Association 2011*, pp. 1521–1526 (Sydney).
24. Moon, H. J., Choi, M. S., Kim, S. K., & Ryu, S. H. (2011). Case studies for the evaluation of interoperability between a BIM based architectural model and building performance analysis programs, *Proceedings of Building Simulation 2011: 12th Conference of International Building Performance Simulation Association 2011*, pp. 1521–1526 (Sydney).
25. Eastman, C., Teicholz, P., Sacks, R., & Liston, K. (2011). *BIM Handbook: A Guide to Building Information Modelling for Owners, Managers, Designers, Engineers, and Contractors*. John Wiley & Sons. Inc., Canada.
26. Cormier, A., Robert, S., Roger, P., Stephan, L., & Wurtz, E. (2011). Towards a BIM-based service oriented platform: Application to building energy performance simulation, *Proceedings of Building Simulation 2011: 12th Conference of International Building Performance Simulation Association*, 2011 (Sydney).
27. Zuo, J., & Zhao, Z.-Y. (2014). Green building research–current status and future agenda: A review. *Renewable and Sustainable Energy Reviews*, 30, 271–281.

28. Wong, J. K. W., & Zhou, J. (2015). Enhancing environmental sustainability over building life cycles through green BIM: A review. *Automation in Construction*, 57, 156–165.
29. Maltese, S., Tagliabue, L. C., Cecconi, F. R., Pasini, D., Manfren, M., & Ciribini, A. L. C. (2017). Sustainability assessment through green BIM for environmental, social and economic efficiency. *Procedia Engineering*, 180, 520–530.
30. Gao, H., Koch, C., & Wu, Y. (2019). Building information modelling based building energy modelling: A review. *Applied Energy*, 238, 320–343.
31. Pezeshki, Z., Soleimani, A., & Darabi, A. (2019). Application of BEM and using BIM database for BEM: A review. *Journal of Building Engineering*, 23, 1–17.
32. Sanhudo, L., Ramos, N. M. M., Poças Martins, J., Almeida, R. M. S. F., Barreira, E., Simões, M. L., & Cardoso, V. (2018). Building information modeling for energy retrofitting – A review. *Renewable and Sustainable Energy Reviews*, 89, 249–260.
33. Kamel, E., & Memari, A. M. (2019). Review of BIM's application in energy simulation: Tools, issues, and solutions. *Automation in Construction*, 97, 164–180.
34. Wong, J. K.-W., & Kuan, K.-L. (2014). Implementing "BEAM Plus" for BIM-based sustainability analysis. *Automation in Construction*, 44, 163–175.
35. Lu, Y., Wu, Z., Chang, R., & Li, Y. (2017). Building Information Modeling (BIM) for green buildings: A critical review and future directions. *Automation in Construction*, 83, 134–148.
36. Azhar, S., Carlton, W. A., Olsen, D., & Ahmad, I. (2011). Building information modeling for sustainable design and LEED rating analysis. *Automation in Construction*, 20, 217–224.
37. Chen, P.-H., & Nguyen, T. C. (2017). Integrating web map service and building information modeling for location and transportation analysis in green building certification process. *Automation in Construction*, 77, 52–66.
38. Ansah, M. K., Chen, X., Yang, H., Lu, L., & Lam, P. T. I. (2019). A review and outlook for integrated BIM application in green building assessment, *Sustainable Cities and Society*, 48, 101576.
39. Cemesova, A., Hopfe, C. J., & Mcleodc, R. S. (2015). PassivBIM: Enhancing interoperability between BIM and low energy design software, *Automation in Construction* 57, 17–32.
40. http://hospital 2020.org/Agreenhospital.html
41. Autodesk. Revit. Available from: https://www.autodesk.com/products/revit/overview, 2019.
42. Bahar, Y. N., Pere, C., Landrieu, J., & Nicolle, C. (2013). A thermal simulation tool for building and its interoperability through the building information modeling (BIM) platform, *Buildings Journal*, 3, 380–398.
43. Klimatoloji – I, Çevre Ve Orman Bakanliği Devlet Meteoroloji İşleri Genel Müdürlüğü, Ankara Mart – 2005 DMI Yayinlari Yayin No: 2005/1.
44. https://www.vaisala.com.
45. Lee, H.-J., Kim, S.-Y., & Yun, C.-Y. (2017). Comparison of solar radiation models to estimate direct normal irradiance for Korea, *Energies*, 10, 594.
46. Roshan, Gh. R., Ghanghermeh, A. A., & Attia, S. (2017). Determining new threshold temperatures for cooling and heating degree day index of different climatic zones of Iran, *Renewable Energy*, 101, 156–167.
47. W.M.O (1991). International Meteorological Vocabulary, W.M.O, No. 182, TP 91, p. 116.
48. Rahimikhoob, A., Behbahani, M. R., & Nazaifar, M. H. (2008). Estimation of cooling degree days from AVHRR data and an MIF neural network, case study: Khoozestan province of Iran, *Canadian Journal of Remote Sensing*, 34, 596–600.

# 4 Indoor Environment and Well-Being

## *The Case of Academic Workplace in Historic Building*

*Lina Seduikyte, Egle Staniskiene, Zivile Stankeviciute, Indre Grazuleviciute-Vileniske, Laura Stasiuliene, Paris Fokaides, and Tomas Sorius*
Kaunas University of Technology

*Paris Fokaides*
Frederick University

**CONTENTS**

## 4.1 INTRODUCTION: BACKGROUND AND DRIVING FORCES

The work experience of every worker in every office, factory, warehouse, shop, or bank is intimately affected by the qualities and organization of the physical work environment. The building they work in and the way it is planned will influence occupants' behavior, such as proximity to colleagues, visibility to a supervisor, the degree of acoustic and visual privacy, the levels of temperature and light they work under, and the quality of the air they breathe. It will very often affect their health (Maddalena et al., 2015; Baldry, 1999).

Different studies (Chan et al., 2016; Maddalena et al., 2015; WHO, 1999) describe the effect of indoor pollutants on health. The sick building syndrome (SBS) causes eye, nose, and throat irritation, dry mucous membrane and dry skin, erythema (skin redness), mental fatigue, headaches, a high frequency of airway infections and coughs, hoarseness and wheezing, itching and nonspecific skin hypersensitivity, and nausea and dizziness. Common symptoms characterized by nasal dryness or congestion, such as a stuffy blocked nose or a runny nose, and also mental effects, such as difficulty to concentrate, can arise (Clements-Croome, 2008; Kortum & Burton, 2010). The largest impact of buildings on the outdoor environment during the whole life cycle comes from the energy used to ensure a comfortable and healthy indoor climate. According to Orme (1998), ventilation accounts for 25%–30% of the total building energy use.

Over the past 20 years, it has been empirically assessed that most building environments have a direct effect on an occupant's personal well-being and performance; however, it is only through more recent studies that a clearer understanding of the occupied environment has been discovered. Bakó-Biró et al. (2008) and Wargocki (2007) have shown that learning in primary school children is affected by $CO_2$ levels and hence ventilation. The UK Building Schools Exhibition and Conference (CIBSE, 2010) asked head teachers if they felt modern buildings affect learning. Around 78% said they felt there was a clear link between the quality of school design and levels of pupil attainment. Williams (2005) reported a similar conclusion for 12 primary schools that he had assessed using the Building Quality Assessment method (Clements-Croome, 2011).

Sick buildings can induce the appearance of psychological symptoms such as stress, anxiety, and aggression in occupants (Kamaruzzaman and Sabrani, 2011; Runeson-Broberg and Norbäck, 2013). These psychological disorders may further result in increasing people's susceptibilities towards other environmental factors, lowering workers' performances, and increasing absenteeism (Realyvásquez et al., 2016; Rydstedt, 2016). These negative psychological effects of being in sick buildings can further result in subsequent issues for occupants such as increasing the possibility of occurrence of hazardous events in workplace. Likewise, SBS can also influence occupants' satisfaction. Gavhed and Toomingas (2007) counted poor illumination levels, besides indoor climate, the air quality, the ambient noise level, as the major contributors leading to dissatisfaction among Swedish employees. In another study, Newsham et al. (2009) studied the effects of physical measurements such as thermal, lighting, and acoustic variables, furniture dimensions, and an assessment of potential exterior view on the employees' performances at an open-plan office building in Michigan, United States. Results highlighted the significant role of window access at the desk in satisfaction with lighting, particularly through its effect on satisfaction with outside view (Ghaffarianhoseini et al., 2018).

Within the boundaries of sustainable design goals and the achievement of the indoor thermal and air quality that are at the basis of well-being, this research encompasses the description and analysis of the indoor features of the Technical College building, an example of Inter-War architectural development (1918–1939) in Kaunas (Lithuania).

## 4.2 INDOOR ENVIRONMENT-RELATED WELL-BEING

Throughout history, air, sunlight, sound, and water have been fundamental to the needs of people. Today, sensitive control of these needs may use traditional or new solutions or integrate these but does not alter the basic truth that the built environment is fundamental to human's sense of well-being and it is the totality of this idea that we need to understand and value (Clements-Croome, 2004).

The literature review allows concluding that there is no general agreement concerning the meaning of well-being. According to Warr (1987), the two halves of the term "well-being" point directly to its meaning, as it involves "being" and "well"—living in a state that is in some sense good. For instance, Danna and Griffin (1999) define the employee well-being as the state of individuals' mental, physical, and general health, as well as their experiences of satisfaction both at work and outside of work. Following this definition, employee well-being is influenced by the pleasure or displeasure derived from the job itself, as well as individuals' interactions with colleagues, teammates, and supervisors (Nielsen et al., 2017). Generally, employee well-being could include both work-related and non-work-related well-being. This chapter limits the focus on indoor environment-related well-being. The holistic definition of well-being is based on health care, philosophy, psychology, and sociology literatures, which converge on three core dimensions of well-being: psychological, physical, and social. The psychological approach defines well-being in terms of subjective experience and functioning, the physical approach defines well-being in terms of bodily health and functioning, and the social approach defines well-being in terms of relational experience and functioning. These are the three key dimensions of well-being because they are valued as ends in and of themselves rather than as means to other ends (Grant et al., 2007).

**Psychological well-being (happiness)** focuses on subjective experiences of individuals. Two views on happiness are evident in the literature: hedonic and eudaimonic (Fisher, 2010). Hedonic views of happiness rely on pleasant feelings and favorable judgments (Fisher, 2010); they are concerned with subjective experiences of pleasure or the balance of negative and positive feelings and thoughts in individuals' judgments (Grant et al., 2007). Meanwhile, eudaimonic views of happiness involve doing what is virtuous, true to one's self, and morally right (Fisher, 2010); moreover, the eudaimonic component of psychological well-being is concerned with the realization of the human potential (Grant et al., 2007).

According to Wright and Cropanzano (2000), psychological well-being has three defining characteristics. First, people are happy when they subjectively believe themselves to be so. Second, well-being involves certain emotional conditions; psychologically well people are particularly more prone to experience positive emotions and less prone to experience negative emotions. Third, well-being refers to one's life as a whole.

Turning to happiness at work, it should be admitted that "happiness" is not the term that has been extensively used in research (Fisher, 2010). However, this does mean that researchers are not interested or even neglect employee happiness at work. On the contrary, many happiness-related constructs have been studied. Such constructs include job involvement, engagement, thriving, vigor, typical mood at work, and effective organizational commitment (Fisher, 2010). However, the most common means of operationalizing the happiness component in organizational research refer to the measurement of job satisfaction (Wright and Cropanzano, 2000; Fisher, 2010).

**Physical well-being (health).** Organizational researchers analyze the linkage between work and employee physical health in at least three ways: work is a potential source of disease and injury; work can be a source of stress; and work is a source of benefits that allows the purchase of health care services (Grant et al., 2007). The relationships between work and health are bidirectional, whereas work has specific impacts on health, health is a determinant of the ability to work and remain in work. The relevant aspect is that the impact of work can have immediate or delayed effects on health outcome (Parent-Thirion et al., 2016). Occupational health care literature differentiates between stressors and strain: stressors (work intensification, workload, etc.) refer to situations or events that raise the stress, meanwhile strain (burnout, stress) refers to responses to stressors (van de Voorde et al., 2012).

WHO defines a healthy workplace as follows (WHO 2010 in Kortum and Burton, 2010): a healthy workplace is one in which workers and managers collaborate to use a continual improvement process to protect and promote the health, safety, and well-being of all workers and the sustainability of the workplace by considering the following identified needs: health and safety concerns in the physical work environment; health, safety, and well-being concerns in the psychosocial work environment including organization of work and workplace culture; personal health resources in the workplace; and ways of participating in the community to improve the health of workers (Ruohomäki et al., 2015).

**Indoor Environmental Quality: Whose Responsibility?**

Improving workers' well-being also means improving their cognitive ability, comfort, mood, motivation, vigor, creativity, work satisfaction, and ultimately health while decreasing depression, tension, and fatigue (Kwallek and Lewis, 1990; Paul, 1997; Lan et al., 2010; Röcker, 2010; Reijula et al., 2011).

According to the recent literature, intelligent buildings should increase well-being by providing users with a favorable multi-sensory experience. An indoor environment must display the following characteristics (Reijula and Sundman, 2004; Clements-Croome, 2011; Bluyssen, 2014): a fresh thermal environment; clean indoor air; ventilation rates to provide fresh air with good distribution and acceptable levels of $CO_2$; good natural lighting and minimal lighting glare from within and outside the space; acoustic quality, low noise levels and spatial planning; and settings to suit various types of work (Ruohomäki et al., 2015).

## 4.3 CASE STUDY OBJECT

### 4.3.1 Case Object—Technical College Building

More than 6,000 buildings from Kaunas' Inter-War intensive construction period have survived till today and constitute valuable historic urban landscape and legacy of the Inter-War Modernism architecture. School buildings are an inseparable part of it and until today serve as important focal points and landmarks of the city. Stasys Kudokas (1893) was one of the most prominent architects of the Inter-War era, Lithuania, who contributed to the development of Kaunas architecture character. In 1934, he designed the building of the Technical College, located in Tvirtoves avenue in Kaunas, which is still used for educational purposes and it is selected for this study.

The monumental building constructed on the small hill became the main landmark in expanding low-rise urban area of Inter-War era. The building was opened in 1938. In 1941, the extension of the building was designed by the same architect, although the project was not implemented and building was expanded in the Soviet era with different architectural stylistics corresponding to this period (Laurinaitis, 2018).

Currently, the building houses the educational institution that is similarly entitled Kaunas Technical College (Figure 4.1); thus, the building maintains its original function. The building is protected by the state as the individual object of regional significance with valuable architectural features, continuity of function, and the links with prominent personalities, such as building's architect S. Kudokas. The architectural style of the building is defined as "modernized historicism" (Laurinaitis, 2018).

Summarizing the review of the history and features of the case study building, its main contributions to sustainability can be distinguished with reference to

**FIGURE 4.1** Kaunas Technical College. (Photograph by the authors.)

Luca et al. (2018): the building certainly contributes to *place-making* as a part of the historic urban landscape of Kaunas, the creation of prominent architecture, and as a living educational institution with historical continuity; it can be seen both as sociocultural and economic *resource* with embodied *energy*; the building contributes to the sphere of *education* both as a part of an educational institution and as a heritage building bearing its own information and significance; *equity* and *well-being* aspects can be related with the continuity of educational institution in the building, this institution as a workplace, and as a place for learning and community development. However, the well-being aspect is also related with the comfort of the indoor environment for people using the building and this aspect is further developed in this study.

### 4.3.2 Case Study Methods

For the case study of Kaunas Technical College building, the following methods were used:

- Field measurements for the measurement of temperature, relative humidity (RH), and $CO_2$ concentration; field measurements reflect the existing state of the indoor climate.
- Computational fluid dynamics (CFD) simulations for the visualization and calculation of the temperature and $CO_2$ distribution in one of the rooms of the college; CFD simulations were used to show indoor climate parameters after possible reconstruction of ventilation system.
- Questionnaire survey.

#### 4.3.2.1 Field Measurements

Field measurements were performed in eight naturally ventilated rooms (three offices, three classrooms, and two computer classrooms) of Kaunas Technical College. Field measurements were made in the period of March 4–22, 2018.

Temperature and RH were measured with HOBO MX11011 data loggers (±0.2°C, ±2% RH accuracy). The temperature and RH were measured in all tested rooms at 1.1 m height (head height of a seated person, Figure 4.2).

$CO_2$ concentration was measured with indoor air quality (IAQ)-CALC 7545 IAQ meters (TSI, United States, with an accuracy of ±3%) at a height of 1.1 m; $CO_2$ concentration was measured in classroom number 3 (classroom for 18 students, 37.25 $m^2$ area).

#### 4.3.2.2 Numerical Method

The commercial CFD tool FloVENT (Mentor Graphics, United States) was chosen to visualize and calculate the temperature and $CO_2$ distribution in the tested classroom. FloVENT is a tool used to investigate the airflow, contamination, and heat performance of both individual thermal zones and whole buildings. In this study, the k-ε turbulence model (LVEL k-ε) was employed as it shows good performance in predicting the behavior of indoor airflows, temperatures, and contaminant

**FIGURE 4.2** Tested classroom.

distribution in buildings (Nielsen, 2015). To achieve steady-state simulation results, the double-precision solver of FloVENT was applied.

CFD model of the classroom number 3 was designed as simplified replicate of the real classroom in Kaunas Technical College building. The geometry of the furniture was built according to the physical form of the real furniture. Simplified geometrical shape models of sitting occupants were used to evaluate the effect of convection heat flow generated by the surface of the occupant. The thermal attribute of 85 W was added to each model of an occupant to simulate dry heat losses through the skin (Zukowska et al., 2012). A heating source of three cuboids with thermal attribute of 500 W was used to represent radiators under the windows. All other boundary conditions (walls, ceiling, floor, windows temperatures) were chosen to replicate parameters achieved during the field measurements.

Two different mechanical ventilation systems were studied (mixing and displacement ventilation), and different CFD models with different boundary conditions were created. These two types of ventilation systems were selected as commonly used air distribution concepts. In a mixing ventilation system, air is supplied in an upper part of the room, that is, above the occupied zone where people are with high initial velocity (approx. 1 m/s). High initial velocity ensures that room air circulates and mixes. Air can be exhausted through the upper or lower part of the room. In displacement ventilation system, air is supplied directly to the occupied zone (at floor level) with low initial velocity (approx. 0.2 m/s) and a few degrees lower air temperature compared with the room temperature. In displacement ventilation convective flows from occupants and warm equipment, for example, computers are the driving force directing contaminated air towards the ceiling zone where exhaust is usually designed.

In the case of mixing ventilation, three square ceiling supply air diffusers and three exhaust air ceiling grilles were added to the geometrical model of the classroom. The supply and exhaust airflows were set to 60 l/s for each supply and exhaust unit. Supply air temperature was set to 16°C. In the case of displacement ventilation, two displacement diffusers (wall type) and three exhaust air ceiling grilles were added to the model. The supply airflow was set to 120 l/s for each supply air diffuser and 80 l/s for exhaust grilles. The location of air exhaust grilles was the same in both mixing and displacement ventilation cases, as their position on the ceiling in this model had no significant influence; 400 ppm of $CO_2$ was added to supply air as it is the background level of $CO_2$ outside.

The volume of air exhaled by occupants was set to 0.15 l/s, temperature at +35°C, and attribute of 40,000 ppm of $CO_2$ was added to each occupant at mouth zone to imitate the breathing process.

The model was divided into regions by using Cartesian grid. In order to get more accurate results, the grid density was increased close to the air supply and exhaust units. The grid sensitivity analysis was performed. It was determined that a 100,000-cell grid was sufficient for this study. Negligible changes in results were seen with denser grid as well as calculation took longer time. The results further will be presented with a grid between 100,000 and 200,000 grid cells.

#### 4.3.2.3 Questionnaire Survey

Students who were studding in the classroom number 3 and employees working in the tested building were asked to fill in questionnaires. The questionnaire was prepared by research team and included personal information, questions related to SBS symptoms, health and environment. The purpose of questionnaire survey was to evaluate if students and employees had complains regarding draft, thermal sensation (too cold, too warm), stuffy air, noise, and identify possible SBS symptoms.

## 4.4 RESULTS

### 4.4.1 Indoor Temperature and Relative Humidity

The average, minimum and maximum temperature, and RH values measured in tested rooms are presented in Table 4.1. Data was extracted and analyzed only for the working hours 8 a.m.–5 p.m.

**TABLE 4.1**
**Indoor Temperature and RH Measured in Tested Rooms**

| | | Temperature T, °C | | | RH, % | | |
|---|---|---|---|---|---|---|---|
| Nr. | Purpose of the Room | Ave. | Max. | Min. | Ave. | Max. | Min. |
| 1. | Classroom | 19.7 | 21.7 | 17.9 | 36 | 50 | 25 |
| 2. | Classroom | 19.7 | 22.1 | 15.2 | 33 | 45 | 23 |
| 3. | Classroom | 20.1 | 22.1 | 18.3 | 35 | 46 | 24 |
| 4. | Computer classroom | 19.7 | 22.1 | 15.2 | 33 | 46 | 23 |
| 5. | Computer classroom | 19.7 | 22.1 | 17.9 | 33 | 59 | 23 |
| 6. | Office | 21.3 | 28.3 | 19.1 | 27 | 35 | 23 |
| 7. | Office | 21.5 | 26.7 | 16.4 | 32 | 49 | 23 |
| 8. | Office | 20.2 | 22.1 | 18.3 | 35 | 46 | 24 |

Results show that average temperature was meeting Lithuanian hygiene standard HN 42:2009 “Microclimate in dwellings and public houses” requirements (18°C–22°C for the heating season). However, the average RH in 63% tested rooms was not reaching the required value (35%–60% for the heating season).

### 4.4.2 $CO_2$ Concentration

$CO_2$ concentration was measured for 24 hours in classroom number 3. During the night, $CO_2$ concentration dropped to 568 ppm. As expected, $CO_2$ concentration was varying, during the classes, $CO_2$ concentration was going up, and during the breaks, it was dropping. The maximum $CO_2$ concentration was 2,790 ppm. The average $CO_2$ concentration during the classes was 1,894 ppm.

The measured $CO_2$ values during the working hours were higher than 1,000 ppm. Therefore, it can be stated that IAQ in terms of $CO_2$ concentration was very poor, and this could influence students’ performance and ability to concentrate.

### 4.4.3 Numerical Results

CFD simulations were performed for mixing and displacement ventilation to evaluate possible temperature and $CO_2$ distribution in the tested classroom. Figures 4.3 and 4.4 present temperature and $CO_2$ distribution for the tested classroom with

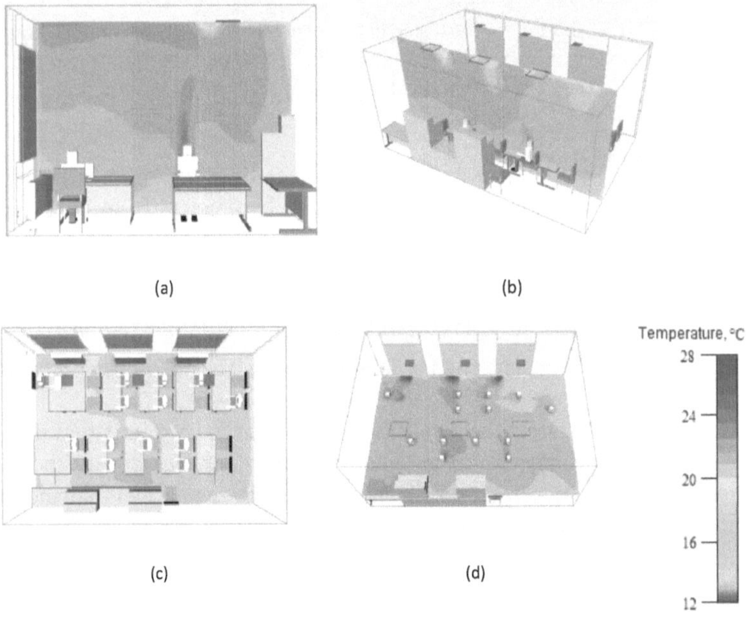


**FIGURE 4.3** Temperature distribution for the tested classroom with mechanical mixing ventilation: (a) section through external wall, diffuser, and a siting person; (b) section across the room; (c) temperature distribution at 0.1 m (at ankles); and (d) temperature distribution at 1.1 m (height of sitting person).

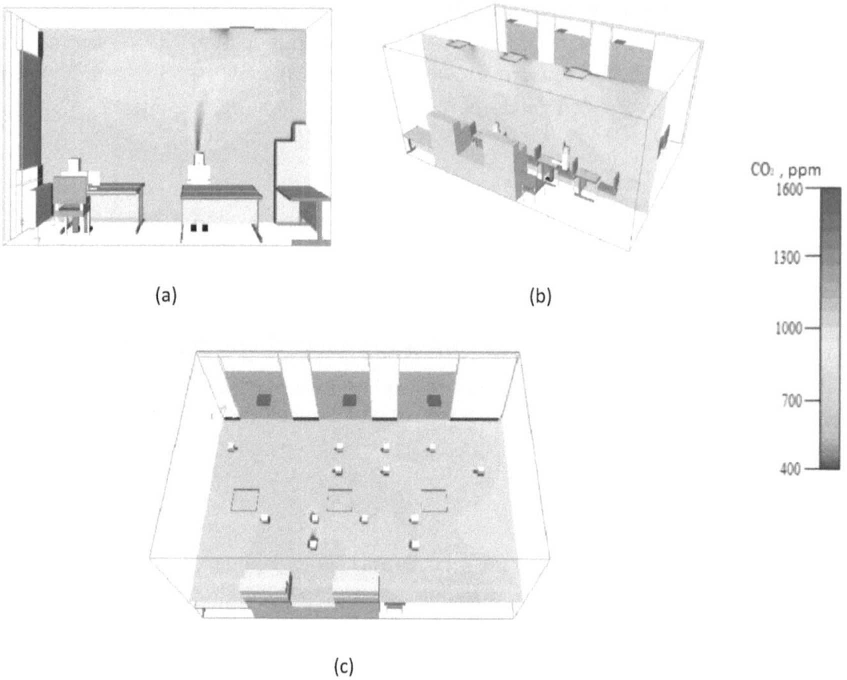


**FIGURE 4.4** $CO_2$ distribution for the tested classroom with mechanical mixing ventilation: (a) section through external wall, diffuser, and a siting person; (b) section across the room; and (c) $CO_2$ distribution at 1.1 m (height of sitting person).

mechanical mixing ventilation when three square ceiling supply air diffusers were used, while Figures 4.5 and 4.6 present the same parameters from the case with mechanical displacement ventilation when two supply wall diffusers were used.

Results of CFD simulation from the case with mixing ventilation show that there is no stratification, both temperature and $CO_2$ concentration spread relatively equally throughout the room. The average temperature in the working zone of the room is about 21.26°C and $CO_2$ concentration is about 788 ppm. The conditions of this model would satisfy requirements of HN 42:2009 norms.

Results of CFD simulation from the case with displacement ventilation show vertical thermal and $CO_2$ concentration gradient. Near the floor, CFD results show lower temperatures and $CO_2$ concentration. Contaminants that are warmer than indoor air, in this case exhaled air, go up and are removed through ceiling grilles. The average temperature in the working zone of the room is about 21.47°C, and $CO_2$ concentration is about 635 ppm. The conditions of this model would satisfy requirements of HN 42:2009 norms.

Summarizing the results, it can be noted that in the current condition, the heritage building of the Technical College only partially meets the requirements for the indoor conditions for its successful continuous use. The average temperature measured in offices and classrooms of tested heritage building during the heating season

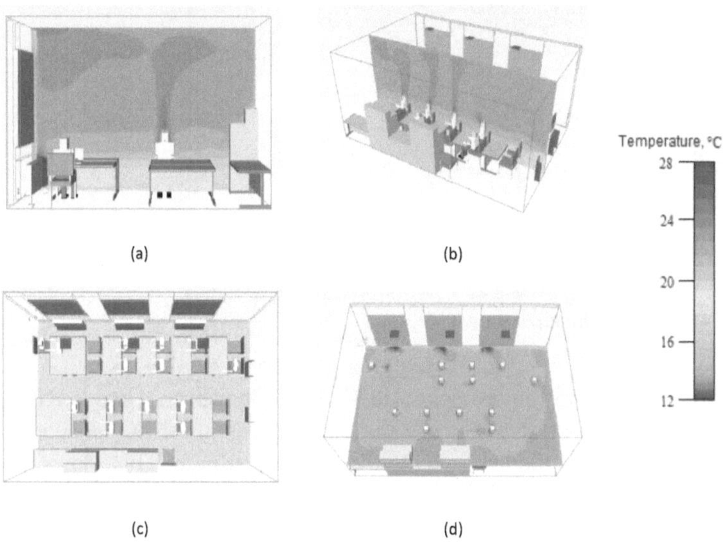


**FIGURE 4.5** Temperature distribution for the tested classroom with mechanical displacement ventilation: (a) section through external wall, diffuser, and a siting person; (b) section across the room; (c) temperature distribution at 0.1 m (at ankles); and (d) temperature distribution at 1.1 m (height of sitting person).

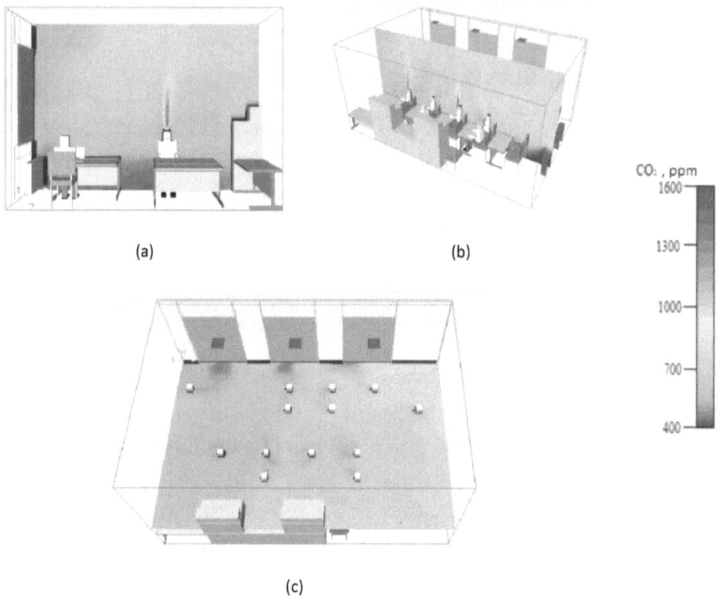


**FIGURE 4.6** $CO_2$ distribution for the tested classroom with mechanical displacement ventilation: (a) section through external wall, diffuser, and a siting person; (b) section across the room; and (c) $CO_2$ distribution at 1.1 m (height of sitting person).

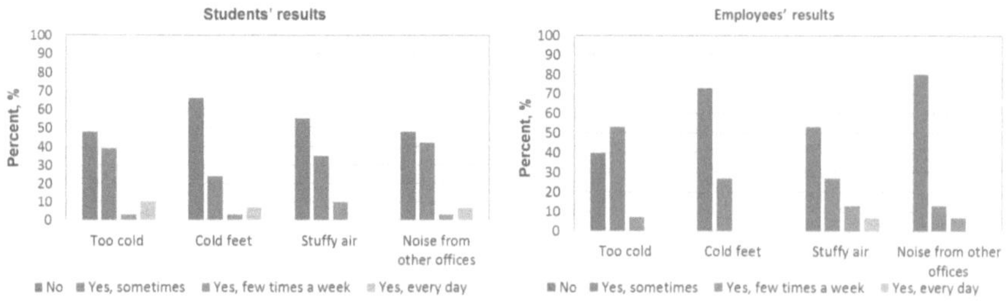


**FIGURE 4.7** Questionnaire results.

was meeting minimum requirements (18°C–22°C); however, average RH was not in compliance with required values (35%–60%). The highest $CO_2$ concentration in the tested classroom of the heritage building was 2,790 ppm. This shows poor IAQ and ventilation, which could influence students' performance and ability to concentrate. The measurement results demonstrate the need of mechanical ventilation in order to have better air quality. CFD results show that both mixing and displacement mechanical ventilation would be able to ensure good thermal and IAQ conditions. In the case of displacement ventilation, $CO_2$ concentration is lower by 19.5% compared with concentration with mixing ventilation. Consequently, these types of forced ventilation could be recommended for improving the indoor comfort in the Technical College building in order to ensure its continuous functioning and contribution to other spheres of sustainability.

### 4.4.4 Questionnaire Survey

The questionnaire survey did not show SBS symptoms (eye, nose, or throat irritation; headache; dizziness; nausea; dry cough; dry or itching skin; difficulty in concentration; fatigue) among people working in tested building.

The main problems which were detected during the questionnaire survey by employees and students were related to insufficient temperature indoors and noise coming from other offices (Figure 4.7).

## 4.5 CONCLUSIONS

As we move more and more into the era of sustainability and smart buildings, we have to bear in mind that a building first of all has to meet the needs of its users. The purpose of this study was to examine the human comfort conditions in an academic building built in the early 20th century in Kaunas, Lithuania. The study handled both field measurements as well as numerical simulations and field research with questionnaires. The main findings of the study can be summarized as follows:

- The measured $CO_2$ values during the working hours were over 1,000 ppm, an indication for poor IAQ, which could affect students' performance and ability to concentrate.

- Results of CFD simulation from the case with mixing ventilation showed that there is no stratification of both temperature and $CO_2$ concentration, and that these two values spread relatively equally throughout the room.
- Also the numerical study underlined the fact that the measurement results demonstrated the need of forced ventilation in order to have better air quality.
- The average temperature measured in offices and classrooms of tested heritage building during the heating season met the minimum requirements (18°C–22°C); however, average RH was not in compliance with required values (35%–60%).
- Poor air quality conditions were observed with the use of questionnaires addressed to buildings occupants.

The findings of this study aspire to support the transition to the sustainable built environment, the cell of which is anticipated to be the smart, sustainable building.

## ACKNOWLEDGMENT

The authors are thankful to administration of Kaunas Technical College for possibility to perform the study in their premises.

## REFERENCES

Bakó-Biró, Z., Kochhar, N., Clements-Croome, D.J., Awbi, H.B., Williams, M., 2008. Ventilation rates in schools and pupil's performance using computerised assessment tests, in: Indoor Air 2008, *The 11th International Conference on Indoor Air Quality and Climate*, Copenhagen.

Baldry, C., 1999. Space: The final frontier, *Sociology*, 33(3), 1–29.

Bluyssen, P.M., 2014. What do we need to be able to (Re)design healthy and comfortable indoor environments? *Intelligent Buildings International*, 6(2), 69–92.

Chan, W.R., Parthasarathy S., Fisk W.J., McKone T.E., 2016. Estimated effect of ventilation and filtration on chronic health risks in U.S. offices, schools, and retail stores. *Indoor Air*, 26, 331–343.

CIBSE, 2010. New schools affect pupil performance, *CIBSE Journal*, February, 6.

Clements-Croome, D.J., 2011. Sustainable intelligent buildings for people: A review. *Intelligent Buildings International*, 3(2), 67–86.

Clements-Croome, D.J., 2008. Work performance, productivity and indoor air. *Scandinavian Journal of Work Environment & Health (Supplement)*, 4, 69–78.

Clements-Croome, D.J., 2004. *Intelligent Buildings: Design, Management and Operation*. London: Thomas Telford.

Danna, K., Griffin, R., 1999. Health and well-being in the workplace: A review and synthesis of the literature. *Journal of Management*, 25, 357–384.

Fisher, C.D., 2010. Happiness at work. *International Journal of Management Reviews*, 12(4), 384–412.

Gavhed, D., Toomingas, A., 2007. Observed physical working conditions in a sample of call centres in Sweden and their relations to directives, recommendations and operators' comfort and symptoms. *International Journal of Industrial Ergonomics*, 37(9), 790–800.

Ghaffarianhoseini, A., AlWaer, H., Omrany, H., Ghaffarianhoseini, A., Alalouch, C., Clements-Croome, D., Tookey, J., 2018. Sick building syndrome: Are we doing enough? *Architectural Science Review*, 61(3), 99–121.

Grant, A.M., Christianson, M.K., Price, R.H., 2007. Happiness, health, or relationships? Managerial practices and employee well-being trade-offs. *Academy of Management Perspectives*, 21(3), 51–63.

Kamaruzzaman, S., Sabrani, N.A., 2011. The effect of indoor air quality (IAQ) towards occupants' psychological performance in office buildings. *Jurnal Rekabentuk dan Binaan*, 4, 49–61.

Kortum, E., Burton, J., 2010. WHO global framework and model on healthy workplaces. In: Anttonen, H., Husman, P., Hussi, T., Leino, T., Ylikoski, M. (eds.) *Proceedings of the International Conference Towards Better Work and Well-Being*, 14–19. Helsinki: Finnish Institute of Occupational Health.

Kwallek, N., Lewis, C.M., 1990. Effects of environmental colour on males and females: A red or white or green office. *Applied Ergonomics*, 21(4), 275–278.

Lan, L., Lian, Z., Pan, L., 2010. The effects of air temperature on office workers' well-being, workload and productivity evaluated with subjective ratings. *Applied Ergonomics*, 42/1(29–36), 1872–9126.

Laurinaitis, P.T., 2018. Aukštesnioji technikos mokykla (dab. Kauno technikos kolegija), http://www.autc.lt/lt/architekturos-objektai/927?id=927 (accessed 11 December 2018).

Luca, E., Sulc, I., Haselsteiner, E., 2018. Regenerative heritage. In: Brown, M., Haselsteiner, E., Apró, D., Kopeva, D., Luca, E., Pulkkinen, K., Vula Rizvanolli, B. (eds.) *Sustainability, Restorative to Regenerative*. Working Group One Report: Restorative Sustainability, COST Action CA16114 RESTORE.

Maddalena, R., Mendell, M.J., Eliseeva, K., Chan, W.R., Sullivan, D.P., Russell, M., Satish, U., Fisk, W.J., 2015. Effects of ventilation rate per person and per floor area on perceived air quality, sick building syndrome symptoms, and decision-making. *Indoor Air*, 25, 362–370.

Newsham, G., Brand, J., Donnelly, C., Veitch, J., Aries, M., Charles, K., 2009. Linking indoor environment conditions to job satisfaction: A field study. *Building Research and Information* 37(2), 129–147.

Nielsen, K., Nielsen, M.B., Ogbonnaya, C., Känsälä, M., Saari, E., Isaksson, K., 2017. Workplace resources to improve both employee well-being and performance: A systematic review and meta-analysis. *Work & Stress*, 31(2), 101–120.

Nielsen, P.V., 2015. Fifty years of CFD for room air distribution. *Building and Environment*, 91, 78–90.

Orme, M., 1998. Energy impact of ventilation. Coventry (United Kingdom): Air Infiltration Ventilation Centre, International Energy AgencyProject; p 43. Technical note, no 49.

Parent-Thirion, A., Biletta, I., Cabrita, J., 2016. *Sixth European Working Conditions Survey (Eurofound): Overview Report*. Luxembourg: Publication Office of the European Union.

Paul, R.D., 1997. Nurturing and pampering paradigm for office ergonomics. *Proceedings of the Human Factors and Ergonomics Society*, 1, 519–523.

Realyvásquez, A., Maldonado-Macías, A.A., García-Alcaraz, J., Cortés-Robles, G., Blanco-Fernández, J., 2016. Structural model for the effects of environmental elements on the psychological characteristics and performance of the employees of manufacturing systems. *International Journal of Environmental Research and Public Health*, 13(1), 104.

Reijula, J., Gröhn, M., Müller, K., Reijula, K., 2011. Human well-being and flowing work in an intelligent work environment. *Intelligent Buildings International*, 3(4), 223–237.

Reijula, K., Sundman-Digert, C., 2004. Assessment of indoor air problems at work with a questionnaire. *Occupational Environmental Medicine*, 61, 33–38.

Röcker, C., 2010. Services and applications for smart office environments – a survey of state-of-the-art usage scenarios, *Proceedings of the International Conference on Computer and Information Technology*, Cape Town, South Africa, 385–401.

Runeson-Broberg, R., Norbäck, D., 2013. Sick building syndrome (SBS) and sick house syndrome (SHS) in relation to psychosocial stress at work in the Swedish workforce. *International Archives of Occupational and Environmental Health*, 86(8), 915–922.

Ruohomäki, V., Lahtinen, M., Reijula, K., 2015. Salutogenic and user-centred approach for workplace design. *Intelligent Buildings International*, 7(4), 184–197.

Rydstedt, L.W., 2016. Psychosocial job strain as a mediator between physical working conditions and symptoms associated with sick building syndrome. *Human Affairs*, 26(4), 440–449.

Van De Voorde, K., Paauwe, J., Van Veldhoven, M., 2012. Employee well-being and the HRM–organizational performance relationship: A review of quantitative studies. *International Journal of Management Reviews*, 14(4), 391–407.

Wargocki, P., 2007. Improving Indoor Air Quality Improves the Performance of Office Work and Schoolwork, Technical University of Denmark, Kgs. Lyngby, Denmark (available at www.inive.org/members_area/medias/pdf/Inive%5CIAQVEC2007%5CWargocki_2.pdf, accessed 10 November 2019).

Warr, P., 1987. *Work, Unemployment, and Mental Health*. Oxford University Press, Oxford.

Williams, B., 2005. Building performance: the value management approach. In: D. Clements-Croome (ed.), *Creating the Productive Workplace*, Routledge, London, 434–457.

World Health Organization (WHO), 1999. Indoor pollutants; exposure and health effects. Copenhagen: WHO, Regional Office for Europe; EUROreports and studies, no 78.

Wright, T.A., Cropanzano, R., 2000. Psychological well-being and job satisfaction as predictors of job performance. *Journal of Occupational Health Psychology*, 5, 84–94.

Zukowska, D., Melikov, A., Popiolek, Z., 2012. Impact of personal factors and furniture arrangement on the thermal plume above a sitting occupant. *Building and Environment*, 49, 104–116.

# 5 Properties and Conversion Technologies of Biomass

*Yaning Zhang, Wenming Fu, Pingfei Xu, Bingxi Li, and Baocheng Jiang*
Harbin Institute of Technology (HIT)

**CONTENTS**

## 5.1 INTRODUCTION

Biomass is any organic material derived directly or indirectly from the process of photosynthesis by plants and algae, except for fossilized materials (Silva et al., 2018), and it mainly includes: (1) oil crops, (2) lignocellulosic biomass, (3) solid waste, and (4) algae (Ruan et al., 2019).

Biomass has the advantages of renewable, availability, neutral $CO_2$/GHG (greenhouse), negligible or zero $SO_x$ emission, low $NO_x$ emission, less combustion ash, safety in processing process, local production, and it is environment-friendly, degradable, sustainable, etc.; biomass is therefore widely studied and used (i.e., making houses as construction materials, feeding animals as forage, cooking as fuels, producing fuels as feedstock, producing chemicals as raw materials) (Ruan et al., 2019).

The properties of biomass influence the design and operation of storage and conversion systems for biomass, they also affect the transportation cost of biomass. Because the properties usually have correlations with each other, one property may help estimate another property. Therefore, full understanding of the properties of biomass is important and essential.

Biomass is a kind of fuel, it can also be converted to the other fuels (biofuels), i.e., biooil, biogas, biochar, through some conversion technologies, i.e., physical technologies, chemical technologies, biochemical technologies, and thermochemical technologies. The biofuel properties and yields are significantly varied by many factors, i.e., biomass properties, unit structures, treating processes, operation conditions (e.g., temperature, pressure, time), knowing the conversion technologies of biomass is therefore helpful for us to acquire the biomass products we need. This chapter initially introduces the properties of biomass, and these include the physical properties, chemical properties, and exergy properties. The biomass conversion technologies are then also detailed, and these include physical technologies, chemical technologies, biochemical technologies, and thermochemical technologies. Hopefully, the contents presented in this chapter will contribute significantly to the present and future of biomass energy.

## 5.2 PHYSICAL PROPERTIES OF BIOMASS

The physical properties of biomass mainly include moisture content, particle size, density, and porosity. The other physical properties mainly include thermal conductivity, specific heat, and so on. In this section, the main physical properties (i.e., moisture content, particle size, density, and porosity) of biomass are introduced.

### 5.2.1 Moisture Content

Moisture content is significant for the utilization of biomass. The moisture content is defined as follows:

$$\text{Moisture content} = \frac{w_{\text{moisture}}}{w_{\text{biomass}}} \times 100\% \quad (5.1)$$

where
$w_{\text{moisture}}$ is the weight of moisture (g or kg)
$w_{\text{biomass}}$ is the weight of biomass (g or kg)

Generally, moisture content of solid biomass is low, for instance, dry straw and dry sawdust have moisture contents of less than 20%. However, for some biomass, i.e., ludge, algae, and residues, the moisture contents would be higher than 95%.

Moisture content plays an important role in the thermochemical reaction processes, i.e., pyrolysis and gasification. During the initial heating process, moisture content in biomass releases when external heat is input. Higher moisture content requires more energy to release the water, and it takes longer time to heat the biomass to a higher temperature, thereby delaying the heating process. On the other hand, it is not proper if the moisture content is too low. For example, for some reactions like wood carbonization, it is very violent at exothermic stage when the biomass has too low a moisture content, which will reduce the yield and strength of charcoal.

Differences in moisture contents could be resulted from using different collection, storage, and drying procedures (Zhang et al., 2012a). Table 5.1 shows the moisture contents of some biomass. It is observed that the moisture contents of biomass vary significantly in the wide range of 3.52%–49.86%, and they can be lowered by

**TABLE 5.1**
**Moisture Contents of Biomass at Different Drying Temperatures (Samuelsson et al., 2006)**

| Biomass | Mean (%, 80°C) | Mean (%, 105°C) | Standard Deviation (%) |
|---|---|---|---|
| Birchbark | 45.94 | 46.49 | 0.146 |
| Birch chips | 40.63 | 40.67 | 0.064 |
| Cork | 3.52 | 4.08 | 0.083 |
| Eucalyptus | 45.99 | 46.21 | 0.047 |
| Hardwood | 14.37 | 14.48 | 0.086 |
| Miscanthus | 37.32 | 37.59 | 0.297 |
| Olive stones | 8.9 | 9.35 | 0.021 |
| Pitchy wood | 12.85 | 13.47 | 0.055 |
| Salix | 28 | 28.2 | 0.035 |
| Spruce chips | 20.04 | 20.4 | 0.038 |
| Spruce needles | 49.86 | 50.27 | 0.072 |
| Triticale | 11.98 | 12.9 | 0.018 |

air temperatures (Samuelsson et al., 2006). When the mean air temperature was increased from 80°C to 105°C (31.25% in increase), the moisture contents released were increased from 3.52%–49.86% to 4.08%–50.27% (0.10%–15.91% in increase).

### 5.2.2 Particle Size

The particle size of biomass is one of the most important properties of biomass, in particular, for cellulose, lignin, and hemicellulose biomass such as straw, wood, and plant shell.

Table 5.2 shows the particle sizes of some biomass. It is observed that the particle sizes are in the range of 0.325–2.0 mm, and different biomass have different particle sizes, and these are mainly caused by the mill methods. The milling methods can be knife milling, hammer milling, ball milling, and so on (Zhang et al., 2013).

Particle size influences the storage and transportation of biomass, and it will cause large space and cost waste if we do not take any procedures to reduce the particle size of biomass. Physical size reduction is most widely applied among kinds of size reduction methods. However, physical size reduction requires energy consumption. Generally, the smaller the size of biomass particles, the more energy it consumes. Table 5.3 shows the energy required for the milling of different biomass (Vidal et al., 2011).

### 5.2.3 Density

Density is another important parameter of biomass, it not only influences the collection and transport of resources but also affects the device size and utilization efficiency. For granular biomass, there are generally three kinds of density: bulk density, apparent density, and true density (Basu, 2018).

Bulk density is defined as the mass per unit bulk volume of biomass which not only includes pores between particles but also pores in the interior of the particle. The definition is expressed by:

$$\rho_{\text{bulk}} = \frac{\text{Total mass of biomass particles or stack}}{\text{Bulk volume occupied by biomass particles or stack}} \tag{5.2}$$

**TABLE 5.2**
**Particle Sizes of Different Biomass (Guedes et al., 2018)**

| Biomass | Particle Size (mm) |
|---|---|
| Rapeseed cake | 2 |
| Palm trunk | 1.5 |
| Palm frond | 1.5 |
| Palm leaf | 1.5 |
| Palm leaf rib | 1.5 |
| Rice husk | 0.6 |
| Cassava rhizome | 0.325 |

**TABLE 5.3**
**Energy Required for Biomass Milling (Vidal et al., 2011)**

| Biomass | Mill Type | Size | Specific Energy (kWh/t) |
|---|---|---|---|
| Wood | | | |
| Hardwoods | Knife, hammer | 1.6 mm | 130 |
| Aspen | Attrition | 1.6 mm | >200 |
| | Hammer | 1.6 mm | 100–150 |
| | Knife | 1.6 mm | 100–150 |
| Poplar | Hammer | <1 mm (95%) | 85 |
| Pine | Hammer | <1 mm (95%) | 118 |
| Pine bark | Hammer | <1 mm (95%) | 20 |
| Hybrid poplar | Attrition | <2.5 mm (>80%) | >130 |
| | Hammer | <3.5 mm (>80%) | 88–112 |
| Herbaceous | | | |
| Straw | Knife | 1.6 mm | 7.5 |
| | Knife | 1.6 mm | <10 |
| | Hammer | 1.6 mm | 42 |
| | Hammer | 1.6 mm | 50 |
| | Hammer | 1.6 mm | 37 |
| | Hammer | 3.2 mm | 28–35 |
| Corn stover | Knife | 3.2 mm | 20 |
| | Hammer | 1.6 mm | 14 |
| | Hammer | 1.6 mm | 15 |
| | Hammer | 3.2 mm | 28–35 |
| Switchgrass | Hammer | 1.6 mm | 52 |
| | Hammer | 3.2 mm | 28–35 |

Zhang et al. (2012a, c) explored the relationship between the bulk density and the average particle size for some biomass (i.e., the rice husks, rice straws, wheat straws). Take rice husk, for example, a negative linear relationship was observed:

$$\mathrm{BD} = 705.52 - 710.20\mathrm{PS} \quad \left(R^2 = 1.00\right) \tag{5.3}$$

where
BD is the bulk density (kg/m$^3$)
PS is the average particle size (mm)

Generally speaking, apparent density is larger than bulk density, and it is defined as the mass per unit apparent volume of biomass which only includes pores in the interior of the particle. The definition is expressed by:

$$\rho_{\text{apparent}} = \frac{\text{Total mass of biomass}}{\text{Apparent volume of biomass including solids and internal pores}} \tag{5.4}$$

True density is the largest among the three densities, and the solid volume is not related to the pores between particles or pores in the interior of the particle. Table 5.4 shows different densities of some kinds of biomass. It is observed that the bulk densities are mainly in the range of 40–450 kg/m$^3$ whereas the apparent densities are mainly in the range of 547.9–630.1 kg/m$^3$ (Zhu and Lu, 2014).

As compared with apparent density, true density is more difficult to be measured. Basu (2018) gives the relationships among the three densities as follows:

$$\rho_{\text{true}} = \frac{\text{Total mass of biomass}}{\text{Solid volume in biomass}} \tag{5.5}$$

$$\rho_{\text{apparent}} = \rho_{\text{true}}\left(1 - \varepsilon_p\right) \tag{5.6}$$

$$\rho_{\text{bulk}} = \rho_{\text{apparent}}\left(1 - \varepsilon_b\right) \tag{5.7}$$

where

$\varepsilon_p$ is the voidage in a biomass particle

$\varepsilon_b$ is the voidage of particle packing

### 5.2.4 Porosity

The porosity of biomass refers to the ratio of pore volume to the total volume of biomass, and the pore volume includes pores between particles and pores in the interior of the particle. There are generally two definitions for porosity. One is effective porosity which is the ratio of interconnected pores volume to bulk volume. The other is absolute porosity or total porosity, and it is defined as the ratio of all pores to bulk volume.

Usually, porosity refers to effective porosity and it's related to the shape, structure, and arrangement of porous materials particles. In general, the porosity of biomass is determined using the water pycnometer method and the porosity of biomass can also be calculated from the following equation (Zhang et al., 2012b):

$$P = \frac{V_i - V_f}{V_s} \times 100 \tag{5.8}$$

**TABLE 5.4**
**Different Densities of Different Biomass (Zhu and Lu, 2014)**

| Biomass | Average Diameter (mm) | Bulk Density (kg/m$^3$) | Apparent Density (kg/m$^3$) |
|---|---|---|---|
| Sawdust | 0.78 | 150–200 | 570.3 |
| Wood chips | – | 200–300 | – |
| Peanut hull | 0.61 | 200–250 | 566.8 |
| Coconut shell | 0.98 | 400–450 | 547.9 |
| Rice husk | 1.50 | 100–140 | 630.1 |
| Broken straw | – | 40–60 | – |
| Straw bale | – | 100–120 | – |

**TABLE 5.5**
**Some Parameters of Corn Residues (Zhang et al., 2012b)**

| Corn[a] | Moisture Content (%) | Average Particle Size (mm) | Bulk Density ($kg/m^3$) | Porosity (%) |
|---|---|---|---|---|
| Cobs | 6.38 | 0.56 | 282.38 | 67.93 |
| Leaves | 7.92 | 0.7 | 81.61 | 86.06 |
| Stalks | 6.4 | 0.49 | 127.32 | 58.51 |

[a] Average of three replicates.

where
$P$ is the porosity of the sample (%)
$V_i$ is the combined volume of the sample plus added water (ml)
$V_f$ is the final total volume of the sample and added water (ml)
$V_s$ is the volume of the sample (ml)

Zhang et al. (2012a, b, c) explored the relationship between porosity and the average particle size for some biomass (i.e., the rice husks, rice straws, corn residues). Table 5.5 shows the average particle size, bulk density, and porosity of corn residues. A positive linear relationship was observed:

$$P = 5.55 + 130.95\text{PS} \quad \left(R^2 = 1.00\right) \tag{5.9}$$

where
$P$ is the porosity (%)
PS is the average particle size (mm)

## 5.3 CHEMICAL PROPERTIES OF BIOMASS

The chemical properties of biomass mainly include elemental compositions, chemical compositions, and heating values. The other chemical properties mainly include ignition temperature, and stoichiometric ratio. In this section, the main chemical properties (i.e., proximate analysis results, ultimate analysis results, and heating values) of biomass are introduced.

### 5.3.1 Proximate Analysis

The proximate analysis of biomass mainly gives the contents of moisture, volatile matter (VM), fixed carbon (FC), and ash of the biomass.

Moisture (M) is the basic analysis index of biomass, and it is crucial not only for transport and storage of biomass but also for pyrolysis and gasification of biomass. Moisture can be divided into two categories; one is crystalline water and the other is free water. For crystalline water in biomass, it is contained within mineral and it is generally counted in the VM. For free water in biomass, it includes external water and inherent water. The external water is easy to evaporate at normal atmospheric temperature depending on the temperature and humidity of air. However, the inherent water is difficult to evaporate unless the temperature is higher than the boiling point of water.

VM represents the quantity of gaseous products of biomass during the thermal decomposition process, and it depends on the material, size, shape of crucible, as well as the heating device and heating conditions (such as the heating rate, temperature, and residence time). Therefore, determination of the VM of biomass should be under certain conditions or according to certain standards.

Ash is the solid residue left after the release of combustible matter in biomass during the complete combustion, and it mainly consists of CaO, $K_2O$, $Na_2O$, MgO, $SiO_2$, $Fe_2O_3$, and $P_2O_5$.

FC refers to the solid residues after removing moisture, VM, and ash (subtracting). It contains mainly carbon; sometimes the other elements such as hydrogen, oxygen, nitrogen, and sulfur may also be included.

The proximate analysis of biomass is mainly performed according to ASTM standard method and thermogravimetric analysis method (Cai et al., 2017). Table 5.6 shows the proximate analysis results of biomass samples under ASTM standards.

**TABLE 5.6**
**Proximate Analysis of Different Biomass Samples (%) (Erol et al., 2010)**

| Biomass | Moisture | Fixed Carbon[a] | Ash[a] | Volatile Matter[a] |
|---|---|---|---|---|
| Broad bean husk | 10.75 | 8.68 | 5.88 | 85.44 |
| Sunflower shell | 10.25 | 11.7 | 3.62 | 84.68 |
| French bean stalk | 9 | 4.67 | 6.32 | 89.01 |
| Sunflower stalk/stover | 8.12 | 5.17 | 8.98 | 85.85 |
| Sourcherry stalk | 6 | 14.1 | 5.05 | 80.85 |
| Walnut shell | 10 | 16.94 | 3.89 | 79.17 |
| Almond shell | 1.25 | 18.73 | 4.05 | 77.22 |
| Peanut shell | 8.12 | 13.6 | 5.99 | 80.41 |
| Cornelian cherry stone | 5.5 | 23.8 | 2.96 | 73.54 |
| Apricot stone | 4 | 17.83 | 1.04 | 81.13 |
| Peach stone | 5 | 20.79 | 1.05 | 78.16 |
| Apricot bagasse | 3.5 | 15.8 | 3.89 | 80.31 |
| Peach bagasse | 6.5 | 6.15 | 1.87 | 91.98 |
| Hybrid poplar | 9 | 6.87 | 3.44 | 89.69 |
| Ash tree | 8.75 | 14.12 | 5.75 | 80.13 |
| Pine cone | 9.25 | 15.15 | 6.89 | 77.96 |
| Soybean cake | 12.5 | 16 | 7.14 | 76.86 |
| Cotton cake | 8.25 | 11.58 | 4.77 | 83.65 |
| Rapeseed | 10.75 | 5.6 | 8.13 | 86.27 |
| Potato peel | 8.5 | 9.56 | 6.29 | 84.15 |

[a] db: dry basis.

It is observed that the contents of moisture, FC (dry basis), VM (dry basis), and ash (dry basis) vary significantly in the ranges of 1.25%–12.5%, 4.67%–23.8%, 73.54%–91.98%, and 1.04%–8.98%, respectively.

### 5.3.2 Ultimate Analysis

The ultimate analysis of biomass mainly pays attention to the five elements of carbon (C), hydrogen (H), oxygen (O), nitrogen (N), and sulfur (S). The other elements are trace elements, they are not reported for the ultimate analysis whereas may be included in the ash analysis. The elemental compositions of biomass are the base for further analysis and calculation like prediction of the heating value and chemical exergy of biomass. Generally, determination of the contents of elements focuses on the organic compound of biomass and it therefore seldom reports the inorganic parts of biomass such as moisture and ash.

Ultimate analysis can be described as follows using as received basis or dry ash free basis:

$$\text{As received basis: } C + H + O + N + S + M + ASH = 100\% \tag{5.10}$$

$$\text{Dry ash free basis: } C + H + O + N + S = 100\% \tag{5.11}$$

where

C is the weight percentages of carbon (%)
H is the weight percentages of hydrogen (%)
O is the weight percentages of oxygen (%)
N is the weight percentages of nitrogen (%)
S is the weight percentages of sulfur (%)
M is the weight percentages of moisture (%)
ASH is the weight percentages of ash (%)

For some woody biomass, the ultimate analysis results of elements may don't change significantly, i.e., about 50% carbon, 6% hydrogen, 44% oxygen, and between 0.1% and 0.5% nitrogen (Nunes et al., 2018). However, in most cases, the ultimate analysis results change significantly, especially for different biomass. Table 5.7 shows the ultimate analysis results of different biomass (dry basis) (Ozyuguran et al., 2018). It is observed that the elements of carbon (C), hydrogen (H), oxygen (O), nitrogen (N), and sulfur (S) vary significantly in the ranges of 37.02%–53.30%, 5.01%–6.79%, 32.01%–47.42%, 0%–8.02%, and 0.31%–0.66%, respectively.

### 5.3.3 Heating Values

Heating value is one of the most important properties of biomass, and it is closely related to the utilization of biomass. It refers to the heat released from the complete combustion of a fuel, and it reflects the ability to convert chemical energy into thermal energy. There are two heating values, one is called higher heating value (HHV) and the other is called lower heating value (LHV). The HHV of biomass is the heat

**TABLE 5.7**
**Ultimate Analysis Results of Different Biomass (Ozyuguran et al., 2018)**

| | Ultimate Analysis (wt%) | | | | |
|---|---|---|---|---|---|
| **Biomass** | C | H | O | N | S |
| Elaeagnus | 44.26 | 6.19 | 46.86 | 1.37 | 0.41 |
| Tea caffeine | 48.59 | 6.43 | 34.78 | 2.59 | 0.46 |
| Tea waste | 45.04 | 6.07 | 40.21 | 3.48 | 0.5 |
| Cornstalk | 42.02 | 5.58 | 43.53 | 1.24 | 0.43 |
| Tobacco waste | 37.02 | 5.01 | 39.95 | 2.2 | 0.45 |
| Apricot stone | 48.07 | 5.99 | 43.89 | 0.05 | 0.39 |
| Apricot pulp | 44.37 | 5.87 | 47.42 | 0.95 | 0.32 |
| Peach pulp | 43.84 | 6.51 | 42.8 | 1.04 | 0.37 |
| Damson plum stone | 50.81 | 6.36 | 40.39 | 1.07 | 0.36 |
| Cornelian cherry stone | 49.03 | 5.86 | 42.67 | 0.05 | 0.34 |
| Sunflower stem and stalk | 37.94 | 5.19 | 46.15 | 0.31 | 0.35 |
| Daphne | 49.03 | 6.4 | 35.9 | 0.94 | 0.42 |
| Thyme | 44.53 | 6.01 | 39.34 | 0.81 | 0.36 |
| Locust bean | 44.31 | 5.7 | 43.1 | 0.92 | 0.42 |
| Flos lavandulae romanae | 45.28 | 5.89 | 36.42 | 0.94 | 0.42 |
| Apple pulp | 47.05 | 6.7 | 42.73 | 0.86 | 0.35 |
| Artichoke husk and waste | 42.08 | 5.92 | 45.88 | 0.83 | 0.36 |
| Sunflower stem | 39.9 | 5.38 | 42.8 | 0.42 | 0.4 |
| Sourcherry stem | 44.78 | 5.75 | 43.22 | 0.5 | 0.4 |
| Soybean residue | 42.96 | 6.21 | 35.8 | 8.02 | 0.57 |
| Black sesame residue | 45.93 | 6.79 | 32.01 | 6.32 | 0.66 |
| Cotton residue | 45.24 | 6.46 | 33.41 | 6.37 | 0.65 |
| Pea stem | 38.97 | 5.45 | 40.31 | 1.79 | 0.42 |
| Grape seed | 50.47 | 6.2 | 35.83 | 2.42 | 0.47 |
| Pine cone | 48.28 | 5.73 | 43.89 | 0.1 | 0.4 |
| Peach stone | 51.98 | 6.13 | 40.41 | 0.02 | 0.48 |
| Sourcherry stone | 53.3 | 6.69 | 37.33 | 1.58 | 0.39 |
| Ash tree wood | 46.72 | 5.95 | 45.32 | 0 | 0.32 |
| Green bean stem and husk | 41.26 | 5.62 | 42.35 | 0.78 | 0.39 |
| Red lentil hull | 43.9 | 6.31 | 42.63 | 1.54 | 0.37 |
| Chickpea husk | 43.8 | 5.81 | 45.67 | 0.38 | 0.35 |
| Broad bean husk | 40.11 | 5.52 | 44.98 | 1.35 | 0.33 |
| Coconut shell | 50.34 | 6.26 | 42.08 | 0 | 0.31 |
| Cacao husk | 43 | 5.69 | 44.38 | 2.1 | 0.41 |
| Peanut husk | 46.89 | 5.9 | 46.07 | 0.61 | 0.37 |
| Broad bean husk | 41.33 | 5.9 | 46.32 | 0.39 | 0.34 |
| Almond shell | 47.7 | 5.88 | 42.58 | 0.05 | 0.31 |
| Walnut shell | 48.23 | 6 | 44.42 | 0.12 | 0.34 |

generated by unit biomass after complete combustion, and the produced water in exhaust gas is in liquid form (latent heat of steam is added to the heat). The LHV of biomass is the heat released by unit biomass after complete combustion. However, the produced water in exhaust gas is in gas form, and no latent heat of steam is added to the heat. The relationship between HHV and LHV can be described by (Basu, 2018):

$$\text{LHV} = \text{HHV} - h_g\left(9\frac{\text{H}}{100} + \frac{\text{M}}{100}\right) \tag{5.12}$$

where

LHV is the lower heating value of biomass (kJ/kg)
HHV is the higher heating value of biomass (kJ/kg)
$h_g$ is the enthalpy difference between stem and liquid water at 25°C, 2,441 kJ/kg

The HHV of biomass can be measured directly, i.e., through an oxygen bomb calorimeter. It can also be estimated based on the proximate analysis results or ultimate analysis results according to the relationships as shown in Table 5.8.

The heating value of biomass not only depends on the types of biomass but also relates to the moisture content. Table 5.9 shows the relationship between LHV and moisture content of some biomass. It is observed that when the moisture content increases in the range of 5%–22%, the lower LHVs of different biomass samples decrease from 14.00%–18.37% to 11.00%–14.61%. Therefore, the higher moisture content of biomass, the lower LHV of biomass.

**TABLE 5.8**
**Correlations Used for Predicting the HHV of Biomass (Yin, 2011)**

| Equation[a] | Based | Unit |
|---|---|---|
| HHV = 19.914 − 0.2324Ash | Proximate analysis | MJ/kg |
| HHV = −3.0368 + 0.2218VM + 0.2601FC | Proximate analysis | MJ/kg |
| HHV = 0.3536FC + 0.1599VM − 0.0078Ash | Proximate analysis | MJ/kg |
| HHV = 0.3259C + 3.4597 | Ultimate analysis | MJ/kg |
| HHV = 354.3FC + 170.8VM | Proximate analysis | kJ/kg |
| HHV = 35430 − 183.5VM − 354.3Ash | Proximate analysis | kJ/kg |
| HHV = −10.8141 + 0.3133VM + 0.3133FC | Proximate analysis | MJ/kg |
| HHV = −0.763 + 0.301C + 0.525H + 0.064O | Ultimate analysis | MJ/kg |
| HHV = 0.4373C − 1.6701 | Ultimate analysis | MJ/kg |

[a] Dry biomass basis (wt.%).
[b] O* is the sum of the contents of oxygen and other elements in the organic matter (O* = 100–C–H–Ash).

**TABLE 5.9**
**LHVs (MJ/kg) of Biomass with Different Moisture Contents (%) (Zhu and Lu, 2014)**

| Moisture | 5 | 7 | 9 | 11 | 12 | 14 | 16 | 18 | 20 | 22 |
|---|---|---|---|---|---|---|---|---|---|---|
| Cornstalk | 15.42 | 15.04 | 14.66 | 14.28 | 14.09 | 13.71 | 13.33 | 12.95 | 12.57 | 12.19 |
| Sorghum stalk | 15.74 | 15.36 | 14.97 | 14.59 | 14.39 | 14.01 | 13.62 | 13.24 | 12.85 | 12.46 |
| Cotton stalk | 15.95 | 15.55 | 15.17 | 14.77 | 14.58 | 14.19 | 13.80 | 13.41 | 13.02 | 12.64 |
| Bean stalk | 15.84 | 15.31 | 14.95 | 14.57 | 14.37 | 13.99 | 13.61 | 13.22 | 12.84 | 12.45 |
| Wheat straw | 15.44 | 15.06 | 14.68 | 14.30 | 14.15 | 13.73 | 13.36 | 12.97 | 12.60 | 12.22 |
| Rice straw | 14.18 | 13.83 | 13.48 | 13.13 | 12.95 | 12.60 | 12.25 | 11.90 | 11.55 | 11.19 |
| Millet straw | 14.79 | 14.43 | 14.06 | 13.69 | 13.51 | 13.15 | 12.78 | 12.46 | 12.05 | 11.69 |
| Willow branch | 16.32 | 15.93 | 15.52 | 15.13 | 14.93 | 14.54 | 14.13 | 13.74 | 13.34 | 12.95 |
| Poplar branch | 14.00 | 13.61 | 13.26 | 12.91 | 12.74 | 12.39 | 12.04 | 11.69 | 11.35 | 11.00 |
| Cow dung | 15.38 | 14.96 | 14.59 | 14.21 | 14.02 | 13.64 | 13.26 | 12.89 | 12.43 | 12.13 |
| Masson pine | 18.37 | 17.93 | 17.49 | 17.05 | 16.83 | 16.38 | 15.94 | 15.49 | 15.05 | 14.61 |
| Birch | 16.97 | 16.42 | 16.13 | 15.72 | 15.51 | 15.10 | 14.69 | 14.28 | 13.87 | 13.46 |
| Linden wood | 16.65 | 16.25 | 15.84 | 15.44 | 15.24 | 14.84 | 14.43 | 14.02 | 13.62 | 13.21 |

## 5.4 EXERGY PROPERTIES OF BIOMASS

Exergy is the maximum work obtainable when a matter is brought to a state of thermodynamic equilibrium with the common components of the natural surroundings by means of reversible processes (Szargut, 1980). It is a very important method for evaluating energy materials and systems and is therefore widely used in the fields of resource accounting, process optimization, economy evaluation, and environment assessment (Dincer, 2002; Daniel and Rosen, 2002; Zhang et al., 2017a). To evaluate and/or study the material characteristics, transfer processes, utilization systems, and environment effects of biomass from exergy aspect, the knowledge and determination of the exergy of biomass are of significant importance.

### 5.4.1 Exergy of Biomass

Generally, there are four forms of exergy for a material: kinetic exergy, potential exergy, physical exergy, and chemical exergy (Figure 5.1). The kinetic exergy is associated with relative motion difference between the material and its surroundings. The potential exergy is associated with the gravitational or electromagnetic difference. The physical exergy is from differences in the pressures and temperatures, and the chemical exergy is from differences in the components and concentrations (Szargut, 2005).

Since the kinetic exergy and potential exergy account for less than 0.001% of the total exergy of a material (mainly fuel), they can therefore be neglected (Zhang, 2012; Zhang et al., 2015b). The physical exergy of a material attributed by pressure and temperature differences is usually also neglected because the material in environmental condition is in equilibrium with the pressure and temperature of the environment.

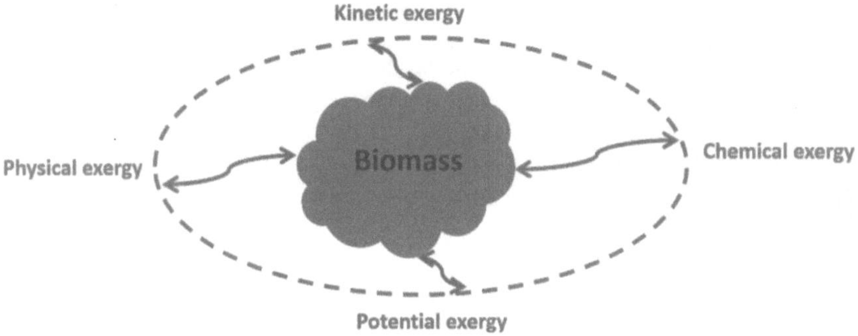


**FIGURE 5.1** Definition of exergy for biomass.

The chemical exergy of biomass, therefore, appears to be a more representative index than the total exergy of a fuel (Rosen and Dincer, 1999).

According to the general definition of exergy for a biomass, the (chemical) exergy of biomass can be calculated based on a multi-process thermodynamic model which is also a comprehensive thermodynamic model (Figure 5.2).

The multi-process thermodynamic model includes three sub-process models: (1) the oxygen separation process, (2) the chemical reaction process, and (3) the products diffusion process.

The oxygen separation process means oxygen ($O_2$) is separated from the environment at the environmental sate ($P_0$, $T_0$), and the exergy involved is oxygen separation exergy. The chemical reaction process requires the biomass react with oxygen at the environmental sate ($P_0$, $T_0$), and the products are the environmental products which are the environmental compositions. The exergy involved in this process is called chemical reaction exergy. The products diffusion process means the environmental products diffuse to the environment and get equilibrium with the environment at the environmental sate ($P_0$, $T_0$). The exergy involved in this process is called products diffusion exergy. The (chemical) exergy of biomass is the sum of oxygen separation exergy, chemical reaction exergy, and products diffusion exergy. This multi-process thermodynamic model would yield accurate results whereas the calculation process is a little complex.

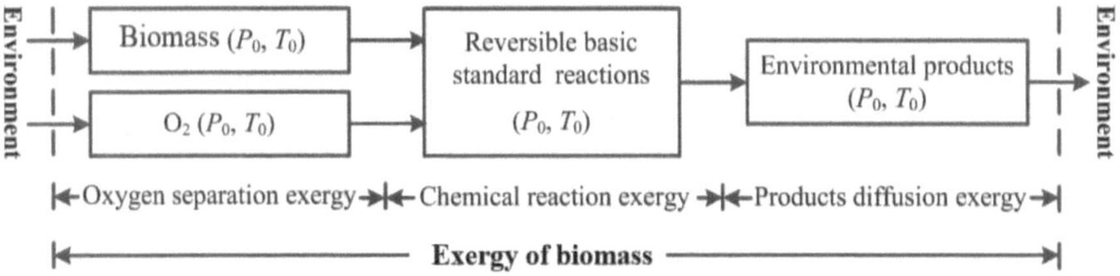


**FIGURE 5.2** Multi-process thermodynamic model for calculating the exergy of biomass.

### 5.4.2 Equations Used

According to the general definition of exergy for a biomass, the multi-process thermodynamic exergy of a biomass includes: (1) the oxygen separation process, (2) the chemical reaction process, and (3) the products diffusion process. The exergy for each process can be formulated.

Oxygen separation exergy ($Ex_S$) can be expressed by:

$$Ex_S = n_{O_2} RT_0 \ln \frac{p^0_{O_2}}{p_0} \tag{5.13}$$

where

$Ex_S$ is the oxygen separation exergy
$n_{O_2}$ is the molar number of separated oxygen
$p^0{}_{O_2}$ is the partial pressure of oxygen at standard or environmental state

The chemical reaction exergy ($Ex_R$) can be expressed by:

$$Ex_R = -\Delta H^0 + T_0 \Delta S^0 \tag{5.14}$$

where

$Ex_R$ is the chemical reaction exergy

Because free enthalpy of standard reaction is defined as:

$$\Delta G^0 = \Delta H^0 - T_0 \Delta S^0 \tag{5.15}$$

where

$\Delta G^0$ is the free enthalpy of standard reaction

Then we have:

$$Ex_R = -\Delta H^0 + T_0 \Delta S^0 = -\Delta G^0 \tag{5.16}$$

The gas diffusion exergy ($Ex_{D,\,gas}$) can be expressed as

$$Ex_{D,\,\text{gas}} = -RT_0 \sum_k n_k \ln y_k^0 \tag{5.17}$$

where

$y^0{}_k$ is the pressure or volume ratio of gas product $k$ at standard or environmental state

The ash diffusion exergy can be expressed by:

$$Ex_{D,\,\text{ash}} = \sum_i m_i ex_i \tag{5.18}$$

where

$Ex_{D,\text{ash}}$ is the ash diffusion exergy
$m_i$ is the mass of ash component $i$
$ex_i$ is the standard chemical exergy of ash component $i$

The ash compositions of biomass mainly include $Al_2O_3$, CaO, $Fe_2O_3$, $K_2O$, MgO, MnO, $Na_2O$, $P_2O_5$, $SO_3$, $SiO_2$, and $TiO_2$.

The products diffusion exergy of biomass is the sum of gas diffusion exergy and ash diffusion exergy:

$$Ex_D = \sum_i m_i ex_i - RT_0 \sum_k n_k \ln y_k^0 \tag{5.19}$$

where

$Ex_D$ products diffusion exergy

The (chemical) exergy of biomass is the sum of oxygen separation exergy, chemical reaction exergy, and products diffusion exergy:

$$Ex = Ex_S + Ex_R + Ex_D \tag{5.20}$$

where

$Ex$ is the (chemical) exergy of biomass

Specifically,

$$Ex = n_{O_2} RT_0 \ln \frac{p_{O_2}^0}{p_0} - \left(-\Delta H^0 + T_0 \Delta S^0\right) + \sum_i m_i ex_i - RT_0 \sum_k n_k \ln y_k^0 \tag{5.21}$$

where

$i$ indicates the ash component $i$
$k$ indicates the gas product $k$

### 5.4.3 Exergy Properties of Biomass

Based on the basic properties (elements of C, H, O, N, and S, contents of moisture and ash, HHV) of biomass as shown in Table 5.10, the exergy properties of different biomass are detailed as shown in Table 5.11.

For the woody biomass, the oxygen separation exergy varies from –190.02 to –71.38 kJ/kg, making up 0.63%–0.96% of the total chemical exergy. The chemical reaction exergy varies from 9.08 to 24.36 MJ/kg, making up 94.42%–97.40% of the total chemical exergy. The products diffusion exergy varies from 356.75 to 1,067.02 kJ/kg, making up 3.24%–6.55% of the total chemical exergy. The total chemical exergy varies from 9.37 to 25.01 MJ/kg.

For the grass and agricultural biomass, the oxygen separation exergy varies from –80.95 to –196.88 kJ/kg, making up of 0.68%–0.91% of the total chemical exergy.

**TABLE 5.10**
**Basic Properties of Biomass (Zhang et al., 2016; Qian et al. 2017)**

| Biomass | C (%) | H (%) | O (%) | N (%) | S (%) | $H_2O$ (%) | Ash (%) | HHV (MJ/kg) |
|---|---|---|---|---|---|---|---|---|
| **Woody Biomass** | | | | | | | | |
| Pine | 50.01 | 5.43 | 37.77 | 0.38 | 0.03 | 4.74 | 1.64 | 19.96 |
| Fir | 45.89 | 5.42 | 38.44 | 0.27 | 0 | 9.53 | 0.45 | 17.80 |
| Conifer | 46.53 | 5.58 | 41.78 | 0.43 | 0 | 4.6 | 1.09 | 24.00 |
| Cardoon | 43.02 | 4.96 | 27.73 | 0.76 | 0.17 | 12 | 11.4 | 15.25 |
| Cytisus | 19.93 | 2.67 | 16.76 | 0.84 | 0.04 | 59.2 | 0.56 | 9.08 |
| Cedar | 46.21 | 5.38 | 35.03 | 0.09 | 0.01 | 12.4 | 0.9 | 18.17 |
| Willow | 48.51 | 5.95 | 41.57 | 0.39 | 0.03 | 2.39 | 1.15 | 19.28 |
| Poplar | 45.91 | 5.67 | 43.36 | 0.74 | 0.03 | 2.66 | 1.64 | 18.87 |
| Oak | 46.81 | 7.45 | 36.95 | 0.53 | 0.02 | 4.7 | 3.54 | 18.23 |
| Citrus | 47 | 6 | 43.2 | 1 | 0.03 | 0 | 2.8 | 18.56 |
| Olive | 48.2 | 5.3 | 44.2 | 0.7 | 0.03 | 0 | 1.5 | 19.10 |
| Pepper | 34.38 | 4.1 | 31.68 | 2.1 | 0.4 | 9.3 | 18.05 | 15.05 |
| Salix | 43.06 | 5.49 | 38.36 | 0.44 | 0 | 11.4 | 1.68 | 17.19 |
| Wood | 41.47 | 4.84 | 34.87 | 1.19 | 0.07 | 15.1 | 2.46 | 16.47 |
| Painted | 44.86 | 4.88 | 37.78 | 0.4 | 0.04 | 10.1 | 1.87 | 17.85 |
| Poplar bark | 52.4 | 6.55 | 38.4 | 0.3 | 0.1 | 0 | 2.2 | 20.90 |
| Willow | 49 | 6 | 42.7 | 0.6 | 0.06 | 0 | 1.6 | 19.60 |
| **Agriculture and Grass Biomass** | | | | | | | | |
| Rice husk | 29.17 | 3.42 | 39.65 | 0.36 | 0.04 | 9.4 | 17.8 | 9.74 |
| Mustard husks | 43.9 | 8.82 | 42.6 | 0.4 | 0.19 | 0 | 4.1 | 19.97 |
| Walnut blows | 53.6 | 6.54 | 36 | 1.4 | 0.11 | 0 | 2.4 | 21.21 |
| Rice straw | 30.2 | 4.56 | 51 | 0.34 | 0 | 0 | 14 | 10.34 |
| Corn straw | 45 | 5.91 | 40.7 | 0.6 | 0.07 | 0 | 7.7 | 18.20 |
| Mint straw | 45.1 | 5.53 | 35.8 | 2.5 | 0.25 | 0 | 10.8 | 17.90 |
| Rape straw | 46.2 | 6.1 | 42.4 | 0.5 | 0.1 | 0 | 4.7 | 18.77 |
| Bana grass | 45.2 | 5.41 | 38.7 | 0.8 | 0.12 | 0 | 9.8 | 17.86 |
| Kenaf grass | 46.6 | 5.78 | 42.9 | 1 | 0.14 | 0 | 3.6 | 18.71 |
| Switchgrass | 47.2 | 5.79 | 41.2 | 0.7 | 0.1 | 0 | 5.1 | 18.82 |
| Max | 46.04 | 5.76 | 43.79 | 0.83 | 0.08 | 12 | 3.49 | 19.62 |
| Vuka | 44.26 | 4.97 | 44.48 | 1.16 | 0.13 | 12 | 4.95 | 19.36 |
| Kadeth | 44.54 | 5.12 | 41.59 | 0.82 | 0.19 | 12 | 7.41 | 18.18 |
| Leger | 46.01 | 5.45 | 44.07 | 0.39 | 0.1 | 15 | 3.97 | 19.38 |
| Shaw | 45 | 6 | 46.52 | 0.42 | 0.05 | 10 | 1.99 | 19.50 |
| Tibor | 44.94 | 5.51 | 46.62 | 1.33 | 0.16 | 10 | 1.39 | 18.98 |
| Kustro | 46.61 | 5.62 | 45.85 | 0.14 | 0.07 | 17 | 1.7 | 19.36 |
| Danko | 45.67 | 6.04 | 45.14 | 0.33 | 0.1 | 18 | 2.71 | 19.25 |
| **Waste Biomass** | | | | | | | | |
| Chicken litter | 37.6 | 4.23 | 15.7 | 3.9 | 0.75 | 0 | 37.8 | 14.05 |
| Wood–agricultural residue | 50.7 | 5.8 | 39.8 | 0.4 | 0.04 | 0 | 3.3 | 19.87 |

(*Continued*)

**TABLE 5.10 (*Continued*)**
**Basic Properties of Biomass (Zhang et al., 2016; Qian et al. 2017)**

| Biomass | C (%) | H (%) | O (%) | N (%) | S (%) | $H_2O$ (%) | Ash (%) | HHV (MJ/kg) |
|---|---|---|---|---|---|---|---|---|
| Wood–almond residue | 47.4 | 5.49 | 39.6 | 0.6 | 0.07 | 0 | 6.9 | 18.63 |
| Wood–straw residue | 47.4 | 5.78 | 38.1 | 0.4 | 0.12 | 0 | 8.2 | 18.80 |
| Currency shredded | 42.9 | 5.95 | 43.6 | 1.8 | 0.3 | 0 | 5.5 | 17.71 |

**TABLE 5.11**
**Exergy Properties of Woody Biomass (Zhang et al., 2016; Qian et al. 2017)**

| | Value (MJ/kg) | | | | Percentage (%) | | |
|---|---|---|---|---|---|---|---|
| Biomass | OSE | CRE | PDX | CE | OSE | CRE | PDX |
| | | Woody Biomass | | | | | |
| Pine | –0.17 | 20.26 | 0.88 | 20.97 | 0.82 | 96.61 | 4.20 |
| Fir | –0.16 | 18.09 | 0.79 | 18.73 | 0.84 | 96.58 | 4.24 |
| Conifer | –0.16 | 24.36 | 0.81 | 25.01 | 0.63 | 97.40 | 3.24 |
| Cardoon | –0.16 | 15.39 | 1.07 | 16.3 | 0.96 | 94.42 | 6.55 |
| Cytisus | –0.07 | 9.08 | 0.36 | 9.37 | 0.76 | 96.91 | 3.81 |
| Cedar | –0.16 | 18.39 | 0.80 | 19.03 | 0.85 | 96.64 | 4.22 |
| Willow | –0.17 | 19.62 | 0.86 | 20.31 | 0.82 | 96.60 | 4.23 |
| Poplar | –0.15 | 19.29 | 0.83 | 19.96 | 0.77 | 96.64 | 4.16 |
| Oak | –0.18 | 18.42 | 0.86 | 19.1 | 0.95 | 96.44 | 4.51 |
| Citrus | –0.16 | 18.97 | 0.86 | 19.67 | 0.82 | 96.44 | 4.37 |
| Olive | –0.16 | 19.55 | 0.86 | 20.25 | 0.77 | 96.54 | 4.23 |
| Pepper | –0.12 | 15.22 | 1.02 | 16.12 | 0.71 | 94.42 | 6.31 |
| Salix | –0.15 | 17.49 | 0.77 | 18.11 | 0.82 | 96.58 | 4.25 |
| Wood | –0.14 | 16.79 | 0.72 | 17.37 | 0.81 | 96.66 | 4.16 |
| Painted | –0.15 | 18.16 | 0.79 | 18.8 | 0.79 | 96.60 | 4.19 |
| Poplar bark | –0.19 | 20.65 | 0.91 | 21.37 | 0.89 | 96.63 | 4.25 |
| Willow | –0.17 | 19.22 | 0.84 | 19.9 | 0.84 | 96.58 | 4.25 |
| | | Agriculture and Grass | | | | | |
| Rice husk | –0.08 | 10.12 | 0.55 | 10.59 | 0.76 | 95.56 | 5.16 |
| Mustard husks | –0.18 | 19.75 | 0.79 | 20.36 | 0.88 | 97.00 | 3.86 |
| Walnut blows | –0.20 | 20.91 | 0.93 | 21.65 | 0.91 | 96.58 | 4.29 |
| Walnut hulls and blows | –0.20 | 20.84 | 0.93 | 21.58 | 0.91 | 96.57 | 4.30 |
| Rice straw | –0.08 | 10.88 | 0.72 | 11.52 | 0.71 | 94.44 | 6.26 |
| Corn straw | –0.16 | 17.86 | 0.78 | 18.48 | 0.85 | 96.65 | 4.22 |
| Mint straw | –0.16 | 17.48 | 0.80 | 18.12 | 0.88 | 96.47 | 4.41 |
| Rape straw | –0.16 | 18.42 | 0.80 | 19.06 | 0.84 | 96.64 | 4.21 |

(*Continued*)

**TABLE 5.11 (*Continued*)**
**Exergy Properties of Woody Biomass (Zhang et al., 2016; Qian et al. 2017)**

| | Value (MJ/kg) | | | | Percentage (%) | | |
|---|---|---|---|---|---|---|---|
| **Biomass** | **OSE** | **CRE** | **PDX** | **CE** | **OSE** | **CRE** | **PDX** |
| Bana grass | −0.15 | 17.52 | 0.79 | 18.15 | 0.85 | 96.53 | 4.33 |
| Kenaf grass | −0.16 | 18.29 | 0.81 | 18.94 | 0.83 | 96.57 | 4.29 |
| Switchgrass | −0.16 | 18.45 | 0.82 | 19.11 | 0.85 | 96.55 | 4.28 |
| Max | −0.15 | 20.04 | 0.86 | 20.75 | 0.74 | 96.58 | 4.12 |
| Vuka | −0.14 | 19.86 | 0.91 | 20.63 | 0.68 | 96.27 | 4.40 |
| Kadeth | −0.15 | 18.59 | 0.95 | 19.39 | 0.75 | 95.87 | 4.89 |
| Leger | −0.15 | 19.79 | 0.84 | 20.48 | 0.74 | 96.63 | 4.11 |
| Shaw | −0.15 | 19.93 | 0.81 | 20.59 | 0.73 | 96.79 | 3.95 |
| Tibor | −0.15 | 19.51 | 0.82 | 20.18 | 0.72 | 96.68 | 4.06 |
| Kustro | −0.15 | 19.78 | 0.83 | 20.45 | 0.75 | 96.72 | 4.06 |
| Danko | −0.15 | 19.65 | 0.83 | 20.32 | 0.76 | 96.70 | 4.09 |
| | | **Waste** | | | | | |
| Chicken litter | −0.15 | 13.85 | 0.72 | 14.42 | 1.02 | 96.05 | 4.99 |
| Wood–agricultural residue | −0.18 | 19.55 | 0.87 | 20.24 | 0.87 | 96.59 | 4.30 |
| Wood–almond residue | −0.16 | 18.29 | 0.82 | 18.94 | 0.85 | 96.57 | 4.31 |
| Wood–straw residue | −0.17 | 18.51 | 0.82 | 19.17 | 0.87 | 96.56 | 4.29 |
| Currency shredded | −0.15 | 17.23 | 0.77 | 17.85 | 0.82 | 96.53 | 4.30 |

OSE, CRE, PDX, and CE indicate oxygen separation exergy, chemical reaction exergy, products diffusion exergy, and chemical exergy, respectively.

The chemical reaction exergy varies from 10.12 to 20.91 MJ/kg, making up 94.44%–97.00% of the total chemical exergy. The products diffusion exergy varies from 546.86 to 948.7 kJ/kg, making up 3.86%–6.26% of the total chemical exergy. The total chemical exergy varies from 10.59 to 21.65 MJ/kg. The last eight biomass are two kinds of wheat straws (max and vuka), two kinds of barley straws (kadeth and leger), two kinds of oat straws (shaw and tibor), and two kinds of rye straws (kustro and danko).

For the waste biomass, the oxygen separation exergy varies from −146.62 to −175.21 kJ/kg, making up of 0.82%–1.02% of the total chemical exergy. The chemical reaction exergy varies from 13.85 to 19.55 MJ/kg, making up 96.05%–96.59% of the total chemical exergy. The products diffusion exergy varies from 718.93 to 869.92 kJ/kg, making up 4.29%–4.99% of the total chemical exergy. The total chemical exergy varies from 14.42 to 20.24 MJ/kg.

We searched for some data of biomass in the reported articles and divided them into three kinds: woody biomass, grass and agricultural biomass, and other biomass. We also calculated their total chemical exergy, oxygen separation exergy, chemical reaction exergy, and products diffusion exergy by the above equations.

## 5.5 CONVERSION TECHNOLOGIES OF BIOMASS

Generally, the conversion technologies mainly include: (1) physical technologies, (2) chemical technologies, (3) thermochemical technologies, and (4) biochemical technologies. Biological technologies such as sequestration and biosequestration can also be used to produce biofuels, e.g., by using algae, bacteria, and cyanobacteria to capture and store the atmospheric greenhouse gas ($CO_2$) via biological processes (Kumar et al., 2018). Sometimes, multi-technologies combining the above technologies are also used.

### 5.5.1 Physical Technologies

The physical technologies used to convert biomass mainly include densification, physical (mechanical) extraction, and distillation.

#### 5.5.1.1 Densification

Solid biofuels like agricultural residues, forestry residues, and other waste materials are often difficult to be directly used because the fuels may have problems, e.g., uneven particle size, low bulk density, and low energy content. These disadvantages can be efficiently overcome by means of densification technologies which can make the solids into compact regular shapes. For some coarse or fluffy materials (e.g., tree branches, cereal straws), chopping, grinding, and pulverizing are efficient approaches to produce high-quality fuels (Champagne, 2008; Zhang et al., 2012c). Sometimes, briquetting or pelletization can be further used to improve the biofuel properties (Taghizadeh-Alisaraei et al., 2017; Jiang et al., 2014).

#### 5.5.1.2 Physical Extraction

Physical extraction is also called mechanical extraction in some publications (Naik et al., 2010). Oil seed crops with high oil content can be physically crushed, and the oils can be extracted and converted to esters which then can be directly used to replace diesel. There is a wide range of oilseed crops that can be used for biodiesel production, and the most commonly used crop is rapeseed. Other feedstock used include cotton, groundnuts, palm, sunflower, soya bean, and recycled frying oil (Champagne, 2008).

Mechanical pressure using screw press (expeller) or extractor can be used to recover crude vegetable oils from the oil seeds. Screw press can be applied in two ways: pre-pressing and full pressing. In the pre-pressing process, only part of the oil is recovered and the partially de-oiled meal (generally contains 18%–20% of oil) can be further extracted through using solvent such as hexane, ethanol, 2-propanol, chloroform, or toluene (Dutta et al., 2014). Combined pre-pressing and solvent extraction is commonly applied for oil seeds with high oil content (30%–40%). Full pressing requires 95 MPa to squeeze out as much oil as possible, preferably up to 3%–5% residual fat for animal materials. Full pressing can also be carried out in a pre-press and a final press (Naik et al., 2010).

#### 5.5.1.3 Distillation

Distillation is an important method for extracting essential oils from oil mixtures, it is the process of separating the components or substances from a liquid mixture by selective boiling and condensation. Distillation may result in essentially complete separation (nearly pure components) or it may be a partial separation that increases the concentration of selected components of the mixture. In either case, the process exploits differences in the volatilities of the mixture components.

For a typical distillation process, oil plants are initially crushed to release the oils, and they are then steam distilled. During this process, the essential oils vaporize and rise up with the steam. The oil vapors are captured and are allowed to condense back into biooil liquids (Champagne, 2008). To achieve a better distillation, high-tech technologies can be used, e.g., vacuum distillation (Rahman et al., 2018) and molecular distillation (Mezza et al., 2018).

### 5.5.2 Chemical Technologies

The chemical technologies used to convert biomass mainly include solvent extraction, supercritical (fluid) extraction, and supercritical water conversion.

#### 5.5.2.1 Solvent Extraction

Extraction refers to a process in which the desired substance is selectively removed from the raw materials by allowing the desired substance to dissolve into the solvent and subsequently recovering the substance from the solvent (Naik et al., 2010). To remove the particular substance from biomass, extraction and separation are both essential. Typically, biomass (wood, wheat straw, aromatic grasses, etc.) contains high volume of macromolecular compounds (polysaccharide, cellulose, hemicellulose, lignin, etc.) which are called primary metabolite. The other low-volume and high-value chemical molecules such as terpenoids, waxes, resins, sterols, and alkaloids are known as secondary metabolites or extractive biomass. In the biorefinery process, these chemicals are initially extracted from biomass by using solvent extraction (Naik et al., 2010).

Solvent extraction generally involves different unit operations: (1) extract the oil from the oil seeds using hexane as a solvent, (2) evaporate the solvent, (3) distil the oil–hexane mixture (also called miscella), and (4) toast the de-oiled meal. In some special cases, the other solvents can be used, e.g., acetone, ethanol, isopropanol, and halogenated solvents (mostly dichloromethane) (Naik et al., 2010).

#### 5.5.2.2 Supercritical Extraction

Supercritical extraction is also called supercritical fluid extraction (SFE). A supercritical fluid is defined as a substance that is under temperature and pressure conditions which are above its vapor liquid critical point (for water it is 644 K and 22 MPa; for $CO_2$ it is 304 K and 7.4 MPa).Under these conditions, the supercritical fluid has some important advantages, e.g., its physical properties (e.g., density, solubility, diffusion coefficient, dielectric constant) can be easily manipulated through slight variations in the temperature and pressure, making the supercritical fluid an excellent medium for extraction and reaction (Lee et al., 2014; Tan and Lee, 2011).

In biorefinery processes, the chemical molecules including terpenoids, waxes, resins, sterols, polycosanol, and alkaloids (known as secondary metabolites or extractive biomass) can be extracted from biomass through using supercritical extraction (Naik et al., 2010). During the process, aromatic woods (cedar wood, sandal wood, pine wood, etc.) are extracted to isolate extractives, the extracted lignocellulosic biomass is further used for hydrolysis and fermentation for producing biofuels. Supercritical extraction is usually performed using $CO_2$ (Naik et al., 2010; Lee et al., 2014).

#### 5.5.2.3 Supercritical Water Conversion

Supercritical water conversion of biomass to biofuels represents an alternative path to acid hydrolysis and enzymatic hydrolysis. Supercritical water can quickly convert cellulose to sugar and convert biomass into a mixture of oils, organic acids, alcohol, and methane. At supercritical (e.g., temperature of 300–644 K, pressure of 200–250 bar) or near critical states (523–573 K), acid ($H^+$) and base components ($OH^-$) of water are separated and dissolved in the biomass. The dissolved supercritical water rapidly breaks the bonds of cellulose and hemicellulose to produce small sugar molecules such as glucose, xylose, and oligosaccharide (Naik et al., 2010). These properties make supercritical water a very promising reaction medium without using any catalyst for the conversion of biomass to value-added biofuels. Supercritical water also breaks the cellulosic bonds and forms gaseous products. The typical overall reaction is (Naik et al., 2010):

$$2C_6H_{12}O_6 + 7H_2O \rightarrow CO_2 + 2CH_4 + CO + 5H_2 \tag{5.22}$$

### 5.5.3 Biochemical Technologies

The biochemical technologies for biomass conversion mainly include digestion, fermentation, and hydrolysis. In some publications, interesterification (Razzak et al., 2013) and transesterification (Suali and Sarbatly, 2012) are also included. In practical applications, the above technologies are usually combined for the conversion of a biomass.

#### 5.5.3.1 Digestion

Biomass digestion utilizes bacteria to facilitate the decomposition of biomass, and this may occur under anaerobic (anaerobic digestion) or aerobic (aerobic digestion) conditions. Compared with aerobic digestion, anaerobic digestion has some advantages: (1) higher organic removal efficiency, (2) higher organic loading rate, (3) lower nutrient requirement, and (4) lower energy requirement. Also, aerobic digestion may not facilitate conversions of biomass to any high value products. Anaerobic digestion is therefore more frequently considered or utilized (Okoro et al., 2017).

Anaerobic digestion is a biochemical conversion of biomass in which numerous species of bacteria participate in the decomposition of organic matter in the absence of oxygen to produce a biogas which contains mainly methane and carbon dioxide. Trace gases such as hydrogen sulfide may also be included (Champagne, 2008).

During an anaerobic digestion process, specialized microorganisms break down the complex organic matter (carbohydrates, proteins, and fats) into smaller molecules (sugars, amino acids, and fatty acids), methane and carbon dioxide are mainly produced through:

$$C_6H_{12}O_6 \rightarrow 3CO_2 + 3CH_4 \tag{5.23}$$

The anaerobic digestion process can be generally divided into four steps (Gould, 2015):

a. Hydrolysis: the large organic polymers (proteins, fats, carbohydrates, etc.) are broken down into smaller molecules (amino acids, fatty acids, simple sugars, etc.).
b. Acidogenesis: acidogenic microorganisms (acidogenic bacteria) further break down the organic matter to produce an acidic environment.
c. Acetogenesis: acetogens catabolize many of the products created in acidogenesis into acetic acid, $CO_2$, and $H_2$.
d. Methanogenesis: methanogens create methane from the final products of acetogenesis, as well as from some of the intermediate products from hydrolysis and acidogenesis mainly through two pathways:

$$CO_2 + 4H_2 \rightarrow CH_4 + 2H_2O \tag{5.24}$$

$$CH_3COOH \rightarrow CH_4 + CO_2 \tag{5.25}$$

Generally, anaerobic digestion is a versatile process, and it can be applied to a wide variety of biomass (Champagne, 2008).

#### 5.5.3.2 Hydrolysis

Chemical hydrolysis and enzymatic hydrolysis are the most successful ones to convert lignocellulosic material into fermentable sugars (Binod et al., 2011; Ruan et al., 2019).

Chemical hydrolysis can be performed in concentrated mineral acids such as $H_2SO_4$ and HCl. It can also be performed in less hazardous and more tractable cellulose solvents such as ionic liquids (ILs). The acid hydrolysis approach is comparatively cheap; however, its application is limited due to the low yields and unfavorable environmental issues involved with the use of strong acids (Ho et al., 2014; Binod et al., 2011).

Enzymatic hydrolysis is usually carried out by highly specific cellulase enzymes, and the products are usually reducing sugars including glucose.

Compared with chemical hydrolysis (performed in strong acids), enzymatic hydrolysis is mainly conducted under milder conditions (e.g., optimum pH of 4.8 and temperature of 45°C–50°C for cellulase enzyme). Also, enzymatic hydrolysis has some advantages: (1) high product yield, (2) less or no by-products, (3) high selectivity, and (4) no corrosion problem (Ho et al., 2014; Binod et al., 2011). However, enzymatic hydrolysis may take a much longer time (several days vs. a few minutes), and the final product may inhibit the enzyme (Binod et al., 2011). At present, a major bottleneck in enzymatic hydrolysis is the high cost of the enzymes (Ho et al., 2014; Binod et al., 2011).

#### 5.5.3.3 Fermentation

The term "fermentation" is derived from the Latin verb "fervere" (means to boil), which describes the action of yeast on the extracts of fruit or malted grain: boiling appears when carbon dioxide bubbles are produced by the anaerobic catabolism of sugar in the extract (Stanbury et al., 2017).

Broadly, fermentation is a metabolic process that consumes sugar in the absence of oxygen. It generally occurs in yeast, bacteria, or oxygen-starved muscle cells, and the products are alcohol, gases, or organic acids (Wikipedia, 2018).

Alcohol and gas are mainly produced through ethanol fermentation (Soccol et al., 2011):

$$C_6H_{12}O_6 \rightarrow 2C_2H_5OH + 2CO_2 \tag{5.26}$$

Organic acids are mainly produced through homolactic fermentation and heterolactic fermentation (Wikipedia, 2018):

$$\text{Homolactic}: C_6H_{12}O_6 \rightarrow 2CH_3CHOHCOOH \tag{5.27}$$

$$\text{Heterolactic}: C_6H_{12}O_6 \rightarrow CH_3CHOHCOOH + C_2H_5OH + CO_2 \tag{5.28}$$

The produced ethanol is then distilled and dehydrated to produce a high-octane, water-free alcohol (Champagne, 2008). Now, fermentation is commercially used on a large scale in various countries to produce bioethanol from biomass. The best known biomass are sugar crops and starch crops. Lignocellulosic biomass such as cereals and woods can also be processed (Champagne, 2008).

### 5.5.4 Thermochemical Technologies

Generally, thermochemical (or thermochemical) technologies mainly include combustion, gasification, pyrolysis, and liquidation. In some publications, some other methods are also included, e.g., incineration (Okoro et al., 2017), torrefaction (Williams et al., 2015), and hydrogenation (Goyal et al., 2008; Suali and Sarbatly, 2012).

#### 5.5.4.1 Combustion

Combustion is a thermochemical process where fuel is burnt in oxygen-excess atmosphere (air or oxygen), and the chemical energy is released through:

$$\text{Biofuel} + O_2 \rightarrow CO_2 + H_2O + \text{heat} \tag{5.29}$$

$$\text{Biofuel} + \text{Air} \rightarrow CO_2 + H_2O + N_2 + \text{heat} \tag{5.30}$$

If the biomass contains ash S or N, ash $SO_x$ or $NO_x$ would also be included in the above equations (Zhang et al., 2017a; Zhang et al., 2018).

Theoretically, the complete combustion would convert C and H to $CO_2$ and $H_2O$, respectively, and the produced heat can result in a high temperature of ~2,000°C

(Ruan et al., 2019). Practically, the incomplete combustion would lead to the production of CO, and this would lead to a low temperature of 800°C–1,000°C (Champagne, 2008). To help a better combustion, excess air or oxygen is usually used (Fokaides and Christoforou, 2016).

#### 5.5.4.2 Gasification

Gasification is also a thermochemical process where the reactions between biomass and gasification agent take place and syngas (also known as producer gas, product gas, synthetic gas, or synthesis gas) is produced. The syngas is mainly composed of CO, $H_2$, $N_2$, $CO_2$, and some hydrocarbons ($CH_4$, $C_2H_4$, $C_2H_6$, etc.). Very small amounts of $H_2S$, $NH_3$ and tars may also be included (Zhang et al., 2015a, b).

Based on the gasification agents used, biomass gasification can be divided into air gasification (Zhang et al., 2017b), oxygen gasification (Niu et al., 2014), steam gasification (Zhang et al., 2012d), carbon dioxide gasification (Gao et al., 2016), supercritical water gasification (Guo et al., 2015), etc. The detailed reactions can be referred to in the literature (Ruan et al., 2019).

Generally, oxygen gasification, steam gasification, carbon dioxide gasification, and supercritical water gasification result in higher HHVs of syngas than air gasification; however, air gasification is the most widely studied and applied because the gasification agent (air) is cheap, the reaction process is easy, the reactor structure is simple, etc.

#### 5.5.4.3 Pyrolysis

Pyrolysis is typically defined as a thermochemical decomposition of biomass feedstock at medium temperatures (300°C–800°C) to high temperatures (800°C–1,300°C) in the inert atmosphere (Zhang et al., 2017c). Some similar definitions are reported in other publications (Du et al., 2016; Naik et al., 2010; Fokaides and Christoforou, 2016; Ho et al., 2014; Champagne, 2008; Goyal et al., 2008; Jayasinghe and Hawboldt, 2012).

The overall reaction of biomass is:

$$\text{Biofuel} + \text{heat} \rightarrow \text{liquid} + \text{syngas} + \text{solid} \tag{5.31}$$

The products of liquid, syngas, and solid are actually all valuable fuels and are defined as biooil (or bio-oil), biosyngas (or bio-syngas), and biochar (or bio-char), respectively. The detailed reactions can be referred to in the literature (Ruan et al., 2019).

Generally, pyrolysis has the following advantages: (1) the main product is biooil and the yield may be as high as 75%, (2) the biooil may have high content of carbon, (3) the biooil may have low contents of nitrogen and sulfur, (4) the HHV of biooil may be very high, e.g., 42 MJ/kg, which is comparable with those of fossil fuels, (5) the residence time is generally short which decreases the operational cost, (6) desired product (biooil, biosyngas, or biochar) could be produced by adjusting the operational parameters, (7) biooil could be easily stored or transported, and (8) biomass feedstock may not need to be processed (Azizi et al., 2018; Ruan et al., 2019).

#### 5.5.4.4 Liquefaction

Liquefaction is a thermochemical process where biomass undergoes complicated chemical reactions in a solvent medium to form mainly liquid products (biooil or bio-oil). Alkali salts (i.e., sodium carbonate and potassium carbonate) act as catalyst for the hydrolysis of macromolecules (i.e., cellulose and hemicellulose) into smaller fragments.

Generally, catalytic aqueous liquefaction may result in higher biooil yield than the non-catalytic aqueous liquefaction. The biooils obtained from liquefaction process generally contain high contents of volatile organic acids, alcohols, aldehydes, ethers, esters, ketones, and non-volatile components, and they can be catalytically upgraded to an organic distillate product which is rich in hydrocarbons and useful chemicals (Naik et al., 2010).

Because the feeding system for liquefaction is complex and expensive, liquefaction has a low interest in applications (Champagne, 2008; Goyal et al., 2008).

### 5.5.5 The Other Technologies

The other technologies mainly include aqueous-phase hydrodeoxygenation (APHDO) and ultrasound technology.

#### 5.5.5.1 Aqueous-Phase Hydrodeoxygenation

APHDO can be used to convert sugar or sugar alcohol derived from biomass to alkanes or mono-functional compounds. Wang et al. (2018) introduced an approach to convert the concentration of sugar alcohol derived from biomass to high alcohol fuels: (1) the biomass (sugarcane bagasse) is processed by hydrolysis/hydrogenation to generate high-concentration (>20 wt%) sugar alcohol, (2) the productions and $H_2$ are fed into APHDO unit where the pressure of $H_2$ was kept at 4 Mpa by pre-controller and rear-pressure controller for partial deoxygenation of sugar alcohol into higher alcohol fuels over RuMo/CNT catalyst.

The alcohol fuel can be blended with gasoline, and it attracts an attention to be an alternative fuel for compression ignition engine, because the combustion performances can be improved and the particle emission can be reduced (Wang et al., 2018).

#### 5.5.5.2 Ultrasound Technology

Although the potential for biofuel production from biological conversion of lignocellulosic biomass and waste materials is huge, it is not explored to its full potential due to the low yields of products obtained. The main reason for the low yields is that the incomplete digestibility of cellulose molecules which are often shielded from hydrolysis into biofuels by a lignin protection and the extent of hydrolysis are also affected by the particle size and moisture content of the biomass or waste materials, crystalline structure of cellulose, as well as the degree of polymerization. Ultrasound irradiation or sonication is a green pre-treatment technique which can be used to enhance the accessibility and digestibility of biomass (Bundhoo and Mohee, 2018).

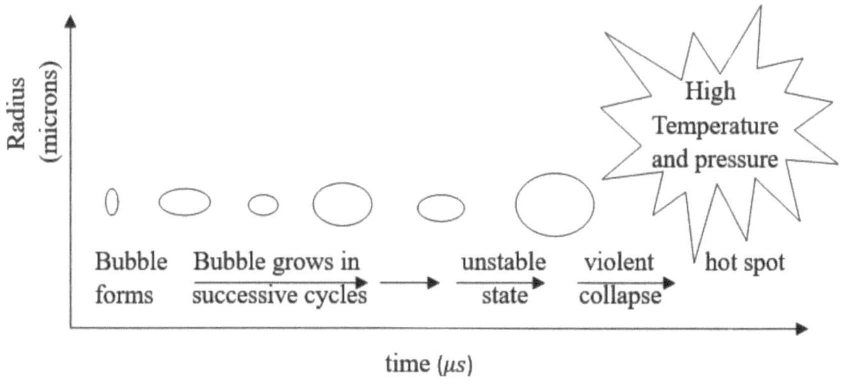


**FIGURE 5.3** Bubble growth and collapse during sonication (Bundhoo and Mohee, 2018).

Ultrasonic waves are sound waves with frequencies of greater than 20 kHz. As waves move in areas of low pressure (rarefaction), gas and steam bubbles are formed and grow until they reach critical diameters when they implode. The collapse of the gas bubbles releases a tremendous amount of energy and increases the temperature and pressure of the conditions (known as "hot spots"). Figure 5.3 shows process of bubble growth and collapse during sonication (Bundhoo and Mohee, 2018). The thermal effect and strong shearing forces may also oxidize organics and assist in solubilization. All these effects of ultrasound irradiation combine and assist in disruption of lignocellulosic structure, breaking down the crystalline nature of cellulose molecules. After the pre-treatment of biomass by sonication, the hydrolysis of biomass improves a lot, subsequently the yields of products increase.

However, ultrasound technology still has some deficiencies, i.e., low energy efficiency and high cost (Bundhoo and Mohee, 2018).

## 5.6 CONCLUSIONS AND FUTURE OUTLOOK

This chapter initially introduces the properties of biomass, and these include the conventional properties of biomass which are physical properties (moisture content, particle size, bulk density, porosity, etc.) and chemical properties (elemental compositions, chemical compositions, heating value, etc.), and also new properties of biomass (exergy properties). Then biomass conversion technologies are also detailed, and these mainly include physical technologies, chemical technologies, biochemical technologies, and thermochemical technologies. The minor conversion technologies including APHDO and ultrasound technology are also included.

The future work mainly falls in the following fields: (1) the other properties of biomass, (2) the relationships among the properties of biomass, (3) the other conversion technologies of biomass, (4) combining and/or improving the conversion technologies of biomass, etc.

## ACKNOWLEDGMENTS

This study was supported by Natural Science Foundation of China (Grant No. 51776049), Harbin Science and Technology Research Funds for Innovative Talents (Grant No. RC2014QN008009), China Postdoctoral Science Foundation (Grant No. 2014M551240), Postdoctoral Scientific Research Development Fund of Heilongjiang Province (LBH-Q18054), Foundation of National & Local Joint Engineering Research Center for Deep Utilization Technology of Rock-salt Resource, Huaiyin Institute of Technology (SF201807), and Externally Collaborative Project from State Key Laboratory of Clean Energy Utilization, Zhejiang University, China (Grant No. ZJUCEU2017011).

## REFERENCES

Azizi K., Moraveji M.K., Najafabadi H.A., (2018). A review on bio-fuel production from microalgal biomass by using pyrolysis method. *Renewable and Sustainable Energy Reviews* 82, 3046–3059.

Basu P., (2018). Chapter 3: *Biomass Characteristics, Biomass Gasification, Pyrolysis and Torrefaction* (3rd edition), Academic Press, London, UK, pp. 49–91.

Binod P., Janu K.U., Sindhu R., Pandey A., (2011). Chapter 10: Hydrolysis of lignocellulosic biomass for bioethanol production. *Biofuels*, Academic Press, Oxford, UK, pp. 229–250.

Bundhoo, Z., Mohee, R., (2018). Ultrasound-assisted biological conversion of biomass and waste materials to biofuels: A review. *Ultrasonics - Sonochemistry* 40, 298–313.

Cai J., He Y., Yu X., Banks S.W., Yang Y., Zhang X., Yu Y., Liu R., Bridgwater A.V., (2017). Review of physicochemical properties and analytical characterization of lignocellulosic biomass. *Renewable and Sustainable Energy Reviews* 76, 309–322.

Champagne P., (2008). Chapter 9: Biomass. *Future Energy*, Elsevier Ltd., London, UK, pp. 151–170.

Daniel J.J., Rosen M.A., (2002). Exergetic environmental assessment of life cycle emissions for various automobiles and fuels. *Exergy, An International Journal* 2 (4), 283–294.

Dincer I., (2002). The role of exergy in energy policy making. *Energy Policy* 30 (2), 137–149.

Du C., Zhao X., Liu D., Lin C.S.K., Wilson K., Luqüe R., Clark J., (2016). Chapter 1: Introduction: an overview of biofuels and production technologies. *Handbook of Biofuels Production* (2nd edition), Elsevier, Duxford, UK, pp. 3–12.

Dutta R., Sarkar U., Mukherjee A., (2014). Extraction of oil from Crotalaria Juncea seeds in a modified Soxhlet apparatus: Physical and chemical characterization of a prospective bio-fuel. *Fuel* 116, 794–802.

Erol M., Haykiri-Acma H., Küçükbayrak S., (2010). Calorific value estimation of biomass from their proximate analyses data. *Renewable Energy* 35(1), 170–173.

Fermentation, From Wikipedia, the free encyclopedia. 2018. https://en.wikipedia.org/wiki/Fermentation.

Fokaides P.A., Christoforou E., (2016). Chapter 3: Life cycle sustainability assessment of biofuels. *Handbook of Biofuels Production* (2nd edition), Woodhead Publishing, Duxford, UK, pp. 41–60.

Gao X., Zhang Y., Li B., Zhao Y., Jiang B., (2016). Determination of the intrinsic reactivities for carbon dioxide gasification of rice husk chars through using random pore model. *Bioresource Technology* 218, 1073–1081.

Gould M.C., (2015). Chapter 18: Bioenergy and anaerobic digestion. *Bioenergy*, Academic Press, London, UK, pp. 297–317.

Goyal H.B., Seal D., Saxena R.C., (2008). Bio-fuels from thermochemical conversion of renewable resources: A review. *Renewable and Sustainable Energy Reviews* 12, 504–517.

Guedes R.E., Luna A.S., Torres A.R., (2018). Operating parameters for bio-oil production in biomass pyrolysis: A review. *Journal of Analytical and Applied Pyrolysis* 129, 134–149.

Guo L., Jin H., Lu Y., (2015). Supercritical water gasification research and development in China. *Journal of Supercritical Fluids* 96, 144–150.

Ho D.P., Ngo H.H., Guo W., (2014). A mini review on renewable sources for biofuel. *Bioresource Technology* 169, 742–749.

Jayasinghe P., Hawboldt K., (2012). A review of bio-oils from waste biomass: Focus on fish processing waste. *Renewable and Sustainable Energy Reviews* 16, 798–821.

Jiang L., Liang J., Yuan X., Li H., Li C., Xiao Z., Huang H., Wang H., Zeng G., (2014). Co-pelletization of sewage sludge and biomass: The density and hardness of pellet. *Bioresource Technology* 166, 435–443.

Kumar M., Sundaram S., Gnansounou E., Larroche C., Thakur I.S., (2018). Carbon dioxide capture, storage and production of biofuel and biomaterials by bacteria: A review. *Bioresource Technology* 247, 1059–1068.

Lee K.T., Lim S., Pang Y.L., Ong H.C., Chong W.T., (2014). Integration of reactive extraction with supercritical fluids for process intensification of biodiesel production: Prospects and recent advances. *Progress in Energy and Combustion Science* 45, 54–78.

Mezza G.N., Borgarello A.V., Grosso N.R., Fernandez H., Pramparo M.C., Gayol M.F., (2018). Antioxidant activity of rosemary essential oil fractions obtained by molecular distillation and their effect on oxidative stability of sunflower oil. *Food Chemistry* 242, 9–15.

Naik S.N., Goud V.V., Rout P.K., Dalai A.K., (2010). Production of first and second generation biofuels: A comprehensive review. *Renewable and Sustainable Energy Reviews* 14 (2), 578–597.

Niu M., Huang Y., Jin B., Wang X., (2014). Oxygen gasification of municipal solid waste in a fixed-bed gasifier. *Chinese J Chemical Engineering* 22, 1021–1026.

Nunes L., Matias J., Catalão J., (2018). Chapter 1: Introduction. *Torrefaction of Biomass for Energy Applications: From Fundamentals to Industrial Scale*, Academic Press, London, UK, pp. 1–43.

Okoro O.V, Sun Z., Birch J., (2017). Meat processing waste as a potential feedstock for biochemicals and biofuels - A review of possible conversion technologies, *Journal Cleaner Production* 142, 1583–1608.

Ozyuguran A., Akturk A., Yaman S., (2018). Optimal use of condensed parameters of ultimate analysis to predict the calorific value of biomass. *Fuel* 214, 640–646.

Qian H., Zhu W., Fan S., Liu C., Lu X., Wang Z., Huang D., Chen W., (2017). Prediction models for chemical exergy of biomass on dry basis from ultimate analysis using available electron concepts. *Energy* 131, 251–258.

Rahman S., Helleur R., Mac Q.S., Papari S., Hawboldt K., (2018). Upgrading and isolation of low molecular weight compounds from bark and softwood bio-oils through vacuum distillation. *Separation and Purification Technology* 194, 123–129.

Razzak S.A., Hossain M.M., Lucky R.A., Bassi A.S., de Lasa H., (2013). Integrated $CO_2$ capture, wastewater treatment and biofuel production by microalgae culturing-A review. *Renewable and Sustainable Energy Reviews* 27, 622–653.

Rosen M.A., Dincer I., (1999). Exergy analysis of waste emissions. *International Journal of Energy Research* 23 (13), 1153–1163.

Ruan R., Zhang Y., Chen P., Liu S., Fan L., Zhou N., Ding K., Peng P., Addy M., Cheng Y., Anderson E., Wang Y., Liu Y., Lei H., Li B., (2019). Chapter 1: Biofuels: Introduction. *Biofuels: Alternative Feedstocks and Conversion Processes for the Production of Liquid and Gaseous Biofuels* (2nd edition), Elsevier, pp. 3–43.

Samuelsson R., Burvall J., Jirjis R., (2006). Comparison of different methods for the determination of moisture content in biomass. *Biomass and Bioenergy* 30, 929–934.

Silva C., Carneiro A., Vital B., Figueiró C., Fialho L., Magalhães M., Carvalho A., Cândido W., (2018). Biomass torrefaction for energy purposes - Definitions and an overview of challenges and opportunities in Brazil. *Renewable and Sustainable Energy Reviews* 82, 2426–2432.

Soccol C.R., Faraco V., Karp S., Vandenberghe L.P.S., Thomaz-Soccol V., Woiciechowski A., Pandey A., (2011). Chapter 5: Lignocellulosic bioethanol: Current status and future perspectives. *Biofuels*, Academic Press, Oxford, UK, pp. 101–122.

Stanbury P.F., Whitaker A., Hall S.J., (2017). Chapter 1: An introduction to fermentation processes. *Principles of Fermentation Technology* (3rd edition), Elsevier, pp. 1–20.

Suali E., Sarbatly R., (2012). Conversion of microalgae to biofuel. *Renewable and Sustainable Energy Reviews* 16, 4316–4342.

Szargut J., (1980). International progress in second law analysis. *Energy* 5, 709–718.

Szargut J., (2005). *Exergy Method, Technical and Ecological Applications*, WIT Press, Boston, MA.

Taghizadeh-Alisaraei A., Assar H.A., Ghobadian B., Motevali A., (2017). Potential of biofuel production from pistachio waste in Iran. *Renewable and Sustainable Energy Reviews* 72, 510–522.

Tan K.T., Lee K.T., (2011). Chapter 14: Biodiesel production in supercritical fluids. *Biofuels*, Academic Press, Oxford, UK, pp. 339–352.

Vidal B.C., Dien B.S., Ting K.C., Singh V., (2011). Influence of feedstock particle size on lignocellulose conversion—A review. *Applied Biochemistry and Biotechnology* 164(8), 1405–1421.

Wang T., Qiu S., Qin Y., Ma Y., Fang Y., (2018). Hydrothermal conversion of biomass to higher alcohol fuels for compression ignition engine. *Energy Procedia* 158, 249–253.

Williams C.L., Dahiya A., Porter P., (2015). Chapter 1: Introduction to bioenergy. *Bioenergy*, Academic Press, London, UK, pp. 5–36.

Yin C.Y., (2011). Prediction of higher heating values of biomass from proximate and ultimate analyses. *Fuel* 90(3), 1128–1132.

Zhang Y., (2012). Thermodynamic Study on Gasification Process of Biomass Fuels. *Dissertation for the Doctoral Degree in Engineering*, Harbin Institute of Technology, Harbin, China (In Chinese).

Zhang Y., Ghaly A. E., Li B., (2012a). Physical properties of rice residues as affected by variety and climatic and cultivation conditions in three continents. *American Journal of Applied Sciences* 9 (11), 1757–1768.

Zhang Y., Ghaly A. E., Li B., (2012b). Physical properties of corn residues. *American Journal of Biochemistry and Biotechnology* 8(2), 44–53.

Zhang Y., Ghaly A. E., Li B., (2012c). Physical properties of wheat straw varieties cultivated under different climatic and soil conditions in three continents. *American Journal of Engineering and Applied Sciences* 5 (2), 98–106.

Zhang Y., Li B., Li H., Zhang B., (2012d). Exergy analysis of biomass utilization via steam gasification and partial oxidation. *Thermochim Acta* 538, 21–28.

Zhang Q., Zhang P., Pei Z.J., Wang D., (2013). Relationships between cellulosic biomass particle size and enzymatic hydrolysis sugar yield: Analysis of inconsistent reports in the literature. *Renewable Energy* 60, 127–136.

Zhang Y., Zhao Y., Gao X., Li B., Huang J., (2015a). Energy and exergy analyses of syngas produced from rice husk gasification in an entrained flow reactor. *Journal of Cleaner Production* 95, 273–280.

Zhang Y., Zhao Y., Gao X., Li B., Huang, J., (2015b). Energy and exergy analyses of syngas produced from rice husk gasification in an entrained flow reactor. *Journal of Cleaner Production* 95, 273–280.

Zhang Y., Zhao W., Li B., Zhang H., Jiang B., Ke C., (2016). Two equations for estimating the exergy of woody biomass based on the exergy of ash. *Energy* 106, 400–407.

Zhang Y., Fan X., Li B., Li H., Gao X., (2017a). Assessing the potential environmental impact of fuel using exergy-cases of wheat straw and coal. *International Journal of Exergy* 23, 85–100.

Zhang Y., Zhao Y., Li B., Gao X., Jiang B., (2017b). Energy and exergy characteristics of syngas produced from air gasification of walnut sawdust in an entrained flow reactor. *International Journal of Exergy* 23 (3), 244–262.

Zhang Y., Chen P., Liu S., Peng P., Min M., Cheng Y., Anderson E., Zhou N., Fan L., Liu C., Chen G., Liu Y., Lei H., Li B., Ruan R., (2017c). Effects of feedstock characteristics on microwave-assisted pyrolysis-a review. *Bioresource Technology* 230, 143–151.

Zhang Y., Gao X., Li B., Li H., Zhao W., (2018). Assessing the potential environmental impact of woody biomass using quantitative universal exergy. *Journal of Cleaner Production* 176, 693–703.

Zhu X., Lu Q., (2014). *Principle and Technology of Biomass Pyrolysis*, Science Press, Beijing, pp. 17–43.

# 6 Wind Resource Forecasting Error in Flat and Complex Terrains

*George Xydis and Evanthia A. Nanaki*
Department of Business Development and Technology, Aarhus University

## CONTENTS



## 6.1 INTRODUCTION

Wind energy is of crucial importance for the energy transition to a sustainable energy future. As a matter of fact, in 2017, 52.5 GW of new wind power installations were added across the globe, bringing the total worldwide capacity up to 539 GW (Global Wind Energy Council, 2017). This was a record year for Europe as well, reaching record numbers with 15.6 GW of new wind power capacity installed (Wind Europe, 2018). This is partially attributed to offshore wind farms and their new 3.1 GW in 2017. It is predicted that this increase will continue during the next few years (Global Wind Energy Council, 2017), establishing wind energy as one of the main ingredients of the worldwide energy mix, which translates into an increasing number of operational wind farms, composed of more powerful wind turbines.

It is a fact that a large percentage of wind farms are developed in mountainous areas with complex terrain, where the estimation of wind potential is complicated. Areas with complex terrain are usually dominated by strong winds; thus, simulating accurately comes along with high uncertainty. Surface roughness and atmospheric instability are some of the factors that drag numerical models and wind energy meteorology to the edge.

The most known wind flow modeling software for wind energy is the linear wind flow model WAsP (Wind Atlas Analysis and Application Program) tool, developed by Risø National Laboratory for Sustainable Energy (Risø). WAsP (Troen, 1990) is widely used to predict the spatial variation of mean wind speed, directional distribution, wind shear, and other boundary-layer features. Most linear models are based on Jackson and Hunt's (1975) theory and were widely used in the 1980s–1990s, where the estimation of wind potential was limited due to the available computational power.

The reputation of the linear models grew mainly because of their ease of use and fast results; their performance has been studied both in flat and complex terrains. However, due to only linear equations used in the WAsP model, users have to be careful when applying this model to intense complex areas.

Computational fluid dynamics (CFD) non-linear models are widely accepted as the next generation wind flow models for long-term wind power forecast applications. Most CFD model the non-linear Navier–Stokes equations and run until convergence is achieved, using a specific wind profile. For special cases, such as two-dimensional or three-dimensional wind flow over obstacles and hills, these models work satisfactorily and give a detailed output on the flow turbulence characteristics (Biatsuamlak et al., 2004). Several research studies have shown that more focused computations, such as the non-linear CFD model, perform better than the WAsP model in many but not in all cases (Berge et al., 2006; VanLuvanee et al., 2009; Xydis, 2012a,2013; Sumner et al., 2010; Periera et al., 2010).

The assessment of the wind resource prediction errors of the WAsP model in areas with flat and complex terrain and the determination of the wind flow simulation errors at these two terrain categories are always important to be clearly estimated. The average wind speed is directly linked to wind power generation, and minimizing the forecasted wind speed errors plays a crucial role in reducing the risk of investment. Decision makers are based on accurate forecasts to move on to wind integration studies, large-scale investments, and inefficient calculations will surely jeopardize the possible investment itself (Xydis, 2015; Xydis & Mihet-Popa, 2016; Xydis & Nanaki, 2015).

Usually, these error insight studies are based on the wind data analysis in two different areas of interest using the WAsP linear model (or other similar models), which is widely used in the wind energy industry. Climatic data from wind field measurements are used in the areas of interest in combination with reanalysis databases (usually more than 20 years of data), applying the methodology proposed based on international standards. To determine the model errors, a comparison between the predicted and actual wind speed values at the wind turbine sites at operating wind farms in the areas of interest is done.

## 6.2 BASIC THEORY

The kinetic energy $KE$ of a gas of mass $m$ (kg) with a velocity $v$ (m/s) is calculated by the equation:

$$KE = \frac{1}{2} m v^2 \tag{6.1}$$

When a mass of air $m$ with air density $\rho$ and velocity $v$ passes through a cross-section $A$, then the mass flow is given by:

$$\dot{m} = \frac{dm}{dt} = \rho A v^3 \quad (6.2)$$

Equations (6.1) and (6.2) show the amount of energy that crosses a cross-section $A$ per time of unit. This action equals the wind power $P$, which is calculated by the equation:

$$P = \frac{dKE}{dt} = \frac{1}{2}\rho A v^3 \quad (6.3)$$

The power produced by a wind turbine is proportional to the wind speed at the engine installation location. The basic equation for calculating the output power of a wind turbine is:

$$P = \frac{1}{2}\rho v^3 A C_p \quad (6.4)$$

where $C_p$ is the power coefficient. The available wind power depends on the wind speed cubed (Di Piazza et al., 2010; Xydis, 2015). Therefore, the average wind speed in different regions plays the most crucial role. A simple example could make that clearer. If in a specific site in country A, the MWh price pays 30% more compared to country B, but in county B in another site, the average wind speed is 1 m/s more (e.g. from 6 to 7 m/s), then we should probably select the site in country B as we will eventually get paid better (since the energy content of the wind power is proportional to the velocity cubed (the third power)). The wind resource potential in an area is estimated based on the distribution of the wind speed, which is determined by the wind intensity and by the frequency of occurrence of wind speed values over a year. However, it is not wind speed the only criterion. Turbulence levels should be low and wind gusts not severe and frequent; otherwise, fatigue problems for wind turbines occur, reducing their lifetime. In order to estimate the wind potential in an area of interest, significant wind intensity information is provided from available mesoscale wind maps, which depict the average annual wind speed at a specific height from the ground surface (Global Wind Atlas, 2019). Figure 6.1 shows the wind potential at 100 m above ground level for a major part of North America.

However, only based on mesoscale analyses (usually 1 km resolution or more), it is impossible to move forward with investments. Investors need to know in advance what exactly is going on regarding wind conditions in the specific area they are interested in, before deciding on further actions. Therefore, micrositing is needed, and a resolution between a few meters up to maximum a hundred meters is required. This is often happening via the installation of wind meteorological masts, which locally measure the wind speed, direction, etc.

Finding and selecting the appropriate site for the mast is decided after a site visit and inspection of the topography of the area, since the morphology of the surface appears to have a significant influence on both wind and turbulence. The location

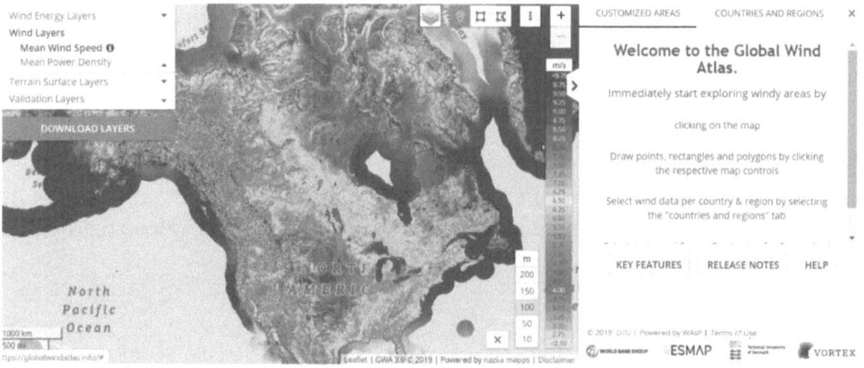


**FIGURE 6.1** Wind power imaging at 100m above ground level for North America (Global Wind Atlas).

of the windmill must be relatively close to the area of interest, away from trees or other obstructions, and above all at the highest hills, so that it is fully exposed to the winds from all directions. The MEASNET standard (MEASNET Site Assessment Procedure) states that in the case of flat terrain, the maximum distance of any wind turbine from the nearest mast is 10km, while in the case of complex terrain, the maximum distance should be maximum 2km (MEASNET Site Assessment Procedure, 2016).

Regarding the measurement periods, if a significant number of data is missing, it is advisable to correlate the data between various wind masts and airport (or other sources) long-term data in order to fill in the blanks and extend the total measurement period. One of the key features of a windmill is its height, which is decided before the installation by the expected wind potential and by the hub height of the wind turbines to be installed. Following the instructions of MEASNET, the main wind speed measurement's height in a wind mast should be at least 2/3 of the proposed hub height. Heights measured near the proposed hub height allow precise determination of the climatic conditions of the wind and at the same time reduce uncertainty in the process of the vertical estimation of these conditions. In addition to the primary anemometer, the masts must have backup anemometers installed in accordance with IEC 61400-12-1 standards to enable comparisons. A common practice is to have anemometers every 10m, exactly as in the left and the middle images in Figure 6.2 (the middle image is more presenting the foundations of a large wind mast). On the right image of Figure 6.2, a small 10m mast is shown.

It should be noted here that except the wind masts, a method used a lot during the last 10–15 years and gains more and more space is Sonic Detection αnd Ranging (SODAR) and Light Detection and Ranging (LIDAR) systems. Both are remote sensing methods with the ability to measure in ore complex terrains and in a more complete way. However, such systems are often too expensive to be installed and are used only in correlation with the set of measurements from the wind masts.

For a given point in space and for period $T$, the average wind speed ($V_{mean}$) of a time series is calculated by the:

**FIGURE 6.2** Risø's 120 m wind mast (left) and a typical example of a 10 m mast (right).

$$V_{\text{mean}} = \frac{1}{N}\sum_{i=1}^{N} V_i \tag{6.5}$$

where $V_i$ is the $i$ speed measurement and $N$ is the total number of measurements. The standard deviation $\sigma$ of the measurements is given by the formula:

$$\sigma = \sqrt{\frac{1}{1-N}\sum_{i=1}^{N}(V_i - V_{\text{mean}})^2} \tag{6.6}$$

The standard deviation has the same dimensions as the variable and it practically measures the dispersion of the values from the mean (Figure 6.3). Specifically, for a case of average wind speed 6.4 m/s at 30 m, the standard deviation of wind speed can be interpreted as significant turbulence levels.

The direction of the wind is constantly changing in space and time and is characterized by the point of the horizon from which the wind blows. It is usually expressed in degrees (values range from 0° to 360°), originating from the earth's magnetic north, in a clockwise direction. The orography (valleys, mountains, hills) and other

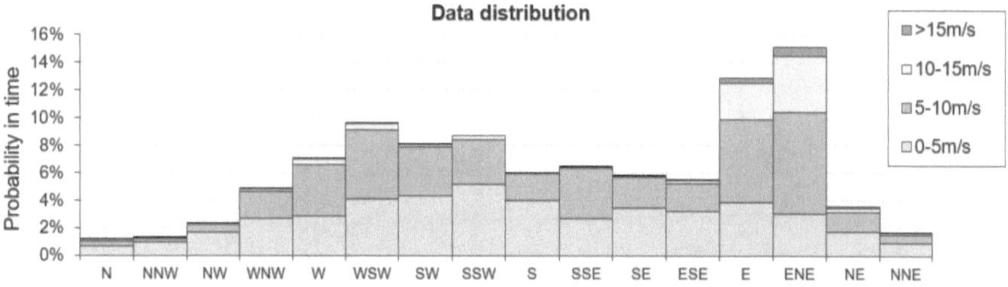


**FIGURE 6.3** Data distribution.

soil features (buildings, vegetation) are some of the factors that influence the main wind direction (Xydis, 2012b; Manwell et al., 2009). Based on the data acquisition of the yearly wind direction, the windrose is created. The windrose is the wind frequency diagram for each direction based on the total number of records. It is a chart, which aims to determine the prevailing and secondary wind directions in a specific site over a period of one year (or more, if possible), which often varies according to the time of year. The circular surface of the chart is divided into 8, 12, 16, or 32 equal sections (in the example of Figure 6.4, the number of sections is 16 of the same example above). Figure 6.4 presents in a different way data distribution presented in Figure 6.3. It is necessary to have long wind measurements to structure and submit a reliable bid. At cases, there are periods of data lost for various reasons. Sometimes the data missing are associated to severe weather phenomena (e.g. thunderstorms, lightnings, etc.), other times with bad management during wind mast erection or at times data storage problems. A typical example of a case with data lost in the middle of the measurement period can be seen in Figure 6.5, where the daily max, the daily mean, and the maximum gust of the day are shown.

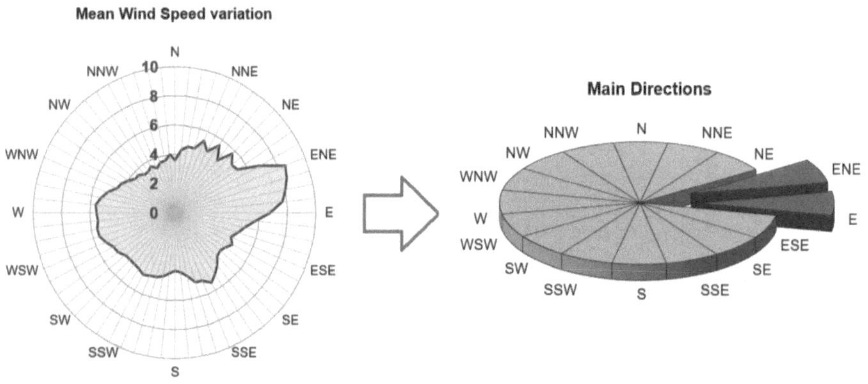


**FIGURE 6.4** A typical 16-sectioned windrose and the translation to the primary and secondary main directions.

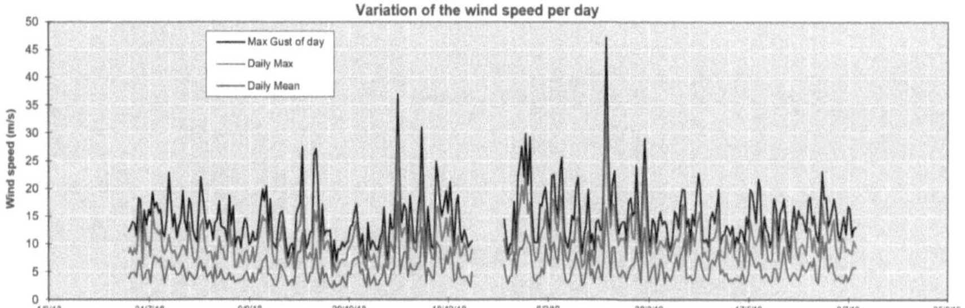


**FIGURE 6.5** Data missing during an annual measurement period between July 2018 and July 2019.

Within the atmospheric boundary layer (ABL), the lowest part of the earth's atmosphere, the wind intensity increases as the height of the ground surface increases. This vertical distribution follows the turbulent flow behaviour. This increase is greater near the ground, while at higher altitudes, the rate of increase falls. As most wind stations are installed and measure the wind speed at lower heights compared to the modern wind turbines' hub height, it is necessary to accurately calculate what the expected wind speed to the hub height is. For this reason, two ways of estimating this have been proposed:

- Exponential law: The calculation of wind speed from one measurement height $Z_r$ to another height $Z$ follows the:

$$V_{(Z)} = V_{(Zr)} * \left( \frac{Z}{Z_r} \right)^{\alpha} \tag{6.7}$$

  where $V_{(Z)}$ is the wind speed at a height $Z$, $V_{(Zr)}$ is the wind speed at a reference height $Z_r$, and $\alpha$ is the shear coefficient dependent on ground roughness. The term shear describes the slope of the wind speed curve with respect to the height of the ground surface. It expresses the height of the wind change within the boundary layer of the atmosphere. This calculation is necessary to estimate the wind speed at the hub height of the wind turbine (Figure 6.6).

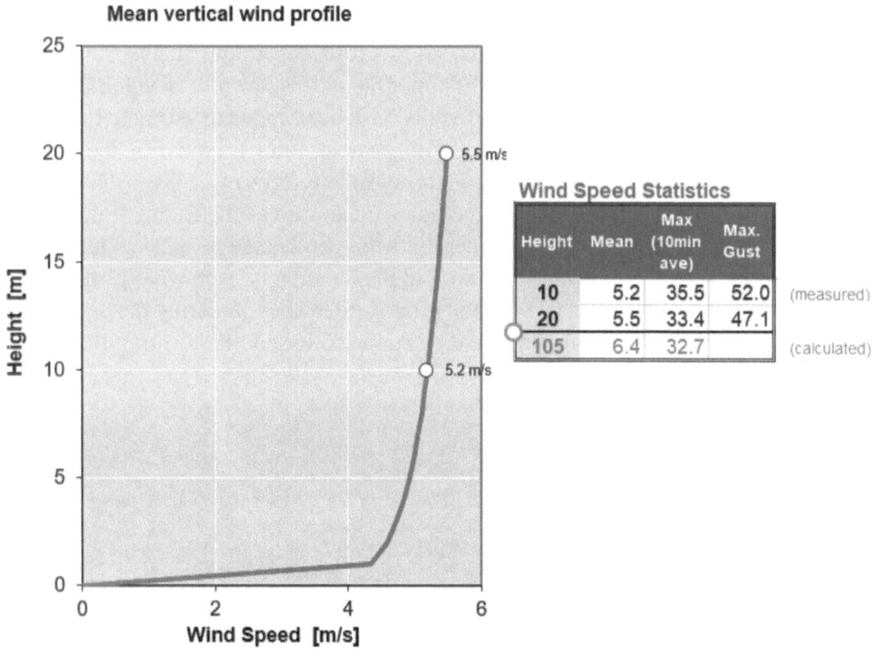


**FIGURE 6.6** Mean vertical wind profile.

- Logarithmic law: To calculate the wind speed from one height $Z_{(r)}$ to another height $Z$, the equation used is:

$$\frac{V(z_1)}{V(z_2)} = \frac{\ln\left(\frac{z_1}{z_0}\right)}{\ln\left(\frac{z_2}{z_0}\right)} \tag{6.8}$$

  where $z_0$ is the roughness length of the surface.

The increase in wind speed is limited by obstacles on the surface of the ground (natural or artificial) by the morphology of the ground around the installation site (vegetation, terrain, and roughness) and by the atmospheric conditions in the wider area, and this is where the roughness length is used, to describe all these variations. Roughness length can be from 0.0002, for ice and the sea, until beyond 300 which could be the roughness length e.g. for megacities (Hansen, 1993). In general, the higher the roughness length, the higher the turbulence levels will be, which subsequently most probably will lead to cancelation of investment plans in this specific area.

## 6.3 DIFFERENT TERRAINS AND ERROR COMPARISON

### 6.3.1 Mixed Topographies

In contrast to the flat terrain, which presents no significant variations in the wide area, the characterization of an area with complex terrain refers to a field with significant slopes and orographic variations, such as hillsides, ridges, canyons, valleys, and mountain basins. The impact of the topography (Figure 6.7) is significant on the flow of wind profile in that area. Attempts have been made to estimate the error involved in model calculations in areas with complex topography. For the classification of surface complexity, the Ruggedness IndeX (RIX) has been introduced (Bowen & Mortensen, 1996). This is an objective measure of the extent of the steep slopes in an area. The RIX indicator identifies the percentage of area with a slope greater than 18.0° over a 3.5 km radius around the location of each wind turbine.

The most common way to deal with the procedure is after gathering the data from the wind masts (via a data logger and then a sim card for communicating the data),

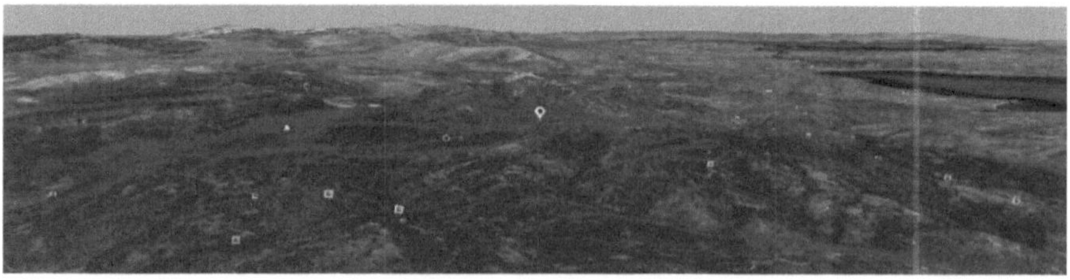

**FIGURE 6.7** A mixed topography (digitally represented here) is often the main reason to high uncertainty.

a collection of the surface data is required. Information on the topography could be extracted from the international DataForWind (2019) database or from the Shuttle Radar Topography Mission (SRTM) 90 m DEM Digital Elevation Database (2019) with digital terrain models. These digital models are converted, with the help of specialized Geographic Information Systems (such as, for instance, the Global Mapper tool), into a specific file format compatible with the wind simulation software tools, such as WindPRO and WAsP. The digital terrain models, used for modeling the wind flow, have often a grid resolution of 15×15 km and the contour curves are usually between 4 and 20 m away aiming for the optimum accuracy.

### 6.3.2 Error Comparison Methodology

The purpose of this work was to suggest a methodology to follow when evaluating the performance of the wind flow simulation model for two specific wind farm sites, located in different (flat and complex) terrains. The theoretical results of the simulation models are meant to be compared to the actual values of wind speed as recorded at the wind turbine sites of each wind station.

The study, in general, includes the following steps:

- Collection and analysis of wind data
- Create a wind flow model using one of the know wind simulation tools (WAsP, WindFarm, WindSim, WindPRO, Meteodyn WT, ZephyCFD, openWind)
- Create a wind map of the wider area
- Predicting wind speed at each wind turbine location and comparing it to the actual wind speed as recorded by the Supervisory control and data acquisition (SCADA) wind turbine system.

Surface roughness was extracted (and digitized) from the existing Corine Land Cover maps (CORINE2000). CORINE2000 is a digital land cover map used for European Union countries (similar databases exist for other continents/countries) available from the European Environment Agency (EEA).

One of the most important calculations in this methodology is the long-term value of the wind potential for each wind measurement site. The use of the "Measure-Correlate-Predict" (MCP) methodology based on input from long-duration wind data (such as the airport or national meteorological agency measurements) allows the uncertainty to be reduced. The MCP module, which is available in the most software tools, is used to correct the wind data by reducing it over a longer period of time, with reference to a correspondingly longer time frame reference point. This section covers the four most common methods of data correlation, such as Linear Regression, Matrix, Weibull Scale, and Wind Index. In the process of correlating the data, when there is high availability in both the reference site and the long-term data point, it is recommended to use the Linear Regression or the Matrix method (Thøgersen et al., 2007). Data from the European Center for Medium-Term Meteorological Forecasts (ECMWF) have been used as a global database boundary, covering all of Europe including most parts of Turkey and Ukraine, except for the northern coast

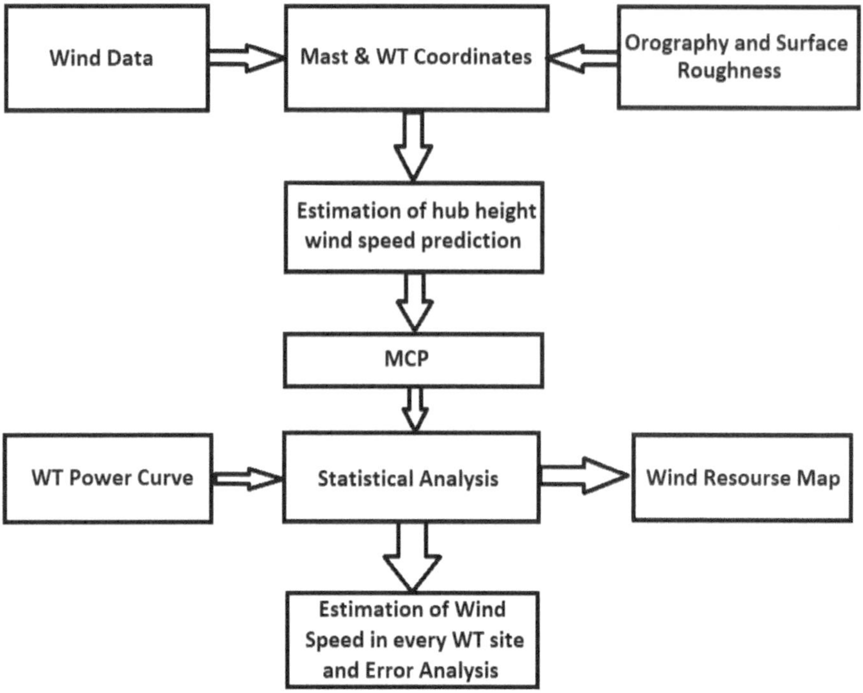


**FIGURE 6.8** Flow chart for estimating wind potential and identifying errors in wind speed forecast at wind turbine sites.

of Scandinavia. The final result of the MCP procedure is a statistical distribution of wind speed based on the surface data and the local climate (Figure 6.8).

In order to exploit the historical period of available wind data and thus reduce the uncertainties in the analysis, wind speed data at the location of the wind mast for one year at the hub height were correlated with long-term wind speed data, taking into account the long-term source of the EMD-ConWx (Europe Mesoscale Data). This procedure is performed through the WindPro tool "MCP," where the wind mast measurements and the EMD-ConWx node reanalysis data are compared. This process goal is to create a single wind speed time series in place of the wind mast at the hub height of the wind turbines (usually around 80 m).

Following the suggested methodology, the wind flow simulation model was solved for the given climatic and roughness conditions, predicting the average annual wind speed at each wind farm engine site. The true value of wind speed in each wind turbine is recorded in the SCADA system of the wind power station, by the anemometers installed in the nacelle of each wind turbine. The deviation of the projected average annual velocity from the actual average annual value, per wind turbine and per year, is calculated by the formula:

$$\sigma = \frac{|ws(\text{forecast}) - ws(\text{real})|}{ws(\text{real})} * 100\% \tag{6.9}$$

where $\sigma$ is the error, *ws*(forecast) is the wind speed predicted, and *ws*(real) is the measured wind speed.

## 6.4 CASE STUDIES IN DIFFERENT TERRAINS

There is a large number of risks associated with each wind project. Among the risks are all kinds of simulations of the development phase. Wind speed estimations compared to real measurements reveal errors that can be crucial in terms of project viability. Risks can be divided into the following broad categories:

- Project management: an experienced team of the appropriate size
- Regulatory: site lease, construction permit/consent, grid connection
- CAPEX: credible projection based on careful and realistically assessed market-based costs, including appropriate contingencies
- Energy yield: data-based assessment with appreciation of all uncertainties
- OPEX: detailed studies, nature of guarantees, experience of service provider
- Power purchase agreements (PPAs): exposure to market fluctuations; political risk
- Other: other important risks will depend on stage and country.

The individual circumstances of the wind farm may provide a good reason to deviate from the above; however, in general, the above schedule would result in minimized risk and hence maximized risk-adjusted return and value for the wind farm. As the project is developed, key risks need to be identified, analyzed, and assessed in terms of the impact on the overall economics and viability of the project. In many cases, the developer is not investing in the required studies and surveys. This could happen for a number of reasons, including lack of financial and managerial resources or conditions at the site were found to differ significantly from what could be reasonably anticipated at the early stages. In that case, the investor will need to make an important decision on the level of risk.

### 6.4.1 Complex Topography

An anonymized (due to confidentiality issues not much could be shared) due diligence, contracted by a bank, was focused on the technical part of the analysis of this wind farm, located in the Dodecanese region, Greece, consisting of a number of wind turbines, each of which has a nominal capacity of approximately 1 MW. The goal of the due diligence of the specific wind farm, based on the data provided by an independent power producer, was to quantify the risks of future operation for the years to come and to recommend a strategy to eliminate all possible risks. Increased roughness levels and orography play a more important role compared to the flat terrain (Papadopoulou et al., 2019), which subsequently influence the wind turbine performance, the repair process, and service and maintenance (Koscis & Xydis, 2019). In Figure 6.9, the net wind farm produced electricity per month is shown for the years 2016–2019. For 2016, 18,976 MWh were produced; for the year 2017, 22,869 MWh; 22,235 MWh for 2018; and finally 22,820 MWh for 2019.

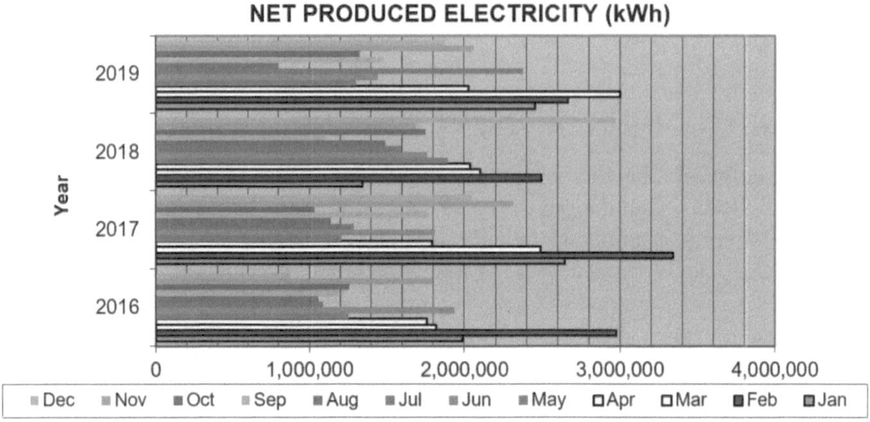


**FIGURE 6.9** Monthly net produced electricity for the years 2016–2019 at the site with complex terrain.

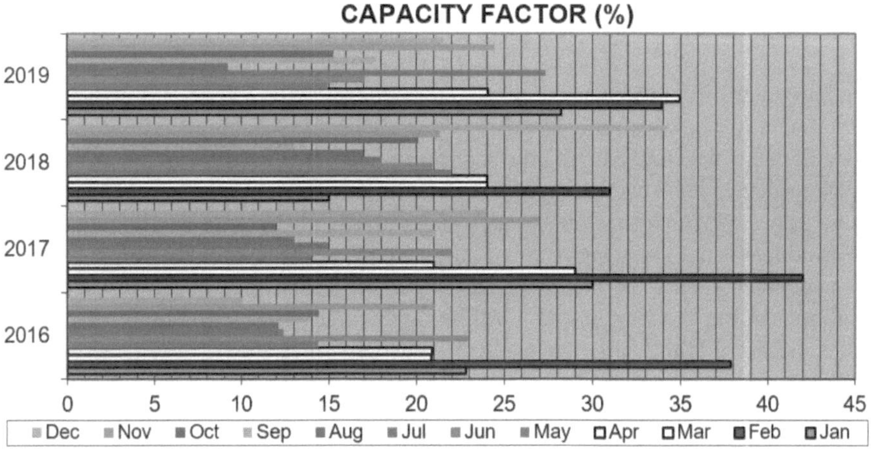


**FIGURE 6.10** Monthly capacity factor for the years 2016–2019 at the site with complex terrain.

In Figure 6.10, the net wind farm capacity factor per month is shown for the years 2016–2019. It is shown that only for the winter months the results are satisfactory. For 2016, the average net CF was estimated 18.64%, while for the years 2017–2019 between 21.72% and 22.5%.

It is clear that 2016 was the worst—in terms of electricity production and productivity—out of the four years. Following the methodology presented and by formula (6.9), the error per month was estimated (Figure 6.11). Table 6.1 includes all the data and the error calculation per month and per year.

The best year in terms of approximation and smaller error is 2018. The average error is 14.77%, while for the year 2016, the calculated error was 26.04%! It should

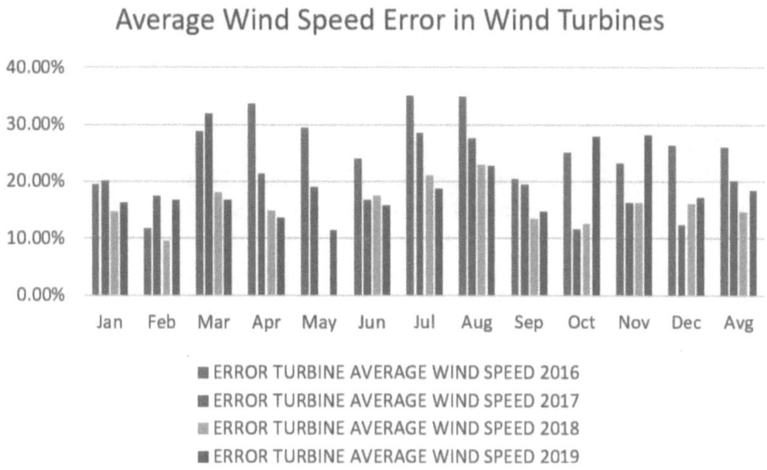


**FIGURE 6.11** The average wind speed error estimated per month throughout 2016–2019.

**TABLE 6.1**
**Error Turbine Average Wind Speed in Complex Terrain**

| | 2016 (%) | 2017 (%) | 2018 (%) | 2019 (%) |
|---|---|---|---|---|
| Jan | 19.53 | 20.20 | 14.65 | 16.28 |
| Feb | 11.77 | 17.53 | 9.57 | 16.73 |
| Mar | 28.88 | 31.91 | 18.13 | 16.79 |
| Apr | 33.63 | 21.32 | 14.82 | 13.70 |
| May | 29.47 | 19.13 | 0.00 | 11.38 |
| Jun | 24.10 | 16.71 | 17.50 | 15.85 |
| Jul | 35.09 | 28.48 | 21.06 | 18.81 |
| Aug | 34.83 | 27.63 | 22.99 | 22.82 |
| Sep | 20.43 | 19.47 | 13.54 | 14.65 |
| Oct | 25.17 | 11.58 | 12.58 | 27.95 |
| Nov | 23.20 | 16.27 | 16.25 | 28.27 |
| Dec | 26.33 | 12.39 | 16.12 | 17.22 |
| **Average** | **26.04** | **20.22** | **14.77** | **18.37** |

be stated here that, according to data, for the year 2016, the wind turbine availability was 94%, while for 2017, 2018, and 2019 was 98.5%, 97.1%, and 96.7% respectively.

### 6.4.2 Flat Topography

The due diligence, contracted by the same bank as in the complex terrain case, was focused on the technical part again. The wind farm was located in the Crete, Greece, consisting of a number of wind turbines, each of which has a nominal capacity of approximately 0.5 MW.

The goal of the due diligence of the specific wind farm, based on the data provided by an independent power producer, was to quantify the risks of future operation for the years to come and to recommend a strategy to eliminate all possible risks. The net wind farm produced electricity per month is shown for the years 2016–2019 in Figure 6.12. For 2016, 14,570 MWh were produced; for the year 2017, 13,491 MWh; 12,400 MWh for 2018; and finally, 13,033 MWh for 2019. In Figure 6.13, the net wind farm capacity factor per month is shown for the years 2016–2019.

It is seen that the results are much better compared to the results of the complex terrain. This has to do not only with the higher capacity factor and the yearly annual speed but also with the lower turbulence and other losses (proximity to the grid, etc.). For 2016, the average net CF was estimated 32.64%, while for the years 2017 31%; for 2018, 28.5%; and for 2019, 29.3%.

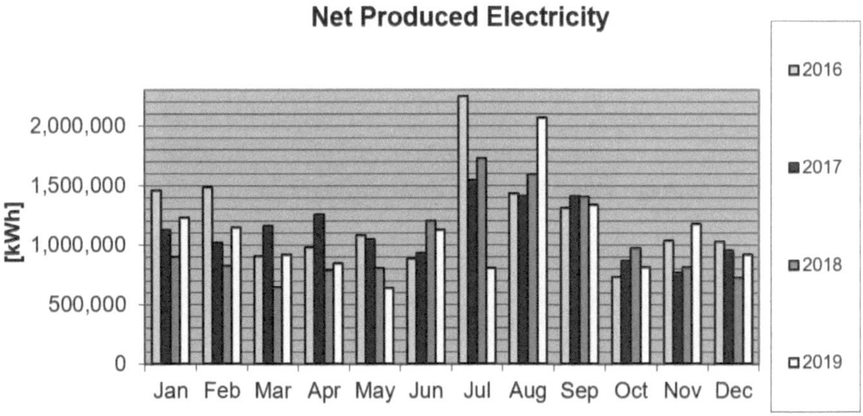


**FIGURE 6.12** Monthly net produced electricity for the years 2016–2019 at the site with flat terrain.

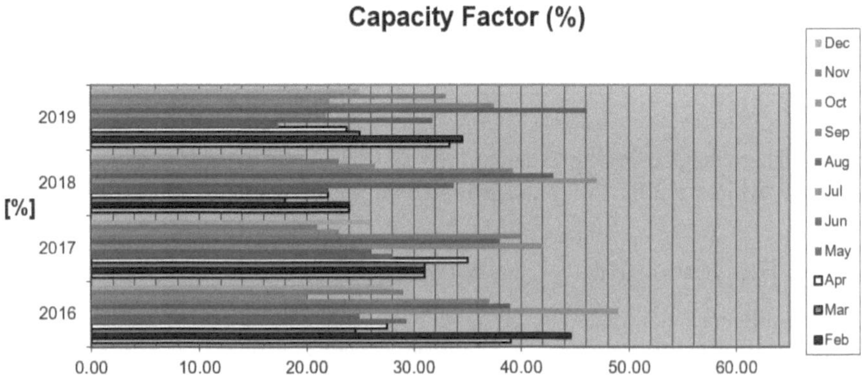


**FIGURE 6.13** Monthly capacity factor for the years 2016–2019 at the site with flat terrain.

Following the methodology presented in this chapter using Equation (6.9), the error per month was estimated (Figure 6.14). Table 6.2 includes all the data and the error calculation per month and per year. It is clear that in the flat terrain—if year 2016 is excluded where error reached 13.83%—in all other years 2017–2019, the error was practically insignificant (Table 6.2).

It should be stated here that, according to data provided, for the years 2016–2019 the wind turbine availability was between 99.73 and 99.9%; therefore, there is no significant impact in this case based on that as there was in the case of the complex terrain.

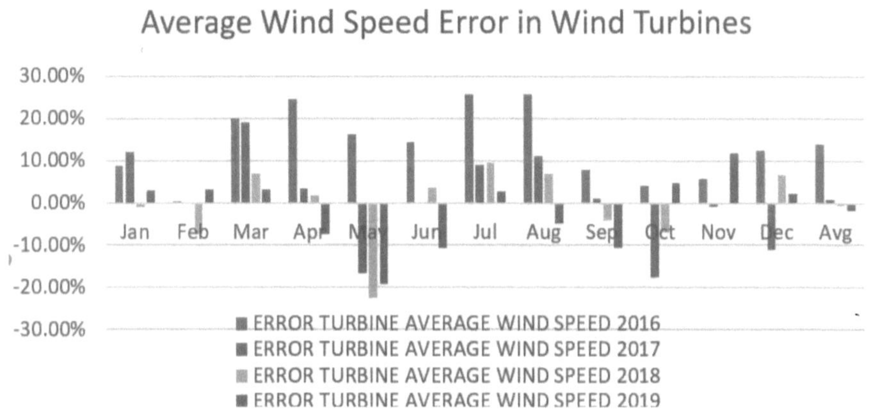


**FIGURE 6.14** The average wind speed error estimated per month throughout 2016–2019.

**TABLE 6.2**
**Error Turbine Average Wind Speed in Flat Terrain**

| | 2016 (%) | 2017 (%) | 2018 (%) | 2019 (%) |
|---|---|---|---|---|
| Jan | 8.82 | 11.92 | –1.00 | 3.03 |
| Feb | 0.27 | 0.15 | –7.29 | 3.05 |
| Mar | 19.87 | 19.09 | 6.88 | 3.15 |
| Apr | 24.64 | 3.39 | 1.81 | –7.22 |
| May | 16.31 | –16.61 | –22.54 | –19.18 |
| Jun | 14.34 | 0.14 | 3.54 | –10.63 |
| Jul | 25.83 | 9.06 | 9.32 | 2.64 |
| Aug | 25.80 | 11.15 | 6.86 | –4.73 |
| Sep | 7.73 | 1.13 | –4.07 | –10.57 |
| Oct | 4.14 | –17.48 | –6.88 | 4.70 |
| Nov | 5.74 | –0.87 | 0.17 | 11.84 |
| Dec | 12.48 | –10.96 | 6.54 | 2.27 |
| **Average** | **13.83** | **0.84** | **–0.56** | **–1.80** |

## 6.5 CONCLUDING REMARKS

In this work, wind measurements were used in two areas, in flat and complex topography, which were qualitatively tested. The wind flow was then simulated in each area using a software simulation tool, and the predicted wind speed was calculated at each wind turbine location. These results were compared against the actual mean annual wind speed values as recorded by the wind turbine SCADA system during the operation of the wind stations.

It was found that the calculations of the simulation tools are more accurate for the prediction of wind in areas with flat terrain. In practice, non-linear models (CFD) can more accurately predict the wind in complex terrains. If you exclude only one year out of the four examined, where the error was rather high, approx. 13%, for the other three years, the error was insignificant. On the contrary, the prediction error of the model for areas with complex terrain was found at least 14% for one of the years and a lot higher (even up to 26%) for the other three years.

## REFERENCES

Berge, E., Gravdahl, A.R., Schelling, J., Tallhaug, L., & Undheim, O. (2006). Wind in complex terrain. A comparison of WAsP and two CFD-models. *Proceedings from EWEC 2006*. Athens, Greece.

Biatsuamlak, G.T., Stathopoulos, T., & Bedard, C. (2004). Numerical evaluation of wind flow over complex terrain: Review. *Journal of Aerospace Engineering*, vol. 17, Issue 4, pp. 135–145.

Bowen, A.J., & Mortensen, N.G., (1996). Exploring the limits of WAsP the wind atlas analysis and application program. *Proceedings of the European Union Wind Energy Conference*, Goteborg, 584–587.

DataForWind database. DataForWind: Services for professionals in wind energy, [Online]. Available: http://www.dataforwind.com/, [Accessed 3 Sept 2019].

Di Piazza, A., Di Piazza, M.C., Ragusa, A., & Vitale, G. (2010). Statistical processing of the wind speed data for energy forecast and planning. *International Conference on Renewable Energies and Power Quality*, 23–25 March 2010, Grenada, Spain.

Global Wind Atlas. Danmarks Tekniske Universitet, [Online]. Available: https://globalwindatlas.info/, [Accessed 30 Oct 2019].

Global Wind Energy Council. Global wind report; *Technical Report*; Global Wind Energy Council: Brussels, Belgium, 2017.

Hansen, F.V., Surface roughness lengths, army research laboratory, 1993, [Online]. Available: https://apps.dtic.mil/dtic/tr/fulltext/u2/a274550.pdf [Accessed 21 Nov. 2019].

Jackson, P.S. & Hunt, J.C.R. (1975). Turbulence wind flow over low hill. *Quarterly Journal of the Royal Meteorological Society*, vol. 101, pp. 929–955.

Koscis, G. & Xydis, G. (2019). Repair process analysis for Wind Turbines equipped with Hydraulic Pitch mechanism on the U.S. market in focus of cost optimization. *Applied Sciences*, vol. 9, p. 3230, DOI: 10.3390/app9163230.

Manwell, J.F., McGowan, J.G., & Rogers, A.L. (2009). *Wind Energy Explained - Theory, Design and Application*. Chichester, West Sussex, United Kingdom: John Wiley & Sons Ltd.

MEASNET site assessment procedure, Version 2.0, 2016, [Online]. Available: http://www.measnet.com/wp-content/uploads/2016/05/Measnet_SiteAssessment_V2.0.pdf [Accessed 9 Aug. 2019].

Papadopoulou, K., Alasis, C., & Xydis, G. (2019). On the wind blade's surface roughness due to dust accumulation and its impact on the wind turbine's performance: A heuristic QBlade-based modelling assessment. *Environmental Progress & Sustainable Energy*, DOI: 10.1002/ep.13296.

Periera, R., Guedes, R., & Santos, C.S. (2010). Comparing WAsP and CFD wind resource estimates for the "regular" user. *Proceedings from EWEC 2010.* Warsaw, Polland.

SRTM 90m DEM Digital Elevation Database, [Online]. Available: http://srtm.csi.cgiar.org/ [Accessed 22 Oct. 2019].

Sumner, J., Sibuet Watters, C., & Masson, C. (2010). CFD in wind energy: The virtual, multiscale wind tunnel. *Energies*, vol. 3, pp. 989–1013.

Thøgersen, M.L., Motta, M., Sørensen, T., & Nielsen, P. (2007, May). Measure-correlate-predict methods: case studies and software implementation. *European Wind Energy Conference & Exhibition* (p. 10), Milan, Italy.

Troen, I. (1990). A high resolution spectral model for flow in complex terrain. *Ninth Symposium Turbulence Diffusion 1990*, Roskilde, 417–420.

VanLuvanee, D., Rogers, T., Randall, G., Williamson, A., & Miller, T. (2009). Comparison of WAsP, MS-Micro/3, CFD, NWP, and analytical methods for estimating site-wide wind speeds. *Presentation from AWEA Wind Resource Assessment Workshop.* Minneapolis, MN.

Xydis, G. (2012a). Effects of air psychrometrics on the exergetic efficiency of a wind farm at a coastal mountainous site – An experimental study. *Energy*, vol. 37, pp. 632–638, DOI: 10.1016/j.energy.2011.10.039.

Xydis, G. (2012b). Wind-direction analysis in coastal mountainous sites. An experimental study within the gulf of Corinth, Greece. *Energy Conversion and Management*, 64, pp. 157–169.

Xydis, G. (2013). The wind chill temperature effect on a large-scale PV plant – An exergy approach, The wind chill temperature effect on a large-scale PV plant—An exergy approach. *Progress in Photovoltaics: Research and Applications*, DOI: 10.1002/pip.2247.

Xydis, G. (2015). A wind energy integration analysis using wind resource assessment as a decision tool for promoting sustainable energy utilization in agriculture. *Journal of Cleaner Production*, vol. 96, pp. 476–485 (Sp. Volume: CP Strategies), DOI: 10.1016/j.jclepro.2013.11.030.

Xydis, G. (2015). Wind energy integration through district heating. A wind resource based approach. *Resources*, vol. 4, pp. 110–127, DOI: 10.3390/resources4010110.

Xydis, G. & Nanaki, E. (2015). Wind energy based electric vehicle charging stations sitting. A GIS/wind resource assessment approach. *Challenges*, vol. 6, Issue 2, pp. 258–270, DOI: 10.3390/challe6020258.

Xydis, G. & Mihet-Popa, L. (2016). Wind energy integration via residential appliances. *Energy Efficiency*, vol. 10, Issue 2, pp. 319–329, DOI: 10.1007/s12053-016-9459-2.

Wind Europe. Wind in Power 2017. Annual combined onshore and offshore wind energy statistics; *Technical Report*; Wind Europe: Brussels, Belgium, 2018.

# 7 Wind Power Forecasting via Deep Learning Methods

*Fazıl Gökgöz and Fahrettin Filiz*
Ankara University

## CONTENTS



## 7.1 INTRODUCTION: BACKGROUND INFORMATION ABOUT WIND POWER FORECASTING

Wind power not only represents a sustainable and non-polluting electricity supply but also is a strongly volatile energy type. The power system is maintained by adjusting electricity supply and demand to stand apart of fluctuating in power systems. Accurate wind power forecasting has great importance in adjusting the supply and demand capacity of the power system.

Wind power is weather-dependent production. In contrast to conventional power stations, wind energy depends on environmental conditions in particular to the wind. Due to the fluctuations in wind, derived power also fluctuates heavily. Wind fluctuations result in questionable power output. This uncertainty has a negative effect on the quality of operations of power systems, such as distribution, dispatch, and peak load management (Athari & Wang, 2018).

The predictability of wind energy is as important as the wind energy potential. Naturally, uncontrollable wind power has effect on the power grid. Accurately forecasted wind power significantly reduces operational costs and provides reliable power grid operations. The accuracy of the wind power forecast helps to determine the power generation cost (Q. Wang, Martinez-Anido, Wu, Florita, & Hodge, 2016).

Finally, forecasting studies on wind power have shown extensive improvement. Those studies have revealed that wind power forecasting is a crucial component of the power system. The volatility of wind power often affects the stability, the quality, and the security of the power system. It is notable important that, healthy and reliable power system can be achieved with the help of accurate wind power forecasting.

This forecasting also helps electric market stakeholders to make important decisions on power either purchasing or generating.

**Nomenclature**

*ARMA* Autoregressive Moving Average Model
*ARX* Autoregressive with Exogenous Input
*ANN* Artificial neural network
*DBN* Deep Belief Network
*DNN* Deep Neural Network
*GARCH* Generalized Autoregressive Conditional Heteroscedasticity
*GRU* Gated Recurrent Units
*LSTM* Long Short-Term Memory
*MA* Moving Average Model
*MAE* Mean Absolute Error
*MSE* Mean Squared Error
*NARX* Nonlinear Autoregressive with Exogenous Input
*RNN* Recurrent Neural Network

Accurate wind power forecasting involves numerous challenges. With the development of machine learning, forecasting models started to receive attention. Intelligence approaches gain dominance on wind power forecasting. Deep learning is a type of machine learning, which has many hidden layers. Besides the other field, deep learning has recently been applied to wind power forecasting problems. Wind power forecasting provides cost savings for utility operations, facilitates plant dispatch scheduling, and optimizes plant maintenance.

This study introduces common deep learning methods applied to wind power forecasting to obtain better forecasting accuracy. For the methodology, RNN and DBN methods are selected to apply in this study. The rationality on selection RNN and DBN methods was mentioned respectively. RNNs are powerful for sequence data modeling. In case of RNN, LSTM and GRU models are applied to forecast wind power. DBN is a stack of Restricted Boltzmann Machines that belongs to the energy-based model. The accuracy of the method's results is compared with MAE and MSE.

In this study as parameter wind power, wind speed and temperature are selected and their data were analyzed. This chapter reviews and compares the most common deep learning methods in the literature that can be applied to the Turkish power market. The performance of deep learning approaches was compared in the mean of accuracy and efficiency. Mean absolute error and mean squared error methods were applied to measure model performance. Finally, challenges and opportunities were the current method's challenges, and potentials were discussed.

This study is organized as follows: Section 7.2 introduces the wind power forecast methods; Section 7.3 overviews deep learning models; Section 7.4 includes the structure of the proposed model whereas experimental methods were implemented and the results were evaluated and compared; and Section 7.5 provides the conclusion and also includes discussion of the study.

## 7.2 DESCRIPTION OF THE WIND POWER FORECAST METHODS

Forecasting can be defined as the prediction of the future results by processing the current data by some special methods. Wind power forecasting methods are grouped in terms of algorithms used and time horizons.

There are three different types of methods that are available in wind power forecasting, respectively, physical, statistical, and hybrid methods. Physical approaches are based on physical laws that govern the atmospheric conditions. Different parameters such as roughness of surface, pressure, temperature, humidity, and obstacles of the atmosphere are evaluated to forecast wind power. The physical methods are limited, need more calculation on performance, and do not have solid results in short-term forecasting (Lei, Shiyan, Chuanwen, Hongling, & Yan, 2009). According to a review study done by H. Liu, Chen, Lv, Wu, and Liu (2019), physical methods are not commonly used in wind power forecasting.

The statistical approaches rely on the historical data. Statistical methods can be divided into time series and artificial intelligence methods. Generally, statistical models are successful in time series approach; however, their ability to process non-linear data and fitting performance is limited (Ouyang, Zha, & Qin, 2017).

Kalman filter (Zuluaga, Álvarez, & Giraldo, 2015), Box–Jenkins models (Yatiyana, Rajakaruna, & Ghosh, 2018), MA, ARMA models (Erdem & Shi, 2011), GARCH models (Zhou & Fang, 2010) are some of the statistical models used for wind power forecasting.

In recent years, deep learning forecasting methods have gained growing popularity with the advancement of machine learning methods. Deep learning methods yield more successful results when compared with physical and statistical methods especially problems that cannot be analytically defined (Marugán, Márquez, Perez, & Ruiz-Hernández, 2018).

Fuzzy logic depends upon the principles of approximate reasoning. Fuzzy logic uses variables for dealing with uncertainty. Fuzzy logic models use soft linguistic variables such as long, medium, and short to explain the relationships (Negnevitsky & Potter, 2006) (Saleh, Moustafa, Abo-Al-Ez, & Abdullah, 2016).

Support vector machines are learning algorithms that can be applied to the classification and regression problems. Support vector-based regression tools have been applied to wind power forecasting (Zeng & Qiao, 2011) (Zhang, Lai, Niu, Wang, & Zhang, 2012).

ANN is an intelligent approach that is generally used to wind energy forecasting (Sideratos & Hatziargyriou, 2012) (Bhaskar & Singh, 2012). Deep learning networks have multiple layers so that they can learn the representation of data with layers of abstraction. Commonly used methods in the wind power forecasting are LSTM (López, Valle, Allende, Gil, & Madsen, 2018), the convolutional neural network (H. Z. Wang et al., 2017), and the Restricted Boltzmann Machine (Xiaosheng et al., 2016). The intelligent models have good wind power forecasting performance according to numerous studies in the literature (Haque, Nehrir, & Mandal, 2014).

The wind power forecasting approach may be classified in different ways. H. Liu et al. (2019) classified wind energy forecasting generically. First, they grouped deterministic and uncertainty classification. There are probabilistic estimation,

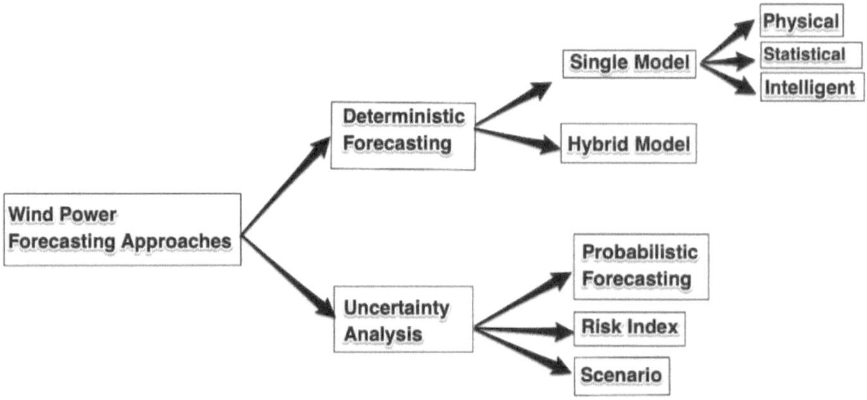


**FIGURE 7.1** Wind power forecasting approaches (Liu et al., 2019).

risk index, and scenario titles under uncertainty analysis. In this study, we examine models in the intelligent group, which are single models under deterministic estimation. Figure 7.1 shows classifications made by H. Liu et al. (2019).

RNNs are able to get abstract features and invariant structures in data. RNN can be extremely challenging. The vanishing and exploding gradient problem is an important problem in deep learning methods. LSTM network can greatly prevent gradient explosion and vanishing in the training process. Therefore, LSTM is used for classifying and forecasting and clustering analyses (Liu, Mi, & Li, 2018).

DBN is composed of RBMs. RBM consists of a two-layer neural network so that the methodology learns a probability distribution of data. DBN methods are popular DNN methods used for wind power forecasts (Qureshi, Khan, Zameer, & Usman, 2017).

Qureshi et al. (2017) proposed DBN for wind power forecasting. Wang et al. (2018) proposed a DBN for wind power forecasting.

Physical and statistical methods each have their own characteristics. Using these features together, hybrid methods are also created. In this way, the strengths of each method can be used together. For example, LSTM is used with other methods. Y. Liu et al. (2019) used LSTM with a discrete wavelet transform for wind power forecasting. López et al. (2018) used LSTM and Echo state networks together for wind power forecasting.

## 7.3 DEEP LEARNING OVERVIEW

During the last decade, artificial learning approaches have gained popularity. Support vector machines, fuzzy logic, deep learning, and artificial neural networks are popular with intelligent approaches. Artificial neural networks have shown significant success, especially we also use DBN and RNN in this study.

The deep learning network can find the data characteristic via its powerful nonlinear mapping capability. Along with these capabilities, deep learning methods are also used in wind power forecasting.

### 7.3.1 LSTM and GRU

LSTM is classified under deep learning methodology which is a modified version of RNN. LSTM is a different type of implementation of RNN which enables long-term learning dependencies.

The RNN is trained using either recurrent learning or back propagation. LSTM compensates for this failure. LSTM is an RNN that provides a model that can store information for time periods, especially those used to overcome the problems of vanishing gradient. LSTM is trying to solve the problem of training over retaining memory and long sequences. LSTMs solve the gradient problem by introducing a more gate. LSTMs have an input gate, output gate, forget gate, and output activation function. LSTM has a memory gate so it can store data information in the long run. LSTM uses a memory gate to memorize the state.

LSTM and GRU both are RNN methods. GRU is different from the LSTM according to utilizing gating information to prevent vanishing gradient problem. GRU has gating units that control the data flow. Furthermore, GRU does not have a specialized memory cell. Figure 7.2 discusses the gates of LSTM and GRU. $i$ corresponds to the input gate, $f$ shows the forget gate, $o$ shows the output gate, and $c$ and $\tilde{c}$ are the memory cell and the cell content. GRU includes reset and update gates, activation, and candidate activation. $r$ and $z$ represent the reset and update gates, whereas $h$ and $\tilde{h}$ denote the activation and the candidate activation, respectively (Chung, Gulcehre, Cho, & Bengio, 2014).

### 7.3.2 DBN

DBN model depends on the RBM. RBMs are a two-layered artificial neural network with generative capabilities which is the basic component of DBN.

RBMs are two-layer artificial neural networks that can find probability distribution from model inputs. RBM's neurons form a bipartite graph. This gives an efficient training capacity for the algorithm when compare Boltzmann Machine approaches.

RBM is an energy-based neural network structure that includes visible and hidden layers. Furthermore, the visible layer collects data whereas the hidden one learns items from the visible layer.

RBMs have a constraint that neurons must establish a bipartite graph. RBM's neurons only have connections with other layers, neurons in the same layer have no

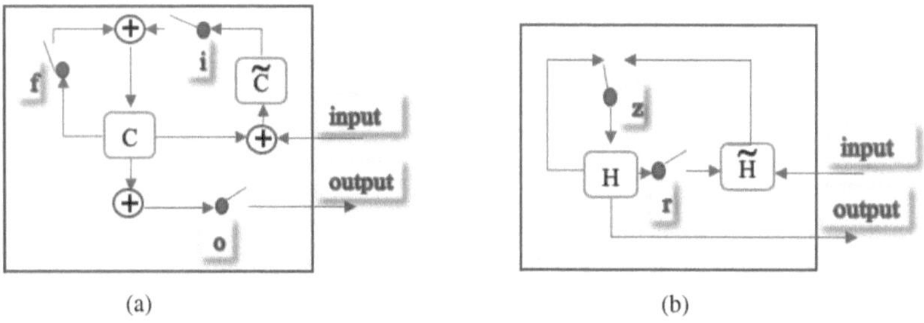


**FIGURE 7.2** (a) LSTM versus (b) GRU (Chung, Gulcehre, Cho, & Bengio, 2014).

connection with each other. Connections between layers are symmetric and bidirectional, in order to allow transfer information in both directions.

DBN is a learning mechanism based on DNN having the capability of unsupervised pre-learning. DBN is composed of multiple RBMs. DBN first performs layer-wise greedy and unsupervised learning, whereby each layer is represented by an RBM. DBN is trained in an unsupervised way (without target labels) to reconstruct its inputs. Each layer of DBN acts as a feature generator and converts the input to more abstract representation. After unsupervised learning, DBN can further be fine-tuned through supervised learning (use target labels) to perform classification or regression. This supervised learning after the unsupervised greedy layered-wise training is mostly performed using gradient descent.

DBN has two phases in training. The first phase, the DBN is trained in an unsupervised way with a contrastive divergence algorithm. In this way, a set of data with reduced characteristics is obtained. The second phase begins with the addition of an artificial neural network layer. Newly added layers improve the model with supervised learning and must be trained with the desired goals to perform forecasting.

## 7.4 EMPIRICAL STUDY

### 7.4.1 Data and Methodology

According to current data, wind power in Turkey in 2019 constitutes 7.40% of the total energy produced and reached about 7,615 MWh. Official records have shown that there are 3,155 turbines and 183 wind farms all around the country. As Figure 7.3 illustrates, wind energy power capacity is constantly increasing in Turkey (Turkish Wind Energy Association, 2019).

Empirical models generally use historical wind speed and wind power data to forecast future values. Meteorological conditions, especially wind speed and temperature, can be counted as a major factor in wind power generations.

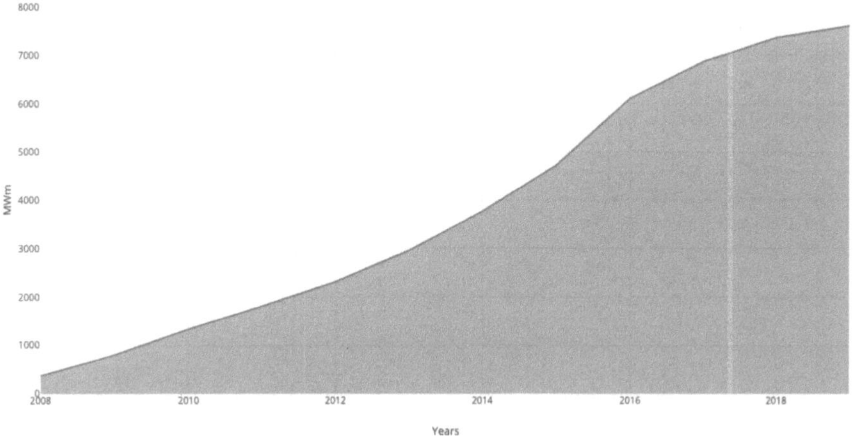


**FIGURE 7.3** Turkey wind energy capacity (Turkish Wind Energy Association, 2019).

In this study, our dataset contains wind speed, wind power, and weather temperature. Wind power measurements have an hourly interval. Temperature and wind speed data of İzmir, Çanakkale, Manisa, and Balıkesir were employed.

However, wind energy capacity of these cities constitutes 49.54% of all capacities (Turkish Wind Energy Association, 2019).

In the study, the temperature and wind speed data of these cities are used as input. Figure 7.4 illustrates the wind power capacities for Turkey.

Wind speed is a crucial parameter for wind power generation. Wind speed is the main source of wind energy, so it is vital to use the wind speed as input for the model. There are different approaches using wind speed. Statistical and intelligence methods have also been used to forecast wind speed. Table 7.1 shows wind speed forecasting methods (Yatiyana et al., 2018). The distance between wind power generation is referred to as distance or spatial resolution. In this study, we have used hourly available wind speed data in this empirical study. Figure 7.5 shows wind power generation in Turkey between March 2019 and December 2019.

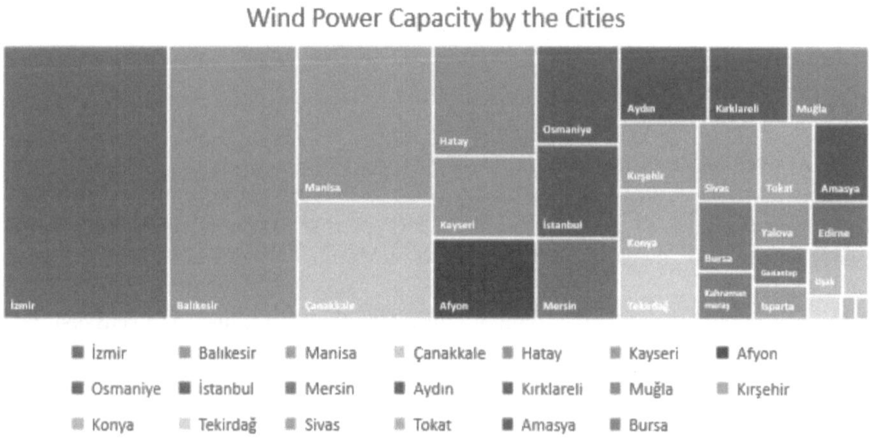


**FIGURE 7.4** Wind power capacity by the cities (Turkish Wind Energy Association, 2019).

## TABLE 7.1
## Wind Speed Prediction Methods

| Models | Time Horizon | Distance (Spatial Resolution) |
|---|---|---|
| AR, ARMA, ARX, NARX | Seconds to hours | Between 0 and ~10 m |
| ANN, SVR, fuzzy theory | Seconds to hours | Between ~10 and ~500 m |
| Sky imagers | Seconds to 30 minutes | Between ~1 and ~2,000 m |
| Satellite imagery | ~15 minutes to ~6 hours | Between ~1,000 and ~10,000 m |
| Mesoscale forecasting | ~4 hours to ~120 hours | Between ~5,000 and ~20,000 m |
| Global forecasting | ~12 hours to days | Between ~10,000 and 90,000 m |

*Source:* Yatiyana et al. (2018).

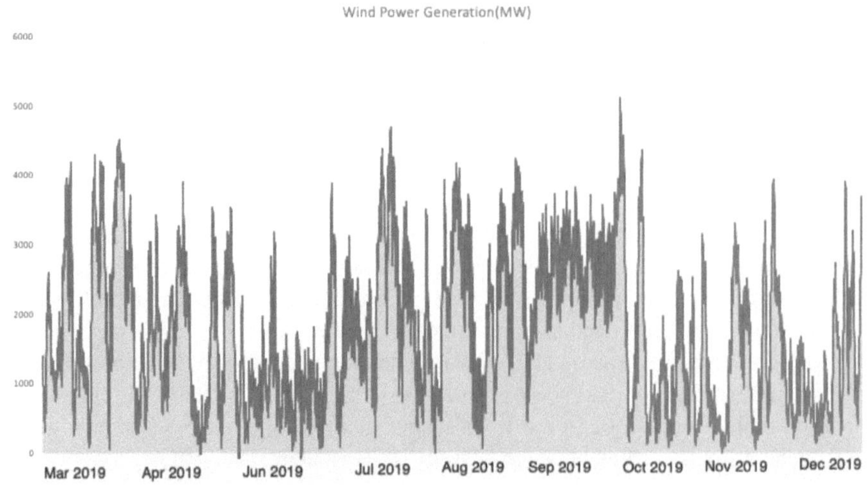


**FIGURE 7.5** Wind power generation with 1-hour resolution.

Many deep learning approaches are established in forecasting the wind power. So as to ensure the accuracy of forecasting wind speed, temperature and wind power data have been utilized as the input of the deep learning approaches. Data contains hourly wind power generation for the period of March 2019 to December 2019 with 1-hour resolution.

Wind speed and temperature data of İzmir, Çanakkale, Manisa, and Balıkesir cities are used as inputs to the models. Each series is first preprocessed using minimax scalar, then preprocessing outputs are used as inputs to deep learning models that forecast wind power forecast. Figure 7.6 shows wind speed data by cities with 1-hour resolution, and Figure 7.7 shows temperature data by cities between March 2019 and December 2019.

### 7.4.2 Empirical Results

This study gives models in wind power forecasting including DBNs and RNNs. Three most widely used DNNs—GRU, LSTM, and DBN—in wind power forecast tasks are employed.

In addition, hidden neuron size and data preprocessing can affect the method's performance very dramatically. Therefore, the optimization of these two parameters is crucial to get better performance of DNNs. Generally, data processing is important because of reducing the impact of abnormal conditions in the data and finding missing data. Data preprocessing effect models performance. Without preprocessing, all models do not give reasonable accuracy results. It was observed that the lack of preprocessing was negatively impacted by the results of all models.

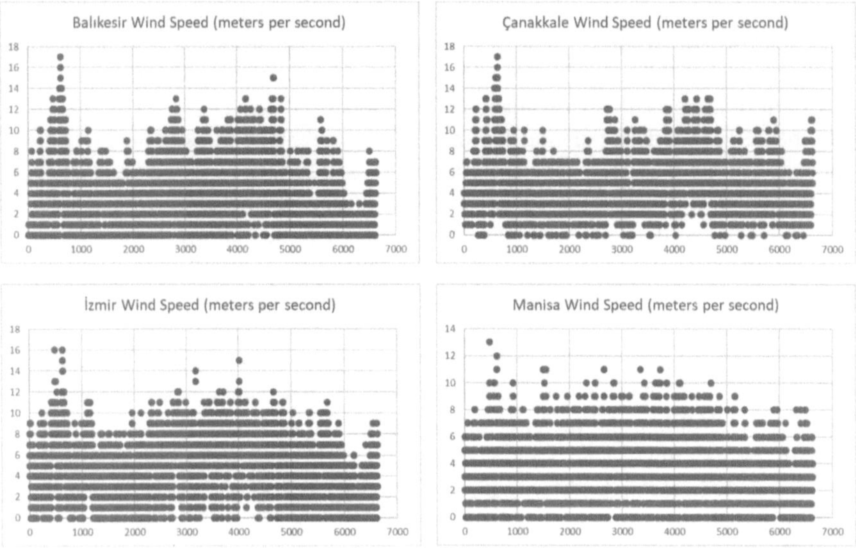


FIGURE 7.6 Wind speed data with 1-hour resolution.

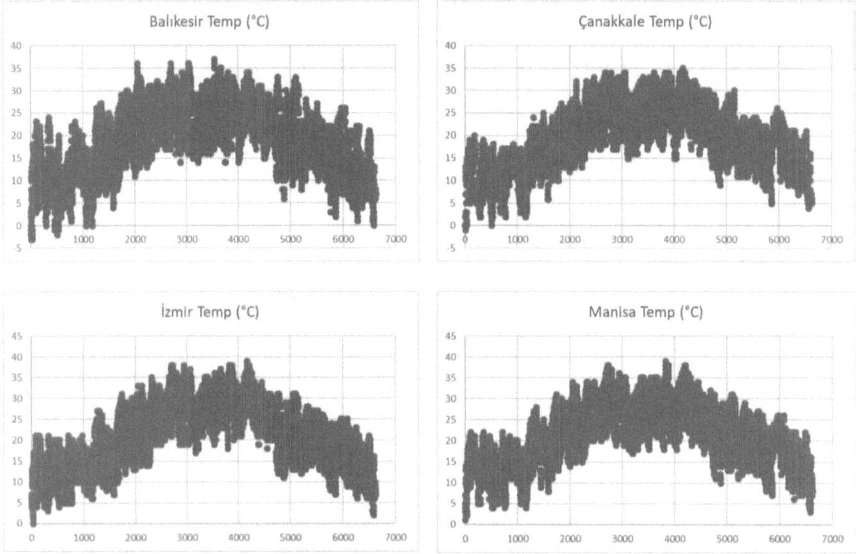


FIGURE 7.7 Temperature data with 1-hour resolution.

Selecting the number of neurons has a significant role in the training time of the models, and the fact that having more neurons does not mean that the model will be more successful.

The number of layers has been changed from 1 to 3, and the number of neurons in the layer has been changed from 5 to 30, and different models have been created.

Among the different models created by changing the number of layers and the number of neurons, the five best results according to the MAE values are given in Tables 7.1–7.3. It is observed that an increasing number of layers and number of neurons do not appear to increase the performance of the model.

In the LSTM models, the model with one layer and 30 neurons in the layer gives the most successful MAE result. The other successful result is the one-layer and ten-neuron model, which gives the successful MAE result. The increase in the number of layers and neurons does not have a positive effect on model performance.

Increasing the number of layers in GRU models has a positive effect on the model's performance. It is also seen that the performances of the top five models are very close to each other.

From the comparison of the methods, it can be said that DBN models give more successful results when the number of neurons is low which is also depicted in Table 7.4. In addition, the increase in the number of neurons also has a negative effect on training time.

GRUs train faster and perform slightly better than LSTMs. GRU has partially achieved more successful results. DBNs performed slightly better than both LSTM and GRU (Table 7.5).

The choice of the type of deep learning models may depend heavily on the dataset and corresponding task. According to the created models, DBN models gave relatively better results than LSTM and GRU models. Among the neural network models, it is interesting to note that DBN outperforms RNNs. Therefore, DBN models can be used in the Turkish wind power forecast.

**TABLE 7.2**
**LSTM Results**

| Model Name (Layer-Neuron Number) | Mean Absolute Error | Mean Squared Error |
|---|---|---|
| 1–30 | 0.119 | 0.022 |
| 1–10 | 0.120 | 0.022 |
| 1–20 | 0.122 | 0.022 |
| 1–20_1–20 | 0.123 | 0.023 |
| 1–30_1–10 | 0.124 | 0.023 |

**TABLE 7.3**
**GRU Results**

| Model Name (Layer-Neuron Number) | Mean Absolute Error | Mean Squared Error |
|---|---|---|
| 1–10_1–30_1–10 | 0.119 | 0.022 |
| 1–20_1–10_1–10 | 0.120 | 0.022 |
| 1–10_1–10_1–10 | 0.121 | 0.022 |
| 1–5_1–30_1–10 | 0.121 | 0.022 |
| 1–5_1–10_1–20 | 0.122 | 0.022 |

**TABLE 7.4**
**DBN Results**

| Model Name | Mean Absolute Error | Mean Squared Error |
|---|---|---|
| 10 | 0.109 | 0.018 |
| 10×10 | 0.113 | 0.019 |
| 50 | 0.109 | 0.018 |
| 50 × 50 | 0.111 | 0.019 |
| 100 | 0.110 | 0.017 |

**TABLE 7.5**
**Best Results of Models**

| Algorithm | Model Name | Mean Absolute Error | Mean Squared Error |
|---|---|---|---|
| LSTM | l–30 | 0.119 | 0.022 |
| GRU | l–10_l–30_l–10 | 0.119 | 0.022 |
| DBN | 10 | 0.109 | 0.018 |

## 7.5 DISCUSSION AND CONCLUSIONS

Accurate wind power forecasts reduce the physical effects of the weather conditions for the wind power systems. Furthermore, wind power forecasting helps to improve the power systems operations. The accuracy of wind power forecasting leads us to determine the power generation cost.

There are many researches released and many academic studies on wind power forecasting and researchers continue to propose many different models. However, it is crucial to evaluate which model is superior since each mathematical model used may have significant dependencies.

It would be inaccurate to conclude that one of the models is better than all other in any case. Therefore, the investigated models can perform well in Turkish wind power data; however, it is difficult to mention that the model will definitely work properly in every case. Our empirical study investigates the accuracy of the wind power forecasting models depending upon different factors and conditions.

Our empirical results have revealed that wind speed and weather data are crucial for analyzing wind power forecasting. Furthermore, LSTM, GRU, and DBN algorithms have found significance and applicable to the Turkish power market.

However, as we analyze the accuracy and efficiency of forecasting approaches, we have found out that deep learning models are superior to neural network models in terms of MEA and MSE figures, which lead the model performances.

In this regard, we may conclude that deep learning models are of great importance in analyzing the power markets and have a significant role for the decision makers especially in the energy forecasting area.

## REFERENCES


Athari, M. H., & Wang, Z. (2018). Impacts of wind power uncertainty on grid vulnerability to cascading overload failures. *IEEE Transactions on Sustainable Energy*, 9(1), 128–137. doi: 10.1109/TSTE.2017.2718518.

Bhaskar, K., & Singh, S. N. (2012). AWNN-Assisted wind power forecasting using feed-forward neural network. *IEEE Transactions on Sustainable Energy*, 3(2), 306–315. doi: 10.1109/TSTE.2011.2182215.

Chung, J., Gulcehre, C., Cho, K., & Bengio, Y. (2014). Empirical evaluation of gated recurrent neural networks on sequence modeling, 1–9. Retrieved from http://arxiv.org/abs/1412.3555.

Erdem, E., & Shi, J. (2011). ARMA based approaches for forecasting the tuple of wind speed and direction. *Applied Energy*, 88(4), 1405–1414. doi: 10.1016/j.apenergy.2010.10.031.

Haque, A. U., Nehrir, M. H., & Mandal, P. (2014). A hybrid intelligent model for deterministic and quantile regression approach for probabilistic wind power forecasting. *IEEE Transactions on Power Systems*, 29(4), 1663–1672. doi: 10.1109/TPWRS.2014.2299801.

Lei, M., Shiyan, L., Chuanwen, J., Hongling, L., & Yan, Z. (2009). A review on the forecasting of wind speed and generated power. *Renewable and Sustainable Energy Reviews*, 13(4), 915–920. doi: 10.1016/j.rser.2008.02.002.

Liu, H., Chen, C., Lv, X., Wu, X., & Liu, M. (2019). Deterministic wind energy forecasting: A review of intelligent predictors and auxiliary methods. *Energy Conversion and Management*, 195(May), 328–345. doi: 10.1016/j.enconman.2019.05.020.

Liu, H., Mi, X. Wei, & Li, Y. F. (2018). Wind speed forecasting method based on deep learning strategy using empirical wavelet transform, long short term memory neural network and Elman neural network. *Energy Conversion and Management*, 156(October 2017), 498–514. doi: 10.1016/j.enconman.2017.11.053.

Liu, Y., Guan, L., Hou, C., Han, H., Liu, Z., Sun, Y., & Zheng, M. (2019). Wind power short-term prediction based on LSTM and discrete wavelet transform. *Applied Sciences (Switzerland)*, 9(6). doi: 10.3390/app9061108.

López, E., Valle, C., Allende, H., Gil, E., & Madsen, H. (2018). Wind power forecasting based on echo state networks and long short-term memory. *Energies*, 11(3). doi: 10.3390/en11030526.

Marugán, A. P., Márquez, F. P. G., Perez, J. M. P., & Ruiz-Hernández, D. (2018, October 15). A survey of artificial neural network in wind energy systems. *Applied Energy*. Elsevier Ltd. doi: 10.1016/j.apenergy.2018.07.084.

Negnevitsky, M., & Potter, C. W. (2006). Innovative short-term wind generation prediction techniques. *2006 IEEE PES Power Systems Conference and Exposition, PSCE 2006-Proceedings*, 60–65. doi: 10.1109/PSCE.2006.296250.

Ouyang, T., Zha, X., & Qin, L. (2017). A combined multivariate model for wind power prediction. *Energy Conversion and Management*, 144, 361–373. doi: 10.1016/j.enconman.2017.04.077.

Qureshi, A. S., Khan, A., Zameer, A., & Usman, A. (2017). Wind power prediction using deep neural network based meta regression and transfer learning. *Applied Soft Computing Journal*, 58, 742–755. doi: 10.1016/j.asoc.2017.05.031.

Saleh, A. E., Moustafa, M. S., Abo-Al-Ez, K. M., & Abdullah, A. A. (2016). A hybrid neuro-fuzzy power prediction system for wind energy generation. *International Journal of Electrical Power and Energy Systems*, 74, 384–395. doi: 10.1016/j.ijepes.2015.07.039.

Sideratos, G., & Hatziargyriou, N. D. (2012). Probabilistic wind power forecasting using radial basis function neural networks. *IEEE Transactions on Power Systems*, 27(4), 1788–1796. doi: 10.1109/TPWRS.2012.2187803.

Turkish Wind Energy Association. (2019). Turkish wind energy statistic report July 2019, 6–7. Retrieved from https://www.tureb.com.tr/turebsayfa/duyurular/turkiye-ruzgar-enerjisi-istatistik-raporu-temmuz-2019.

Wang, H. Z, Li, G. Q., Wang, G. B., Peng, J. C., Jiang, H., & Liu, Y. T. (2017). Deep learning based ensemble approach for probabilistic wind power forecasting. *Applied Energy*, 188, 56–70. doi: 10.1016/j.apenergy.2016.11.111.

Wang, K., Qi, X., Liu, H., & Song, J. (2018). Deep belief network based k-means cluster approach for short-term wind power forecasting. *Energy*, 165, 840–852. doi: 10.1016/j.energy.2018.09.118.

Wang, Q., Martinez-Anido, C. B., Wu, H., Florita, A. R., & Hodge, B. M. (2016). Quantifying the economic and grid reliability impacts of improved wind power forecasting. *IEEE Transactions on Sustainable Energy*, 7(4), 1525–1537. doi: 10.1109/TSTE.2016.2560628.

Xiaosheng, P., Lei, X., Wen, J., Yuan, X., Wenhan, F., Shuanglei, F., & Bo, W. (2016). A very short term wind power prediction approach based on Multilayer Restricted Boltzmann Machine. *Asia-Pacific Power and Energy Engineering Conference, APPEEC*, 2016-December, 2409–2413. doi: 10.1109/APPEEC.2016.7779917.

Yatiyana, E., Rajakaruna, S., & Ghosh, A. (2018). Wind speed and direction forecasting for wind power generation using ARIMA model. *2017 Australasian Universities Power Engineering Conference*, AUPEC 2017, 2017-November, 1–6. doi: 10.1109/AUPEC.2017.8282494.

Zeng, J., & Qiao, W. (2011). Support vector machine-based short-term wind power forecasting. *2011 IEEE/PES Power Systems Conference and Exposition*, PSCE 2011, 0511, 1–8. doi: 10.1109/PSCE.2011.5772573.

Zhang, Q., Lai, K. K., Niu, D., Wang, Q., & Zhang, X. (2012). A fuzzy group forecasting model based on least squares support vector machine (LS-SVM) for short-term wind power. *Energies*, 5(9), 3329–3346. doi: 10.3390/en5093329.

Zhou, H., & Fang, J. (2010). Wind power forecasting based on econometrics theory. *EPEC 2010- IEEE Electrical Power and Energy Conference: "Sustainable Energy for an Intelligent Grid*," 1–6. doi: 10.1109/EPEC.2010.5697193.

Zuluaga, C. D., Álvarez, M. A., & Giraldo, E. (2015). Short-term wind speed prediction based on robust Kalman filtering: An experimental comparison. *Applied Energy*, 156, 321–330. doi: 10.1016/j.apenergy.2015.07.043.

# 8 Green Energy
## *Solar, Wind, Geothermal, Tidal Storage*

*Manju Santhakumari*
CSIR-National Geophysical Research Institute

## CONTENTS



## ABBREVIATIONS

**AC** Alternating current
**A-CAES** Adiabatic compressed air energy storage
**CAES** Compressed air energy storage
**DC** Direct current
**D-CAES** Diabetic compressed air energy storage
**DLCs** Double-layer capacitors
**DWPT** Dynamic wireless power transfer
**ESSs** Energy storage systems

| | |
|---|---|
| **I-CAES** | Isothermal compressed air energy storage |
| **LHS** | Latent heat storage |
| **MATLAB** | Matrix laboratory |
| **PCM** | Phase change material |
| **PHS** | Pumped hydro storage |
| **PV** | Photo voltaic |
| **SEM** | Smart energy matrix |
| **SHS** | Sensible heat storage |
| **SMES** | Superconducting magnetic energy storages |
| **TCES** | Thermochemical energy storage |
| **TES** | Thermal energy storage |
| **UPS** | Uninterruptible power supply |
| **USA** | United States of America |
| **WTES** | Wind-powered thermal energy storage |

**Nomenclature**

| | |
|---|---|
| $/kW | Dollar per kilowatt |
| % | Percentage |
| °C | Degree centigrade |
| $CO_2$ | Carbon dioxide |
| E | Energy |
| I | Current |
| J | Joule |
| K | Kelvin |
| kW | Kilowatt |
| L | Inductance of the coil |
| m | Meter |
| $m^3$ | Cubic meter |
| mm | Milli meters |
| $MnO_2$ | Manganese dioxide |
| MPa | Mega Pascal |
| MW | Megawatt |
| MWh | Megawatt hour |
| $NO_x$ | Nitrogen oxides |
| rpm | Rotations per minute |
| $SO_2$ | Sulfur dioxide |
| V | Volt |
| W/kg | Watt per kilogram |
| W/L | Watt per liter |
| Wh/kg | Watt-hour per kilogram |
| Wh/L | Watt-hour per liter |
| Zn | Zinc |

## 8.1 INTRODUCTION

The necessities of life, such as air, water, food and energy, are inevitable for the survival of living beings in this universe. A secure, sustainable, and economical strategy to conserve the energy is necessary for the betterment of society. As the world goes behind the low carbon systems, renewable energy storage technologies have evolved substantially. Electricity generated from fossil fuels provides a stable form of energy where the stockpiled fuel is immediately accessed. However, for non-conventional energy sources, such as solar, wind, wave, and tidal, energy cannot be stockpiled due to its non-availability and intermittent nature. Adequate storage technologies replace the systems based on fossil fuels with intermittent non-conventional energy systems. Energy storage system (ESS) improves the efficiency of the electrical system by providing a balance between supply and demand, thereby ensuring the stability and reliability of the electrical systems.

A significant transition from the current status of the power systems is essential to tackle the intermittent renewable energy sources to meet the demand. Efficient, cheap, and reliable techniques to store the energy provide value addition to the non-conventional energy generation. The ongoing revolution in the renewable energy sector highlights the demand for advanced storage technologies and integration of energy in the power sector. The fabrication and the commercialization of storage technologies have a considerable impact on the power sector in terms of modeling. Researchers have concentrated on enhancing the reliability, safety, power quality, efficiency, optimization of the grid planning, and reducing the construction cost of the EESs. Storage technologies enhance the flexible distribution of electricity and transmission grids, thereby improving the quality of power and reducing greenhouse gas emissions.

Development and installation of advanced storage technologies empower the nation's electrical grid and efficient utilization of renewable energy sources. These progress in the energy sector decreases the peak power demand, reduces the electricity bills, and decreases the greenhouse gas emissions. This chapter addresses the five major classifications of the storage technologies for grid-scale energy use based on the energy forms.

## 8.2 STORAGE TECHNOLOGIES

With global acceptance of renewable energy sources, storage technologies for these intermittent sources have shown greater momentum in recent years. These storage technologies mitigate the variations in power supply, improve the flexibility of the system, enhance the storage capability with advanced technology, and supply the generated electricity from non-conventional energy sources. EES system stores energy as potential, kinetic, thermal, and chemical energy and supplies to the customers based on the demand in the form of electricity. These systems consist of a power conversion unit, storage facility, and a balance of plant. Based on the form of energy stored, researchers have classified the EES into mechanical, electrochemical, chemical, electrical, and thermal storage technologies as shown in Figure 8.1 (Achkari & Fadar, 2018).

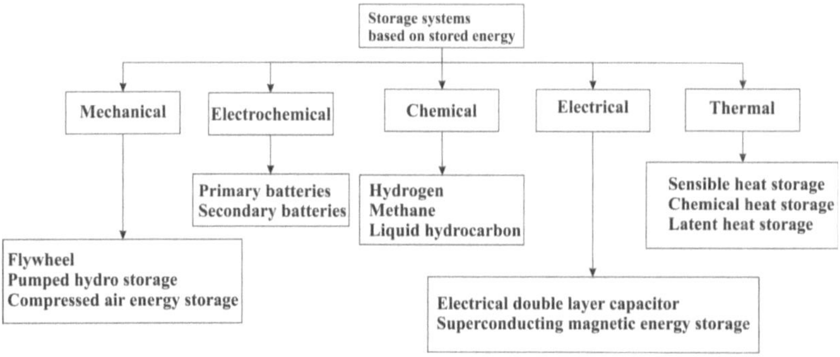


**FIGURE 8.1** Storage technologies used in non-conventional energy systems (Achkari & Fadar, 2018).

## 8.2.1 Mechanical Energy Storage Systems

The mechanical ESS consists of a pumped hydro storage (PHS) system, compressed air ESSs, and flywheel energy storage. The flywheel system stores kinetic energy, whereas other systems store potential energy.

### 8.2.1.1 Flywheel

The flywheel is a mechanical storage energy system, which converts the stored energy to kinetic energy in a rotating mass. The modern flywheel storage system (Figure 8.2) (Díaz-González, Sumper, Gomis-Bellmunt, & Villafáfila-Robles, 2012) consists of rotors with one or two shapes, bearings to support the flywheel rotor, motor/ generator for charging and discharging, power converter, and a containment

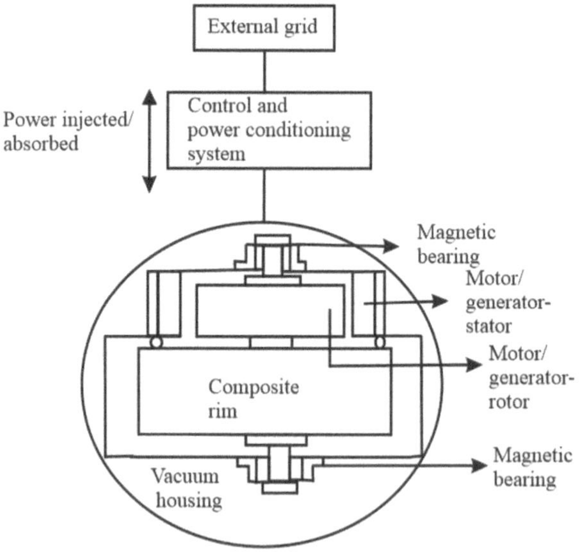


**FIGURE 8.2** Schematic of flywheel storage technology (Díaz-González et al., 2012).

chamber. The system operates in a high-vacuum environment of about $10^{-6}$ to $10^{-8}$ atmospheric pressure to reduce the windage losses and to protect the rotor from the external environment. Electrical power from the grid or another source of electrical energy supplies the input energy to the flywheel. The motor/generator assembly drives the rotating flywheel as it is coaxially connected and helps the exchange of electrical energy to mechanical energy and vice versa. A bidirectional power converter, together with the electrical machine, does the energy conversion processes such as direct current (DC)–alternating current (AC), AC–AC, and combination of these. The mechanical or the magnetic bearing system helps in giving a support mechanism to the flywheel by fixing the rotor in the appropriate position with lower friction. The containment chamber is made of thick steel and maintains the rotor in the vacuum environment. This stationary part reduces the aerodynamic drag losses so that the system can withstand the rotor failures. For solar systems, flywheels integrated with batteries enhance the power output of the system and increase the battery life, whereas, for the wind power system, flywheels reduce the wind oscillations and improve the frequency of the system (Bolund, Bernhoff, & Leijon, 2007).

Based on the rotational speed, the classification of flywheel technologies includes low-speed flywheel (speed less than 10,000 rotations per minute (rpm)) and high-speed flywheels (speed greater than 10,000 rpm). Steel rotors and conventional bearings are used in low-speed flywheels, whereas composite rotors and low-friction bearings were integrated with high-speed flywheels. Flywheels could deliver higher power based on the generator ratings and the power electronics, whereas the lifetime depends on the bearings used in the system. The energy stored in a flywheel depends on the square of the rotational speed, and hence, high-speed flywheels are suitable for energy–mass ratio optimization. Low-speed flywheels are used for uninterruptible power supply (UPS) for internet services, whereas the high-speed flywheels find application in the transport sector. The Piller's POWERBRIDGE is a popular low-speed flywheel (3,600 rpm) UPS with a range of 250–1,300 kW and can deliver 1.1 MW for 15 seconds (Cimuca, Saudemont, Robyns, & Radulescu, 2006). Beacon Power Corporation from Massachusetts (Lazarewicz & Rojas, 2004) has developed multiple paralleled flywheels housing (named as smart energy matrix (SEM)) ten high-energy flywheels which can deliver 1 MW for 15 minutes. By connecting several SEMs, it can deliver 10+ MW for a lifespan of 20 years.

Key merits of the flywheel technology include its higher cyclic ability of $10^5$ cycles with deep discharge, larger power density, higher efficiency, and lower maintenance. However, these systems lag with a higher self-discharge rate of about 20% per hour, making it suitable only for short-term applications. The higher price due to the incorporation of advanced materials while manufacturing, and lower production, makes the flywheel applicable for smoothing power fluctuations up to 15 minutes. The longer life and higher charging–discharging cycles make the flywheel systems more preferable than the batteries.

#### 8.2.1.2 Pumped Hydro Storage (PHS)

The PHS technology is the dominant, efficient, and high-capacity form of grid storage energy system which can store energy during non-peak hours and use it during peak hours. The storage system (Figure 8.3) (Zhou, Benbouzid, Frédéric

Charpentier, Scuiller, & Tang, 2013) consists of two reservoirs where one reservoir locates in a higher elevation than the other. A water tunnel and a piping system separate the two water reservoirs. During non-peak hours and when electricity is cheap (power is available), the system supplies water from the lower reservoir to the upper reservoir, where the supplied energy is set aside as potential energy in the water. Similarly, during peak hours and when electricity is expensive (non-availability of power), upper reservoir water is released to the lower reservoir, thereby releasing energy to produce the electricity. The study shows that the PHS plants stabilize the electrical system with greater penetration of renewable energy sources (van Meerwijk, Benders, Davila-Martinez, & Laugs, 2016).

The PHS systems decrease the harmonic distortions, reduce the voltage sags, and surge. It also protects the electrical system from outages. Despite the technical advantages, the system provides, these facilities lag with some disadvantages. The deployment of the system requires appropriate terrain with adequate elevation for fixing the reservoirs and a huge quantity of water resource. Despite the lower operation and maintenance cost, the facility needs huge capital investment for the civil construction work, which last for one decade. The categorization of pumped storage capacity includes long-term storage (period of one day or more) and short-term storage (period up to 24 hours) (Dursun & Alboyaci, 2010). The storage period of PHS varies from several hours to days or even years based on the evaporation and the penetration. Considering the evaporation and convection losses, almost 71%–85% of the electrical energy can be regained while pumping the water to the reservoir. The storage capacity of the PHS depends on the height of the waterfall, the quantity of water, and the round-trip efficiency covers 75%. Locations with two natural water bodies are suitable for implementing the PHS technology. Open sea, underground storage using flooded mine shafts, or cavities can be used as a lower reservoir. Introduction of pumped storage in isolated island grids having technical limitations enhances the maximum utilization of wind energy for national and community-level usage. Site selection of PHS is a significant part and requires regular consciousness of the scientists for selecting a technically feasible and socially acceptable location. As years go on, the availability of technically feasible and commercially available sites has

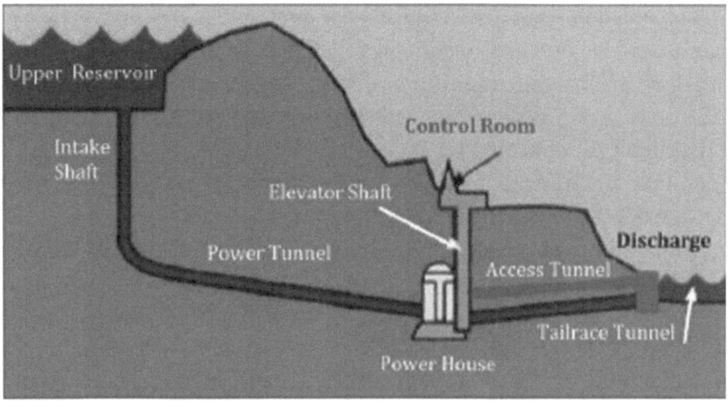


**FIGURE 8.3** Schematic of a PHS system (Zhou et al., 2013).

become scanty, and hence, it is essential to develop innovative methods for the site selection. Connolly et al. (2010) developed a computer program to identify a beneficial location for PHS based on capacity and economics. The developed program scans a particular terrain and traces out the availability of any feasible PHS site in it.

The photo voltaic (PV) based PHS system uses the power obtained from the panels to transmit through the grids and pumps the water from lower to upper reservoir using the extra energy produced during off-peak hours (Javanbakht, Mohagheghi, & Simoes, 2013; Ma, Yang, Lu, & Peng, 2015; Margeta & Glasnovic, 2012). Integration of PHS with intermittent wind power is an economically and technically competitive technology to fill the wind power gaps and to exploit the vast wind potential (Anagnostopoulos & Papantonis, 2012; Brown, Peças Lopes, & Matos, 2008; Kapsali, Anagnostopoulos, & Kaldellis, 2012). A seawater PHS located at the northern coast of Okinawa Island, Japan, generates 30 MW of power with a gross storage capacity of $10^6 m^3$. The octagonal planar shape plant has a maximum depth of 25 m and a crest length of 848 m (Hiratsuka, Arai, & Yoshimura, 1993).

#### 8.2.1.3 Compressed Air Energy Storage (CAES)

The energy-storing technology such as CAES is a large-scale ESS which stores energy in the form of high-pressured air. This technology is applicable for longer duration and large utility-scale at lower costs. The CAES facility (Figure 8.4) (Díaz-González et al., 2012) uses compressed air and stores it under higher pressure in an underground cavern. The storage system consists of a motor/generator, air compressors, turbine train, a container for storing compressed air, and fuel storage and heat exchanger units (Desai, Gonzalez, Pemberton, & Rathjen, 2005). The electrically driven compressors charge the storage and convert the electrical energy to potential energy or pressurized air. Based on the demand to generate electricity, the release of stored pressurized air and expansion takes place through air turbine. The compressed air released from the CAES plant is heated through a recuperator before mixing it with natural gas. The mixture is later expanded through a turbine to produce electricity. The analysis of the hybrid CAES system found that it saves the emission costs and reduces the total suspended particulate, $CO_2$, $NO_x$, and $SO_2$, emissions per year to a significant amount (Chen et al., 2017).

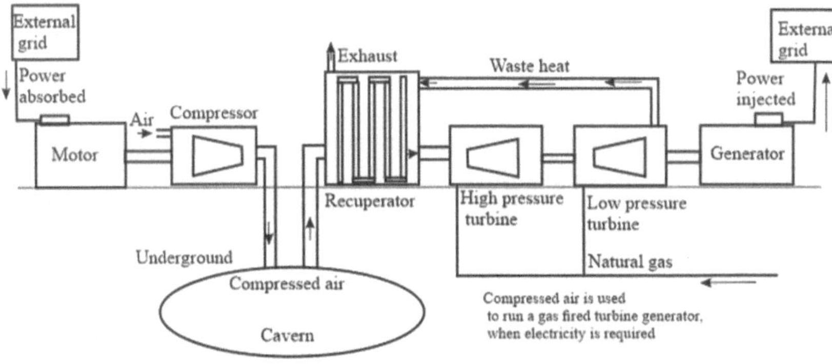


**FIGURE 8.4** Schematic of a compressed air ESS (Díaz-González et al., 2012).

According to the process in which energy is retained from storage to generation, the classification of CAES systems includes diabetic (D)-CAES, adiabatic (A)-CAES, and isothermal (I)-CAES systems (Budt, Wolf, Span, & Yan, 2016). Diabetic CAES systems need a high source of energy through a combustion chamber to allow high-enthalpy air inside the gas turbine. The efficiency of the adiabatic CAES is higher than the diabetic CAES and uses heat exchangers to retain the generated heat to heat the compressed air. Isothermal CAES consists of a series of compressors and cooling systems to compress the air with lower temperature. During the expansion process, the rejected heat from the cooling system is stored and released. Usually, the storage of CAES is kept at underground, which reduces land usage and investment cost. While comparing this with natural structures, it has geological limitations and restricted pressure differences to guarantee stability. Estimation of the critical parameter such as sealing capacity is not an easy task due to the generation of compression power with air leakage.

The first installed CAES plant in the world has cavern size of 3,10,000 $m^3$ integrated with 60 MW compressors at 10 MPa pressure. The unit located at 600 m underground in Huntorf, Germany, can generate 290 MW for 2 hours. The CAES system possesses 90% availability and 99% starting reliability since its installation in 1978. The second CAES unit installed 450 m below the surface in McIntosh, Alabama, USA, compresses air up to 7.5 MPa. The system generates 110 MW energy for 26 hours and holds a storage capacity of 5,00,000 $m^3$. When compared to the Huntorf CAES plant, the McIntosh plant reuses the heat from the gas turbine using a recuperator, thereby reducing the consumption of fuel by ~25%.

Similar to PHS, installation of CAES plant is possible only in some geographical locations which are in proximity to salt caverns, rock mines, or depleted gas fields. Incorporation of a gas turbine is essential, and hence, the unit is not an independent system. These systems highlight the necessity of combusting fossil fuels and associated emissions resulting in being less eco-friendly. Also, the CAES systems cannot be used with the wind turbine, solar PV, or nuclear plants. CAES units have larger storage duration, reduced capital costs, and higher efficiency.

### 8.2.2 Electrochemical Energy Storage Systems

Electrochemical storage systems (Figure 8.5) (Díaz-González et al., 2012) consist of several electrochemical cells connected in between and produce electricity by electrochemical reaction. The cell includes cathode, anode, and an electrolyte kept at solid, liquid, or viscous state. The chemical reaction took place at the cathode and the anode while discharging, and electrons flow from the anode to the cathode. The reverse actions took place while charging, and the battery charges with an applied voltage at the two electrodes. These storage systems consist of primary batteries, secondary batteries, batteries for grid-scale, fuel cells, and electrochemical capacitors.

#### 8.2.2.1 Primary Batteries

Primary batteries are single-use batteries which cannot be recharged and need to be discarded after it discharges completely. Based on the electrolyte present, the

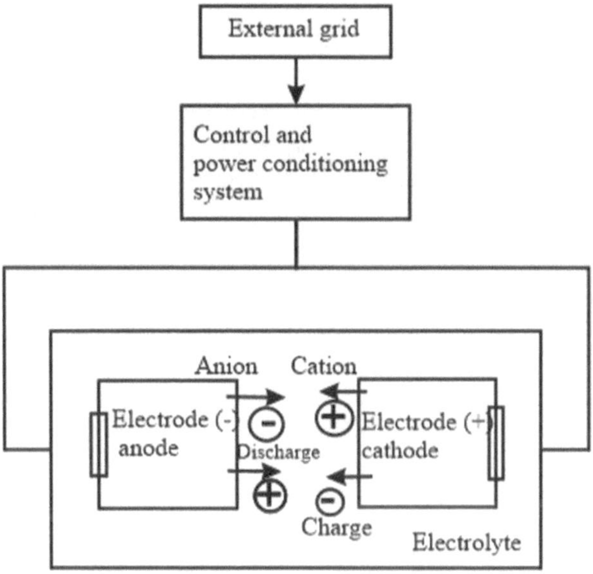


**FIGURE 8.5** Schematic of battery ESS (Díaz-González et al., 2012).

categorization of primary batteries includes aqueous and non-aqueous batteries. These batteries are widely available in the market based on the size such as AA, AAA, and C type; the common form being lithium-primary cells, alkaline battery, and the zinc–carbon battery.

#### *8.2.2.1.1 Lithium-Primary Cells*

The non-aqueous electrolyte-based lithium metal anode batteries were discovered in the early 1970s and had become a commonly available battery among the public in the early 1990s. This battery is one among the low-maintenance batteries which do not need cycling to extend the age of the battery. Apart from this, the self-discharging of this battery is very less compared to other batteries and causes less damage when discarded. Despite the advantages offered by these batteries, it contains hazardous materials such as lithium metals and flammable solvents, causing thermal runaway and exothermic activities. Misuse of these batteries during operation and maintenance causes short-circuit, overheating, over pressurization, and explosion of the battery packs (Lisbona & Snee, 2011).

#### *8.2.2.1.2 Alkaline Battery*

In $Zn\text{-}MnO_2$ alkaline battery, zinc and manganese dioxide act as the electrodes and potassium or sodium hydroxide acts as the electrolyte. These primary batteries have dominated the market for almost six decades (Kordesch, 1981) due to the steady voltage output and higher energy density. Despite the advantages, these batteries face a threat in the local market with primary lithium batteries owing to its higher performance.

*8.2.2.1.3 Zinc–Carbon Battery*

In 2003, the overall weight of zinc–carbon batteries sold in East and West Europe accounted for about 50,197 tonnes which were equivalent to 30.50% of the total annual sale (Sayilgan et al., 2009). These batteries use zinc as the anode, manganese dioxide as cathode, and ammonium chloride/zinc chloride as the electrolyte. The recovery of valuable metals such as zinc, manganese, carbon, plastics, and electrolyte after discarding the batteries was well studied using acid and alkaline leaching processes (Shin, Kang, Yang, Kim, & Sohn, 2007).

### 8.2.2.2 Secondary Batteries

The oldest form of electrochemical energy storage, the secondary batteries come as different sizes in the market and encounter faster development in the recent past. These batteries have larger applications in laptops, computers, ignition systems, and cell phones. The following section discusses the commonly available rechargeable secondary batteries.

*8.2.2.2.1 Lead-Acid Battery*

The lead-acid battery has greater utilization in energy storage applications such as stand-alone PV systems, battery unit for wind power, emergency power units, and starter batteries in automobiles. These batteries offer lower manufacturing cost, 65%–80% efficiency, and good surge capabilities. Despite the advantages, these batteries lag behind shorter life cycle, acid leaking, lower energy density, poor thermal management, the release of poisonous gases, and acid fumes (Chen et al., 2009). Identification of an enormous amount of airborne and blood lead in the lead-acid battery recycling and manufacturing unit in Kenya has raised concern about the health, hygiene, and work culture of the industry (Were, Kamau, Shiundu, Wafula, & Moturi, 2012).

*8.2.2.2.2 Nickel–Cadmium Battery*

Based on the age and availability in the market, the nickel–cadmium batteries rank similar to the lead-acid batteries, where nickel hydroxide and cadmium hydroxide act as the electrodes separated by an alkaline electrolyte. The expensive manufacturing procedure results in a higher cost for these batteries, and it causes severe environmental threat due to landfilling after the usage. These batteries observe memory effect due to reduced discharge voltage in the nickel electrode, and proper operation and maintenance decrease this effect to some extent (Sato, Ito, Arakawa, & Kobayakawa, 1996).

*8.2.2.2.3 Lithium-Ion Battery*

The commercialization of the lithium-ion batteries took place in 1991 by the multinational company Sony (Chen et al., 2009). Lithiated metal oxide acts as the cathode, graphitic carbon as the anode, and lithium salts dissolved in organic carbonates as the electrolyte in this battery. During charging, lithium atoms migrate to the carbon anode through the electrolyte and combine with external electrons and accumulate between the carbon layers as lithium ions. The process will reverse while discharging of the battery. These batteries possess higher efficiency, longer life cycle, higher energy density, power density, and lower self-discharge rates (Chen et al., 2009). Lithium-ion batteries use different materials for its manufacture based on the size, application, and the type of cathode used. The study shows that hydrometallurgical processes recovered materials such as copper, cobalt, and nickel while recycling the lithium-ion batteries (Boyden, Soo, & Doolan, 2016).

##### *8.2.2.2.4 Sodium Sulfur Battery*

Sodium sulfur batteries developed in the 1960s by the Ford Motor Company (Energy storage Council, 2015) consist of liquid sulfur and liquid sodium as two electrodes separated by solid beta alumina ceramic electrolyte. During the discharge process, the positively charged sodium ions migrate through the electrolyte and combine with sulfur to form sodium polysulfides. During the charging period, the entire process took place in reversible mode. These batteries operate at a temperature of around 300°C–350°C (Energy storage Council, 2015) and possess some advantages such as higher life cycle, energy density, and lower cost. Due to the high-operating temperature range, these batteries face operation and maintenance problem along with safety issues.

##### *8.2.2.2.5 Flow Batteries*

The basic difference between a flow battery (redox flow battery) and a conventional battery is that the electrode stores the energy in case of a conventional battery, whereas the electrolyte stores the energy for flow battery. In this battery (Figure 8.6) (Díaz-González et al., 2012), the dissolved electroactive species inside the electrolyte flows through the reactor where the conversion of chemical energy to electrical energy takes place. The cells of the reactor help in pumping the electrolyte stored in the tank and the irreversible reaction helps the battery to charge, discharge, and recharge. These batteries separate power and energy and have the capability to release the energy at a discharge rate of 10 hours. There are two broad flow battery technologies such as redox flow battery and hybrid flow battery. Systems, where the electroactive materials are stored internally, are called hybrid flow batteries (Chen et al., 2009).

### 8.2.3 Chemical Energy Sorage Systems

The storage of electricity from non-conventional energy resources is challenging, and a promising approach to store energy is in the form of chemicals. The main advantage is that the energy density of the chemicals is much higher than that of the energy density of the battery technologies. The chemicals help not only for the storage of energy but also in the transportation sector (as fuel) and chemical industry (as raw material). The following section discusses the various chemicals used as storage compounds.

#### 8.2.3.1 Hydrogen

Hydrogen is an ideal energy carrier owing to its ease of production, the capability to convert to electricity using fuel cell or gas turbine and higher weight-based energy content. The technology such as water electrolysis helps in the production of electricity

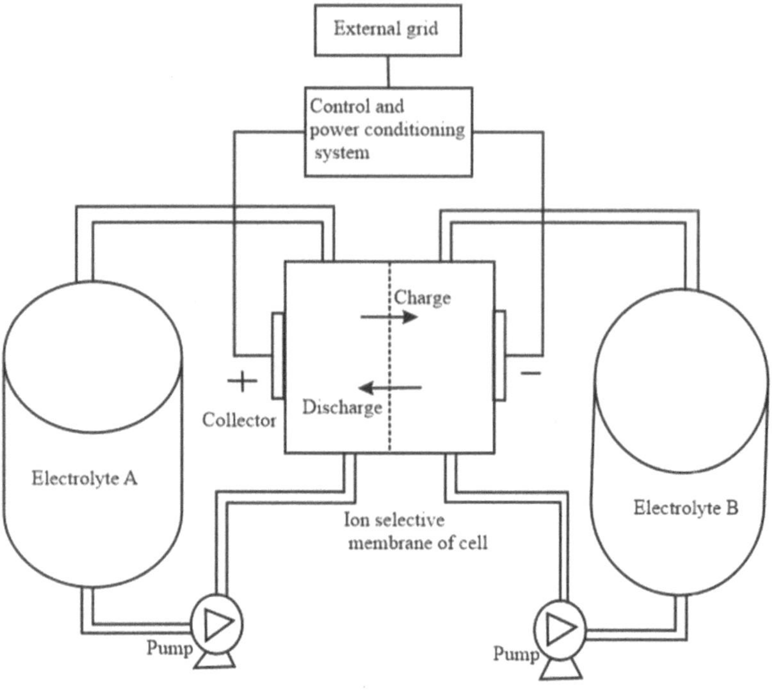


**FIGURE 8.6** Schematic of the flow battery system (Díaz-González et al., 2012).

from non-conventional energy sources. At present, the global hydrogen production has reached around 500 billion cubic meters per year (Shiva Kumar & Himabindu, 2019). Hydrogen production from renewable energy sources uses wind, biomass, and solar energy as the primary source. Electrolysis and gasification technologies were involved in generating hydrogen from the non-conventional sources. Solar hydrogen plants supply hydrogen to the vehicle fueling station using PV panels and use electrolyzers to split the water. The study found that a solar-powered proton exchange membrane electrolyzer enhances the hydrogen generation capacity by 12% so that it could supply optimum hydrogen to fuel cell vehicle (Gibson & Kelly, 2008). A work reported analyzing the hydrogen generation in Argentina found the wind energy generated in a single department could produce ten times the hydrogen energy to power the entire vehicular transportation in the province (Rodríguez et al., 2010).

Generation of free hydrogen is the advantage of electrolysis technology as the end product is free from sulfur and carbon contamination. However, the process lags behind the higher operating cost and the larger requirement of energy than fossil fuel technology. As the research on alternative hydrogen storage solution is progressing with metal hydrides, and complex hydrides (Eberle, Felderhoff, & Schüth, 2009), the technology seems like a cost-effective way to produce hydrogen. Biomass resources use municipal and animal waste, wood and forest residues, and crops and agricultural leftovers to produce hydrogen. Despite the clean energy produced from agricultural and domestic waste, the biomass gasification process lags behind the incapability to produce hydrogen on a larger scale. The photosynthesis (during biomass growing process) absorbs the carbon dioxide released during oxidation of the biomass, which is an added advantage over fossil fuels. The process requires vast land area and huge resources, which is a significant concern for the development of this method.

#### 8.2.3.2 Methane

Methane is another storage molecule with triple volumetric energy density (gaseous methane) than gaseous hydrogen. Storing the non-conventional electricity in molecules reduces the environmental impact due to carbon dioxide and also increases the capacity to save energy for a longer duration (Vogt, Monai, Kramer, & Weckhuysen, 2019). In Germany, the "Power to Gas Platform" has attained much attention after the inauguration in 2012 by the German Energy Agency (Winkler-Goldstein & Rastetter, 2013). The technology explains about the combustion of stored methane to generate electricity during the intermittent period when non-conventional energy productions are not available. Methane combustion produces carbon dioxide, and the by-product later underwent electrolysis and hydrogenated to form methane with an energy efficiency of 75%–80% (Trost, Horn, Jentsch, & Sterner, 2012). The salt cavern and the natural gas grid later store the methane. A downside of this storage method is the involvement of safety aspects due to its higher level of greenhouse gas potential than carbon dioxide.

#### 8.2.3.3 Liquid Hydrocarbon

Liquid hydrocarbon produced from renewable sources forms the backbone of the existing energy scenario owing to the higher energy density. In future, these hydrocarbons might supply fuel to the aviation sector as well (Kivits, Charles, & Ryan, 2010). This chemical ESS is of special interest for the renewable sector as it would help in estimating the future energy scenarios. A study was conducted to model the fluctuating power from wind system and carbon dioxide using a flowsheet simulation to evaluate the economic performance (König, Freiberg, Dietrich, & Wörner, 2015). The authors found that economic viability is possible to some extent by reducing the cost of the electrolyzer and electric feedstock.

### 8.2.4 Electrical Energy Storage Systems

A current-carrying coil saves the electromagnetic energy reflected from the objects in the form of electrical and magnetic waves. The two technologies which store this electrical energy consist of double-layer capacitors (DLCs) and superconducting magnetic energy storages (SMES). These storage systems possess higher efficiency, power density, lower energy density, and lesser degradation after frequent charging and discharging.

#### 8.2.4.1 Electrical Double-Layer Capacitors (DLCs)

The electrochemical DLCs or supercapacitors store energy between two charged electrodes separated by an insulating material in an electric field. These storage devices employ larger surface area electrodes and thinner dielectrics to accumulate more charge carriers and capacitance up to 5,000 Farads. These capabilities help the supercapacitors to possess higher energy density (than conventional capacitors) and higher power density (than batteries). Based on the storing charges, the classification of supercapacitors includes faradaicnon-faradaic and hybrid supercapacitors. Faradaic or pseudo-capacitors charge faradically due to migration of charges between the electrode and electrolyte through electrosorption, oxidation–reduction, and intercalation processes. Non-faradaic capacitors consist of carbon-based electrodes, electrolyte, and a separator which store charge electrostatically. Hybrid supercapacitors provide better performance, higher energy, and power densities than electrical DLCs without compromising the cost and cycle life.

The researchers integrated a high-voltage photo capacitor consisting of four serially connected cells into a dye-sensitized solar module and electrical DLCs in a compact harvesting storage device. The study rendered a competing technology to satisfy the storage technology as the system achieved a remarkable photo charging potential of 2.45 volts (V) as it has never achieved for any storage device (Scalia, Varzi, Lamberti, Jacob, & Passerini, 2018). To produce hydrogen from wind power, the wind power-generating station in a stand-alone radio base station implemented electric DLCs as a power buffer medium. The study found that the incorporation of electric DLCs ensures stabilization of the quantity of generated hydrogen and decreases the hydrogen load capacity (Carton & Olabi, 2010). The usage of DLCs reduces the long charging times and power shortages faced by dynamic wireless power transfer (DWPT) system. Ando et al. (2020) developed a motorcycle system with electric DLC for a DWPT scooter as a power supply unit. As per the study, the transmitted energy to the scooter is 120 Joules (J), which is enough for 100 J necessary for 2 m run of the scooter.

#### 8.2.4.2 Superconducting Magnetic Energy Storages (SMES)

The SMES works on an electrodynamic principle where it stores energy in the magnetic field when a current flows around the coil. The coil must be a superconducting material with negligible electrical resistance, and the storage system possesses short reaction time so that the system releases power within a fraction of cycle for the loss in the line power. Maintaining the coil temperature below the critical temperature of the superconducting material is essential and depends on the type of material. The conductor consists of niobium–titanium, and the coolant could be fluid helium maintained at 4.20 K or super liquid helium at 1.80 K (Chen et al., 2009). A typical SMES consists of superconducting coil, cryogenic system, power conditioning unit, and protection systems. The energy ($E$) stored in the current-carrying coil is $E = 0.5LI^2$; where $L$ is the inductance of the coil and $I$ is the current. The equation applies to all coils; however, coils made from superconducting materials can only withstand the higher currents for useful energy storage capacity. The SMES possesses higher energy storage efficiency (97%), rapid response, less dependence of energy output on discharge rate, higher life cycle, and hence suitable for industrial applications where voltage stability and power quality is significant (Chen et al., 2009). The SMES system finds application in the storage of intermittent energy sources to constant electricity. Despite the advantages offered by the SMES system, the unit faces several problems while implementation such as higher cost and environmental harm linked with the magnetic field.

The researchers (Zhang et al., 2016) proposed an advanced superconducting power-conditioning unit consisting of hydrogen and ESS to resolve the fluctuating output power of the solar power generator. The work highlighted the continuous operation of the SMES system by applying a threshold voltage to the SMES coil in the 1 kilowatt (kW) hybrid ESS. Kalman filter resolved the fluctuating power from the renewable source, where the electrolyzer compensated the longer period component and SMES system compensated the shorter period component. The output power leveling of a 100 megawatt (MW) wind farm (Nomura et al., 2005) uses the SMES system connected to the electrical grid. The optimal selection of the charging or discharging rate of the SMES system reduces the capacity of the converter system and levels the output power of the wind farm based on the power demand. Depending on the wind speed, the SMES system compensated the inertia of the turbine blades by using the stored energy. Integration of large power grid consisting of wind and marine current farm with SMES unit results in power flow control and damping enhancement (Wang, Chen, Lee, & Chen, 2009). The authors simulated the characteristics of the wind farm and marine current farm with 80 MW and 60 MW induction generator, respectively. The model used frequency as well as the time-domain approach to validate the effectiveness of the proposed control unit. The simulated results found that the SMES unit was effective in stabilizing the wind as well as the marine current farm under variable wind speeds. Qais et al. (2020) applied a continuous mixed p-norm algorithm to tune SMES units to smoothen the output power of grid-connected wind power plants and to reduce the power fluctuations. The implemented control scheme shows outstanding performance under various operating conditions and is evaluated based on wind speed variability and uncertainty conditions.

### 8.2.5 Thermal Energy Storage Systems

Thermal energy can be stored by heating or melting or vaporizing a material so that the stored energy can be used for future applications by heating, cooling, and reversing the processes. Thermal energy storage (TES) system balances the seasonal variation on demand and supply of energy system (storage of heat during summer using underground ESSs for use in winter) and reduces the carbon dioxide emission and costs by improving the overall efficiency of energy systems.

Based on the properties of the materials, the classification of TES includes sensible, chemical, and latent heat storage (LHS) systems (Khan, Saidur, & Al-Sulaiman, 2017). In sensible heat storage (SHS), storage of thermal energy takes place by increasing the temperature of solid or liquid and without changing the phase of the material. The temperature of the material increases while absorbing and releasing energy during the charging and discharging process. Due to the variation in internal energy of the material, the accumulation of supplied energy takes place in the form of sensible heat. SHS medium includes solid storage materials such as rocks, bricks, stones, wood, and iron and liquid storage materials such as water, organic liquids, molten salts, and liquid metals. Thermochemical energy storage (TCES) systems provide long-term storage capability with lower energy loss, high-storage density, and consistent temperature output. The principle of TCES includes endothermic and exothermic chemical reaction and stores energy in the form of potential chemical energy. During the endothermic chemical reaction, thermal energy from a source is absorbed, thereby segregating the material into two products. During the exothermic discharging process, these products combine, recover the stored energy, and regenerate the material to reuse. The system releases water vapor while supplying heat and liberates heat energy while absorbing water vapor using sorption materials such as ammoniates, hydrates, and metal hydrides. The stored heat depends on the quantity of storage material, endothermic heat of reaction, and the extent of conversion. In LHS systems, the material changes its phase from solid to liquid by increasing the temperature as latent heat of fusion by storing the energy and changes its phase from liquid to vapor as latent heat of vaporization by releasing the heat. LHS system consists of storage material undergoing phase change such as phase change materials (PCMs), heat exchanging surface, and a container enclosing the material.

Hamed et al. (2017) developed a mathematical and experimental model of charging and discharging behavior of flat-plate solar collector integrated with PCM having a melting temperature of 42°C–44°C. The authors accomplished their experiments during the summer solstice in Gabes (33°51 North, 10°03 East), Tunisia. The conduction and convection processes transmit the thermal energy from the sunlight to the PCM attached below the absorber plate in the collector. The model uses the finite difference method to discretize each node and also MATLAB programming language for modeling the different PCM nodes. The authors also evaluated the effect of PCM thickness from 5 millimeters (mm) to 25 mm on the stored energy, total exergy destroyed during the heat transfer process, outlet water temperature, melt fraction during melting, and solidification. The results show that as the PCM thickness increases from 5 mm to 25 mm, the LHS rises from 69.60% to 73.04% whereas the SHS reduces to 26.96% from 30.40%. As the PCM thickness improves, the energy efficiency of the collector enhances from 76.98% to 82.42%, whereas the exergy efficiency decreases to 2.40% from 2.62%. The variation between the solidification and the melting time increases with PCM thickness, due to convection motion enhancement. Also, PCM lowers the temperature during charging mode and increases its temperature during the discharging mode. The temperature variation is because of the increase in storage volume as thickness increases, and hence, space heating requirements must be reduced during sunshine hours and increased during night-time compared to the conventional solar water heater.

Okazaki et al. (2015) studied and compared the wind-powered thermal energy storage (WTES) systems with battery ESS. The authors found that WTES is more attractive and absorbs surplus energy when connected to the grid. Integration of WTES enhances the energy harvest from wind farms, creates lesser environmental impact, and is less economical (due to low-tech materials for construction and lower noise and vibration of blades).

Incorporation of thermal ESSs enhances the operation of the thermal equipment, helping the consumers in shifting the usage of energy from high to low demand periods, effective operation of the thermal equipment, and facilitates the storage of intermittent energy sources between periods of supply and demand. TES systems also decrease the emission of pollutants and decrease environmental impacts. These systems also provide considerable economic gains during their lifetime by adequate operation of the thermal equipment and shifting of energy usage during lower demand period.

## 8.3 CONCLUSION

Table 8.1 (Chen et al., 2009) provides the comparison of the various technologies based on energy density, efficiency, and cost. Figure 8.7 (Zhou et al., 2013) shows the capacities of various energy storage and power output for different technologies.

**TABLE 8.1**
**Comparison of Various Storage Technologies (Díaz-González et al., 2012)**

| | Energy and Power Density | | | | Efficiency | Lifetime |
|---|---|---|---|---|---|---|
| Technology | Wh/kg | W/kg | MWh | MW | % | Years |
| Flywheel | 5–100 | 11,900 | 0.02–5 | 0.10–20 | 85 | 20 |
| PHS | - | - | 500–8,000 | 10–1,000 | 65–80 | 30–50 |
| CAES | 3.2–5.5 | - | 580 | 50–300 | 70–73 | 30–40 |
| Lead acid battery | 30–50 | 180–200 | 0.001–40 | 0.05–10 | 70–80 | 5–15 |
| Nickel–cadmium battery | 30–80 | 100–160 | 6.75 | 45 | 72 | 13–16 |
| Lithium-ion battery | 80–200 | 245–2,000 | 0.5–50 | 0.1–50 | 78 | 14–16 |
| SMES | 10–75 | - | 0.001 | 1–100 | 80–95 | 20 |

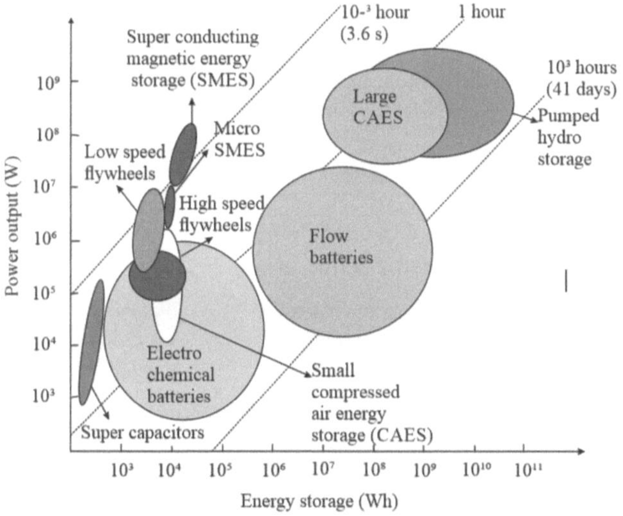


**FIGURE 8.7** Comparison of energy storage technologies (Zhou et al., 2013).

The figure illustrates that most of the technologies are in conjunction with each other. Despite the similarities, to choose a final solution, the cost can be treated as a significant parameter, as shown in Table 8.2 (Zhou et al., 2013). Mechanical storage systems are the widely adopted energy storage units with PHS accounting for 99% of global energy storage capacity. However, due to lesser efficiency and lower commercial progress, CAES and flywheels have not attained much global acceptance. Despite the challenges faced by thermomechanical storage systems, several innovative ideas have come up, which has the potential to be used in future to match the integration of intermittent sources.

Solutions for electrochemical storage such as batteries require advanced research and development work in lithium-ion and next-generation batteries. Existing battery technologies need improvements in energy/power density, charge/discharge rates, cost, and durability. The government must provide adequate funding in material development and fundamental research in new battery technologies for electric vehicles and other storage systems.

With higher energy density in chemicals, the produced chemicals are useful for electricity generation as raw material for the chemical industry and as fuel in the transportation sector. Most of the technologies based on the chemical storage system are at the verge of technological modifications; a comparison is difficult at this stage with the different storage systems.

By storing the energy in the form of an electric or magnetic field, the electrical storage systems increase its efficiency and reduce the reaction time and volumetric energy density. ElectricDLCs are robust technology with a higher power density and negligible cycle degradation. Despite the short reaction time, SMES require sophisticated types of equipment for the operation. The commercialization of SMES is possible with advanced research work on superconductors at room temperature.

Market development and prices are significant barriers that TES technologies face. Each storage application requires a special design to meet the boundary conditions. For TES, most of the research work has concentrated on materials for storage and thermal insulation development. The PCM requires attempts to enhance the reacting materials and system integration processes.

**TABLE 8.2**
**Costs for Energy Storage Systems (Zhou et al., 2013)**

| Storage Systems | Power Cost ($/kW) | Total Power Cost ($/kW) | Energy Cost ($/kWh) |
|---|---|---|---|
| Sodium sulfur (6-hour storage) | 550 | 2,150 | 400 |
| Lead acid (4-hour storage) | 660 | 2,580 | 480 |
| Flow battery (4-hour storage) | 1,300 | 3,100 | 450 |
| Flywheel (15-minute storage) | 3,920 | 4,313 | 1,570 |
| Supercapacitor (2-minute storage) | 350 | 450 | 30,000 |

## ACKNOWLEDGMENT

The author would like to thank the Director, CSIR-NGRI (Dr VM Tiwari) for permitting to publish this book chapter (NGRI/Lib/2020/Pub-10). The author thanks Dr EVSSK Babu for his encouragement and support and also thank him for providing the necessary facilities to complete this book chapter from institutional projects: MLP-6406–28 (EVB) and GEOMET (MLP-0002-FBR-2-EVSSK).


## REFERENCES


Achkari, O., & Fadar, A. El. (2018). Renewable Energy Storage Technologies-A Review. In *Conférence Internationale en Automatique & Traitement de Signal (ATS*-2018) (Vol. 35, pp. 69–79).

Anagnostopoulos, J. S., & Papantonis, D. E. (2012). Study of pumped storage schemes to support high RES penetration in the electric power system of Greece. *Energy*, *45*(1), 416–423. doi:10.1016/j.energy.2012.02.031.

Ando, T., Omori, H., Kimura, N., & Morizane, T. (2020). A Novel Type of Dynamic Wireless Power Transferred Scooter System with EDLC Storage Unit, 1–5. doi:10.1109/appeec45492.2019.8994349.

Bolund, B., Bernhoff, H., & Leijon, M. (2007). Flywheel energy and power storage systems. *Renewable and Sustainable Energy Reviews*, *11*(2), 235–258. doi:10.1016/j.rser.2005.01.004.

Boyden, A., Soo, V. K., & Doolan, M. (2016). The environmental impacts of recycling portable lithium-ion batteries. *Procedia CIRP*, *48*, 188–193. doi:10.1016/j.procir.2016.03.100.

Brown, P. D., Peças Lopes, J. A., & Matos, M. A. (2008). Optimization of pumped storage capacity in an isolated power system with large renewable penetration. *IEEE Transactions on Power Systems*, *23*(2), 523–531. doi:10.1109/TPWRS.2008.919419.

Budt, M., Wolf, D., Span, R., & Yan, J. (2016). A review on compressed air energy storage: Basic principles, past milestones and recent developments. *Applied Energy*, *170*, 250–268. doi:10.1016/j.apenergy.2016.02.108.

Carton, J. G., & Olabi, A. G. (2010). Wind/hydrogen hybrid systems: Opportunity for Ireland's wind resource to provide consistent sustainable energy supply. *Energy*, *35*(12), 4536–4544. doi:10.1016/j.energy.2010.09.010.

Chen, H., Cong, T. N., Yang, W., Tan, C., Li, Y., & Ding, Y. (2009). Progress in electrical energy storage system: A critical review. *Progress in Natural Science*, *19*(3), 291–312. doi:10.1016/j.pnsc.2008.07.014.

Chen, J., Liu, W., Jiang, D., Zhang, J., Ren, S., Li, L., … Shi, X. (2017). Preliminary investigation on the feasibility of a clean CAES system coupled with wind and solar energy in China. *Energy*, *127*, 462–478. doi:10.1016/j.energy.2017.03.088.

Cimuca, G. O., Saudemont, C., Robyns, B., & Radulescu, M. M. (2006). Control and performance evaluation of a flywheel energy-storage system associated to a variable-speed wind generator. *IEEE Transactions on Industrial Electronics*, *53*(4), 1074–1085. doi:10.1109/TIE.2006.878326.

Connolly, D., MacLaughlin, S., & Leahy, M. (2010). Development of a computer program to locate potential sites for pumped hydroelectric energy storage. *Energy*, *35*(1), 375–381. doi:10.1016/j.energy.2009.10.004.

Desai, N., Gonzalez, S., Pemberton, D., & Rathjen, T. (2005). *The Economic Impact of CAES on Wind in TX, OK, and NM. AWEA project database.* Retrieved from http://scholar.google.com/scholar?hl=en&btnG=Search&q=intitle:The+Economic+Impact+of+CAES+on+Wind+in+TX+, +OK+, +and+NM#0

Díaz-González, F., Sumper, A., Gomis-Bellmunt, O., & Villafáfila-Robles, R. (2012). A review of energy storage technologies for wind power applications. *Renewable and Sustainable Energy Reviews*, *16*(4), 2154–2171. doi:10.1016/j.rser.2012.01.029.

Dursun, B., & Alboyaci, B. (2010). The contribution of wind-hydro pumped storage systems in meeting Turkey's electric energy demand. *Renewable and Sustainable Energy Reviews*, *14*(7), 1979–1988. doi:10.1016/j.rser.2010.03.030.

Eberle, U., Felderhoff, M., & Schüth, F. (2009). Chemical and physical solutions for hydrogen storage. *Angewandte Chemie - International Edition*, *48*(36), 6608–6630. doi:10.1002/anie.200806293.

Energy storage Council. (2015). *Global Energy Storage Market Overview & Regional Summary Report*. Mawson ACT.

Gibson, T. L., & Kelly, N. A. (2008). Optimization of solar powered hydrogen production using photovoltaic electrolysis devices. *International Journal of Hydrogen Energy*, *33*(21), 5931–5940. doi:10.1016/j.ijhydene.2008.05.106.

Hamed, M., Fallah, A., & Ben Brahim, A. (2017). Numerical analysis of charging and discharging performance of an integrated collector storage solar water heater. *International Journal of Hydrogen Energy*, *42*(13), 8777–8789. doi:10.1016/j.ijhydene.2016.11.179.

Hiratsuka, A., Arai, T., & Yoshimura, T. (1993). Seawater pumped-storage power plant in Okinawa island, Japan. *Engineering Geology*, *35*(3–4), 237–246. doi:10.1016/0013-7952(93)90012-2.

Javanbakht, P., Mohagheghi, S., & Simoes, M. G. (2013). Transient performance analysis of a small-scale PV-PHS power plant fed by a SVPWM drive applied for a distribution system. *2013 IEEE Energy Conversion Congress and Exposition, ECCE 2013*, Denver, CO, 4532–4539. doi:10.1109/ECCE.2013.6647307.

Kapsali, M., Anagnostopoulos, J. S., & Kaldellis, J. K. (2012). Wind powered pumped-hydro storage systems for remote islands: A complete sensitivity analysis based on economic perspectives. *Applied Energy*, *99*, 430–444. doi:10.1016/j.apenergy.2012.05.054.

Khan, M. M. A., Saidur, R., & Al-Sulaiman, F. A. (2017). A review for phase change materials (PCMs) in solar absorption refrigeration systems. *Renewable and Sustainable Energy Reviews*, *76*, 105–137. doi:10.1016/j.rser.2017.03.070.

Kivits, R., Charles, M. B., & Ryan, N. (2010). A post-carbon aviation future: Airports and the transition to a cleaner aviation sector. *Futures*, *42*(3), 199–211. doi:10.1016/j.futures.2009.11.005.

König, D. H., Freiberg, M., Dietrich, R. U., & Wörner, A. (2015). Techno-economic study of the storage of fluctuating renewable energy in liquid hydrocarbons. *Fuel*, *159*, 289–297. doi:10.1016/j.fuel.2015.06.085.

Kordesch, K. (1981). *Comprehensive Treatise of Electrochemistry*. Plenum Press, NewYork. doi:10.1016/b978-044451645-9/50009-1.

Lazarewicz, M. L., & Rojas, A. (2004). Grid frequency regulation by recycling electrical energy in flywheels. In *IEEE Power Engineering Society General Meeting* (pp. 2038–2042). IEEE, Denver, CO. doi:10.1109/pes.2004.1373235.

Lisbona, D., & Snee, T. (2011). A review of hazards associated with primary lithium and lithium-ion batteries. *Process Safety and Environmental Protection*, *89*(6), 434–442. doi:10.1016/j.psep.2011.06.022.

Ma, T., Yang, H., Lu, L., & Peng, J. (2015). Pumped storage-based standalone photovoltaic power generation system: Modeling and techno-economic optimization. *Applied Energy*, *137*, 649–659. doi:10.1016/j.apenergy.2014.06.005.

Margeta, J., & Glasnovic, Z. (2012). Theoretical settings of photovoltaic-hydro energy system for sustainable energy production. *Solar Energy*, *86*(3), 972–982. doi:10.1016/j.solener.2012.01.007.

Nomura, S., Ohata, Y., Hagita, T., Tsutsui, H., Tsuji-Iio, S., & Shimada, R. (2005). Wind farms linked by SMES systems. *IEEE Transactions on Applied Superconductivity, 15*(2 PART II), 1951–1954. doi:10.1109/TASC.2005.849343.

Okazaki, T., Shirai, Y., & Nakamura, T. (2015). Concept study of wind power utilizing direct thermal energy conversion and thermal energy storage. *Renewable Energy, 83*, 332–338. doi:10.1016/j.renene.2015.04.027.

Qais, M. H., Hasanien, H. M., & Alghuwainem, S. (2020). Output power smoothing of wind power plants using self-tuned controlled SMES units. *Electric Power Systems Research, 178*, 106056. doi:10.1016/j.epsr.2019.106056.

Rodríguez, C. R., Riso, M., Jiménez Yob, G., Ottogalli, R., Santa Cruz, R., Aisa, S., … Leiva, E. P. M. (2010). Analysis of the potential for hydrogen production in the province of Córdoba, Argentina, from wind resources. *International Journal of Hydrogen Energy, 35*(11), 5952–5956. doi:10.1016/j.ijhydene.2009.12.101.

Sato, Y., Ito, K., Arakawa, T., & Kobayakawa, K. (1996). Possible cause of the memory effect observed in nickel-cadmium secondary batteries. *Journal of the Electrochemical Society, 143*(10), 3–6. doi:10.1149/1.1837152.

Sayilgan, E., Kukrer, T., Civelekoglu, G., Ferella, F., Akcil, A., Veglio, F., & Kitis, M. (2009). A review of technologies for the recovery of metals from spent alkaline and zinc-carbon batteries. *Hydrometallurgy*, 97(3–4), 158–166. doi:10.1016/j.hydromet.2009.02.008.

Scalia, A., Varzi, A., Lamberti, A., Jacob, T., & Passerini, S. (2018). Portable high voltage integrated harvesting-storage device employing dye-sensitized solar module and all-solid-state electrochemical double layer capacitor. *Frontiers in Chemistry, 6*, 1–8. doi:10.3389/fchem.2018.00443.

Shin, S. M., Kang, J. G., Yang, D. H., Kim, T. H., & Sohn, J. S. (2007). Comparison of acid and alkaline leaching for recovery of valuable metals from spent zinc-carbon battery. *Geosystem Engineering, 10*(2), 21–26. doi:10.1080/12269328.2007.10541267.

Shiva Kumar, S., & Himabindu, V. (2019). Hydrogen production by PEM water electrolysis – A review. *Materials Science for Energy Technologies*, 2(3), 442–454. doi:10.1016/j.mset.2019.03.002.

Trost, T., Horn, S., Jentsch, M., & Sterner, M. (2012). Erneuerbares Methan: Analyse der CO2-Potenziale für Power-to-Gas Anlagen in Deutschland (Renewable Methane: Analysis of CO2 Potentials for Power-to-Gas in Germany). *Zeitschrift Für Energiewirtschaft, 36*(3), 173–190. doi:10.1007/s12398-012-0080-6.

van Meerwijk, A. J. H., Benders, R. M. J., Davila-Martinez, A., & Laugs, G. A. H. (2016). Swiss pumped hydro storage potential for Germany's electricity system under high penetration of intermittent renewable energy. *Journal of Modern Power Systems and Clean Energy, 4*(4), 542–553. doi:10.1007/s40565-016-0239-y.

Vogt, C., Monai, M., Kramer, G. J., & Weckhuysen, B. M. (2019). The renaissance of the Sabatier reaction and its applications on Earth and in space. *Nature Catalysis*, 2(3), 188–197. doi:10.1038/s41929-019-0244-4.

Wang, L., Chen, S. S., Lee, W. J., & Chen, Z. (2009). Dynamic stability enhancement and power flow control of a hybrid wind and marine-current farm using SMES. *IEEE Transactions on Energy Conversion, 24*(3), 626–639. doi:10.1109/TEC.2008.2008877.

Were, F. H., Kamau, G. N., Shiundu, P. M., Wafula, G. A., & Moturi, C. M. (2012). Air and blood lead levels in lead acid battery recycling and manufacturing plants in Kenya. *Journal of Occupational and Environmental Hygiene*, 9(5), 340–344. doi:10.1080/15459624.2012.673458.

Winkler-Goldstein, R., & Rastetter, A. (2013). Power to gas: The final breakthrough for the hydrogen economy? *Green, 3*(1), 69–78. doi:10.1515/green-2013-0001.

Zhang, Z., Miyajima, R., Sato, Y., Miyagi, D., Tsuda, M., Makida, Y., … Hirano, N. (2016). Characteristics of compensation for fluctuating output power of a solar power generator in a hybrid energy storage system using a Bi2223 SMES coil cooled by thermosiphon with liquid hydrogen. *IEEE Transactions on Applied* Superconductivity, *26*(4), 1–5. doi:10.1109/TASC.2016.2529565.

Zhou, Z., Benbouzid, M., Frédéric Charpentier, J., Scuiller, F., & Tang, T. (2013). A review of energy storage technologies for marine current energy systems. *Renewable and Sustainable Energy Reviews*, *18*, 390–400. doi:10.1016/j.rser.2012.10.006.

# 9 New Energy Mining
## *Compressed Air Energy Storage in Abandoned Mines*

*Bernardo Llamas, Belén Vallespir, Marcelo F. Ortega, and Pedro Mora*
Universidad Politécnica de Madrid

**CONTENTS**



## 9.1 INTRODUCTION

The relationship between energy and subsurface has always existed, after the invention of the steam engine, which led to the industrial revolution (IEA, 2019). Since then, fossil fuels have remained the main source of primary energy for the development of society, and according to the International Energy Agency forecast, they will continue to maintain this position in the coming decades (IEA, 2019).

However, the concern about climate change and sustainable development has intensified the promotion of new energy sources whose environmental footprint is reduced, especially in terms of $CO_2$ emissions. The Kyoto Protocol was a milestone in raising awareness about the need to reduce greenhouse gas emissions. However, despite the efforts made in the development of new energy sources, these have not achieved a definitive implementation, halted by one of the three pillars of sustainability: environment, competitiveness, and security of supply. Therefore, greenhouse gas emissions have not yet been reduced worldwide (IEA, 2019). The Paris Protocol is the last international treaty that aims to limit the concentration of $CO_2$ by no more than 450 ppm during the present century. However, the concentration of $CO_2$ continues to rise, and this year will close with a concentration higher than 410 ppm, with

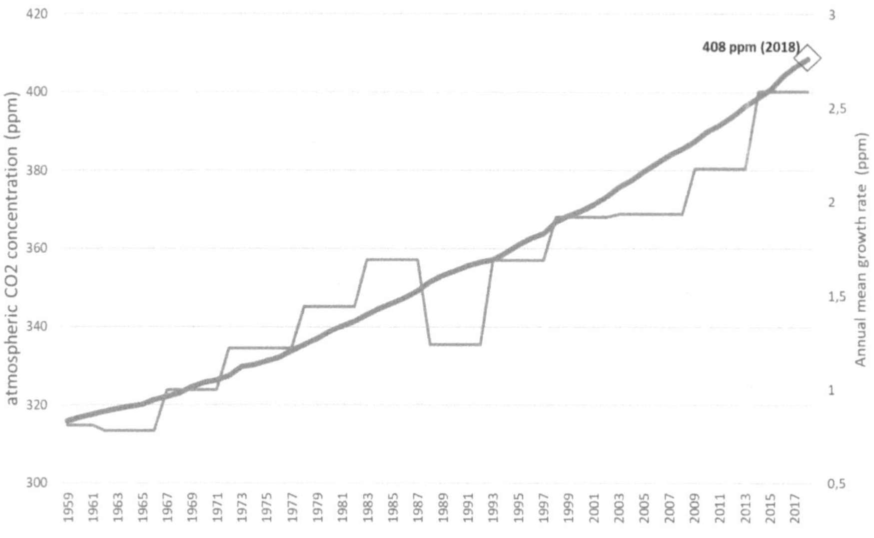


**FIGURE 9.1** Trends in atmospheric carbon dioxide concentration. Mauna Loa Observatory. (UPM).

annual growth rates of 2–3 ppm/year: According to Figure 9.1, $CO_2$ concentration in the atmosphere will reach 450 ppm in less than 20 years, as annual growth rate is higher than 2 ppm $CO_2$/year (NOAA Research webpage, 2020).

Fossil fuels remain as main primary energy consumed, as well as coal and natural gas are the most fuels used to produce electricity (IEA, 2019). For this reason, electricity and heat producers are responsible for more than 13,500 $MtCO_2$ (2017) followed by transport sector (slightly more than 8,000 $MtCO_2$, 2017) (IEA webpage, 2020). Renewable energies are considered one of the options in the urgent need to reduce $CO_2$ emissions. To increase the deployment of these intermittent energy sources (wind and photovoltaic energy), it is necessary to develop energy storage technologies. In this way, it will be possible to increase the load factor and be able to guarantee supply without the need for costly backup systems and that is traditionally based on fossil sources. This research chapter will evaluate the potential for using abandoned coal mines as an ideal infrastructure to store energy in the subsurface. Thus, it could be possible to define a new energy mining resource in the 21st century: abandoned coal mines as compressed air energy storage (túnel-CAES).

### 9.1.1 Electricity Production: From Coal to Renewable Energy

According to Figure 9.2, it is possible to observe the changes in the energy model at the European level. These slight variations in the trend lines that represent the consumption of resources for the electricity production are a reflection of the policies adopted by the member countries of the European Union (EU) and the objective to (1) decrease $CO_2$ emissions (Directive 2003/87/CE, 2003), (Directive 2009/29/CE, 2009), and (2) promote renewable energy (Directive 2009/28/CE, 2009).

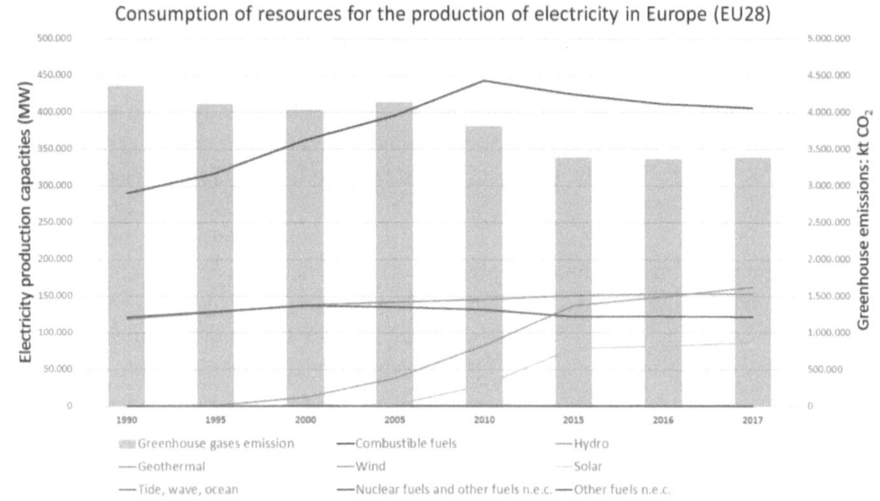


**FIGURE 9.2** Energy resources to produce electricity in Europe (Eurostat).

Since 2006, fossil resources began to decrease their presence in the energy mix, giving way to technologies such as thermal and photovoltaic solar and other renewable sources. Even so, the energy generated by these technologies reached 43.65% in 2017 (Eurostat, 2019). On the contrary, renewable resources have increased their presence in the market in the last 30 years. This increase is very evident in the case of the wind energy which produced 17.47% of the consumed electricity (Eurostat, 2019). The reduction of dependence on fossil fuels and the increase in the presence of renewable energy in the European generation mix have also been reflected in the reduction of greenhouse gas (GHG) emissions. It has been remarkable since 2010, according to the transformation of the mix of generations in several EU members.

A very important parameter for the implementation of renewable energies is the load factor. It indicates the real energy that the plant may generate with respect to the total theoretical energy that this plant could generate if it had generated at full load during the annual period. This parameter may be reduced by many factors such as maintenance operations, the entry of the plant into the market or the intermittency of the primary energy (i.e. solar radiation, wind). In the case of renewable generation plants, the load factor has values between 20% and 40% in wind generation and 10% and 30% in the case of solar generation. Thermal plant or nuclear get higher load factor, 70%–90% and 60%–98%, respectively. Electric storage is a demanded solution to guarantee the main disadvantage of renewables, facilitating the availability of the energy generated by these systems.

### 9.1.2 Electrical Energy Storage

Electrical energy storage (EES) technologies range from small scale, such as a mobile device, to large storage systems such as a hydraulic pumping plant. In the scenario under study, the objective is to: (1) reduce the dependence on fossil fuels which are responsible for the greenhouse emissions and (2) increase renewable and intermittent energies with a low load factor. In order to remain in good conditions of supply and energy quality, it is essential to include EES technologies which may be defined as a process of converting energy from one form (mainly electrical energy) to a storable form and reserving it in various mediums; then the stored energy can be converted back into electrical energy when needed (Luo, Wang et al., 2015).

In addition, EES adds value to the generation systems, facilitates the integration of intermittent generation sources, contributes to distributed generation, and thereby reduces the need for excessive oversizing of the voltage lines in addition to improving the quality of energy and security of supply. As for the demand curve, storage increases the regulation range and avoids deviation in production from what is agreed in the market, which allows us to decouple the generation of electricity consumption. The EES market will multiply by 13 globally by 2024 (Mackenzie, 2019). With this data, many of the technologies listed in Figure 9.1 will be considered, depending on the different characteristics (principle) and application areas (function). Thus, EES may be classified according to two key features of energy: The first is the **principle** to store energy (Figure 9.3).

The second is the **function** that the storage system can play according to the parameters that define it. These are the response time, the power capacity they can accumulate, and the amount of time during which they can operate. Based on the above, they can be classified as follows:

- Energy management: Superior response time, up to hours. Look for the decoupling of demand and energy production.
- Network support: Short response time. They are technologies capable of correcting the fall of a generation source the time it takes to contemplate another energy source.
- Focused on quality improvement: Very low response times, improve by smoothing the wave and avoiding variations in frequency.

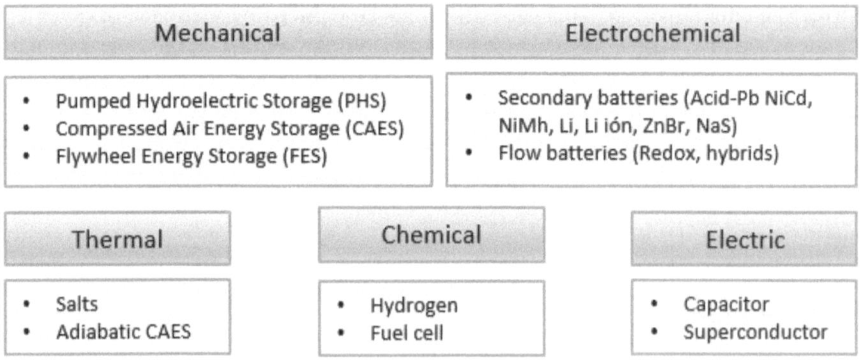


**FIGURE 9.3** Different options to store energy. Classified according to principles to store energy (UPM).

EES technologies to manage energy may be considered pumping hydroelectric storage (PHS), CAES, and hydrogen. Lately, batteries are being introduced in this classification, thanks to the large facilities that are being developed in the world (Yang et al., 2009; Beaudin et al., 2010; May et al., 2018). However, Figure 9.4 represents commercial EES facilities: even if batteries are implemented in several facilities, their capacity is not enough to be considered in the energy management field.

**PHS** is the most mature and commercially implemented technology (Caralis et al., 2019). With the ability to store large amounts of energy into the grid, its cumulative production represents around 3% of the world's electricity production (Caralis et al., 2019). The form of energy storage is potential energy, due to the difference in heights between two points, where water storage tanks are located. When it is necessary to store electrical energy, water is pumped from the lower tank to the upper tank. This water is stored until it is necessary to use the plant as a generator, which, thanks to the difference in height between deposits, the water directed through the forced pipe will drive the turbines. When the upper tank has a natural supply of water, it is called mixed pumping, in case of not having it is a pure pumping.

The main advantages of PHS are that it has a useful life of between 40 and 60 years (Yang et al., 2009), a large installed capacity, and large energy efficiency, between 65% and 80% (Klumpp, 2016). The main disadvantages are the high cost of installation and the geomorphology and topography.

With less commercial development, it is possible to define **CAES**. These stored energy uses electrical energy to compress air and store it underground, in large mining cavities. Currently, the installed capacity worldwide is 440 MW (Kapila et al., 2017). Disadvantage of CAES is energy efficiency which is around 50%. Recent improvements defined Adiabatic CAES (ACAES). In this case, thermal energy generated in the compression process is also stored. Compressed air and heat will be

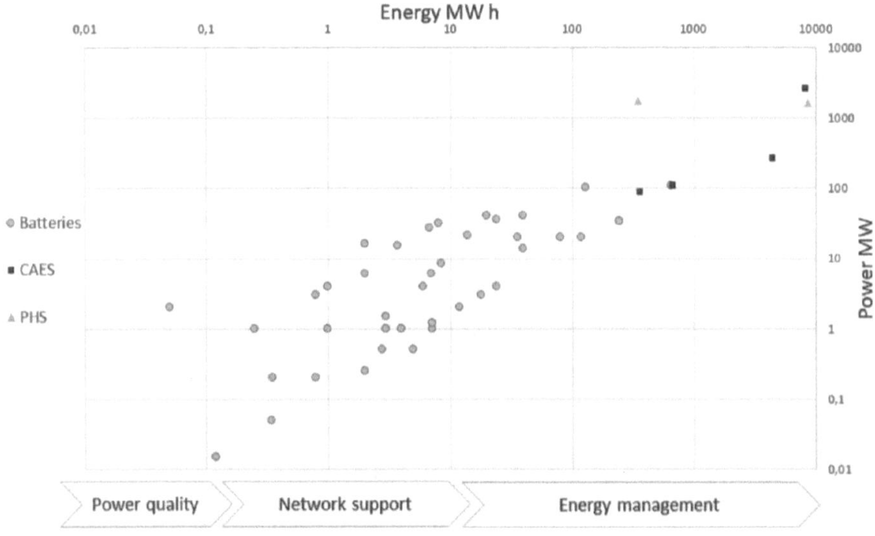


**FIGURE 9.4** Energy storage technologies according to the area of application (UPM).

used later in the expansion process (turbine system). Defining a smart heat management system, non-fossil resources will be necessary to heat the compressed air in the electricity generation stage. According to the latest studies, the overall energy efficiency of the system could reach up to 70% (Díaz-González et al., 2012).

Table 9.1 compares the only two energy storage alternatives with enough capacity to manage energy. Previous research publications indicate that the investment requirement for the CAES technology is most cost-effective than in the case of PHS.

**Hydrogen** may be considered by itself as a source of energy, but in this case, it will be defined as a chemical storage system. The excess energy to be stored is used in a process of electrolysis of water through which said molecule is separated into oxygen, $O_2$ and hydrogen, $H_2$. The latest is a high energetic molecule and it could be used to produce electricity (fuel cell) or to generate natural gas (power to gas).

Hydrogen storage can be done both in solid form, absorbed by another material, in a liquid, cryogenized, or in a gas state. Electrical energy is recovered from hydrogen by combustion of hydrogen or in a fuel cell. Both processes produce a clean final

**TABLE 9.1**
**Comparison of Commercial Electrical Energy Storage Technologies (Luo et al., 2015)**

| | | Pumped Hydroelectric Storage (PHS) | Compressed Air Energy Storage (CAES) |
|---|---|---|---|
| Investment cost (CAPEX) according to energy capacity | $/kWh | 5–100<br>10–12<br>50–100<br>80–200 | 2–50<br>3<br>15–100<br>50–110 |
| Maintenance and operating cost (OPEX) | $/kWh | 3 | 6 |
| Nominal capacity | MW | 100–5,000<br>1–4,000<br>1,000–3,000 | 8–300<br>220<br>100–3,000 |
| Discharge duration | H | 1–24<br>>12 | 1–24<br>>24 |
| Response time | - | Minutes | Seconds/minutes |
| Efficiency | % | 71–85<br>80<br>65–80 | 70–89<br>75<br>60–79 |
| Space requirement | $m^2$/kWh | 0.02 | 0.01 |
| Technological maturity | | Industrial use 250 facilities:<br>99% storage capacity | Industrial use 2 facilities |
| Infrastructure execution time | | High | Medium |

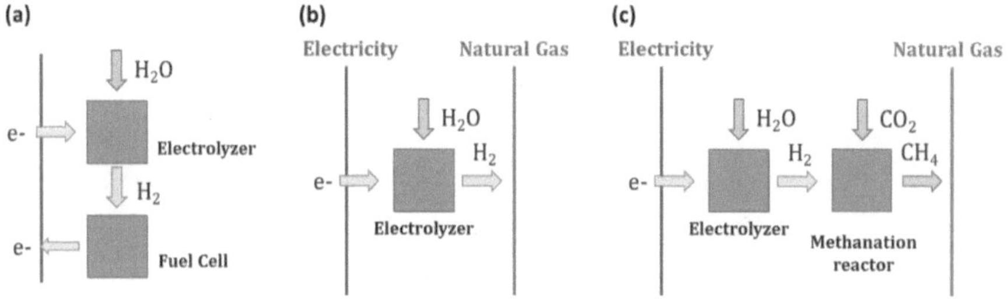


**FIGURE 9.5** Options to store energy with hydrogen: (a) electrolyzer and fuel cell technologies; (b) grid injection into the natural gas network; and (c) power to gas technology (UPM, SLwH2 project).

fuel without GHG emissions (if the electricity consumed in the electrolysis process is renewable). The main barrier to its development is the high cost, both in installation and maintenance and operation (Andersson & Grönkvist, 2019). Some other options are considered (Figure 9.5) to store hydrogen: hydrogen injection (and storage) in the natural gas network and transformation of the hydrogen (methanation process) in combination with $CO_2$ to produce methane (power to gas technology). One of the most relevant advantage for the latest technology is that it could be a solution to use anthropogenic $CO_2$.

## 9.2 NEW ENERGY MINING (21ST CENTURY)

As described in the previous section, the transformation in the use of electricity production technologies is a direct decline in the demand for coal in Europe. This leads to lower production and a singular crisis in that sector that is without the first user for this resource: the electricity sector.

Figure 9.6 shows the reduction of coal production in Europe: it has decreased up to 50% compared to the year 1990. According to the data published by EUROSTAT (Eurostat webpage, 2019), the production of coal through underground mining is only 20% in 2017 compared to the production achieved in 1990. Working conditions and health and safety in operation have been the main causes for the abandonment of this type of exploitation in Europe. In this way, the traditionally considered link between energy and subsoil is weakening.

But the infrastructure developed in the last century could be a nice approach to define a new energetic mining in the 21st century: abandoned galleries could be isolated, and it may be transformed into a container to store compressed air. This storage energy should be defined, but it could achieve some characteristic to be considered as an EES technology to manage significant amount of energy. In addition, the investment cost would be lower than traditional PHS or CAES (Table 9.1). In addition, regions that have traditionally exploited coal could find a new economy based on abandoned infrastructure and thus define a sustainable link between energy and subsurface.

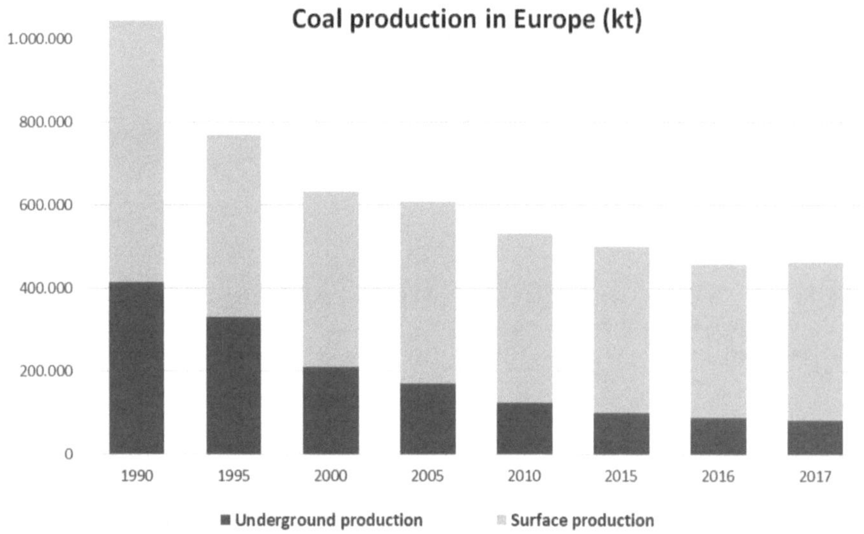


**FIGURE 9.6** Evolution of the coal production in EU28. Data: EUROSTAT, 2019 (UPM).

### 9.2.1 Underground Energy Storage

CAES technologies store electricity in the form of potential energy by introducing high-pressurized air into an underground geological structure. Train of compressors would consume non-demanded electricity to pressurize air up to 70 bar, which is the maximum pressure considered in large CAES at 700m deep. During this process, the air increases its temperature, so it is necessary to insert each compression stage with a heat exchanger. In other words, around 50% of the consumed electricity is transformed into heat. In diabatic CAES, the heat is released into the atmosphere, whereas current research focus on how to store or manage this heat: isothermal or ACAES are under study.

As underground structure, it is possible to differentiate salt caverns, saline aquifers, or underground infrastructures (Figure 9.7). When the plant needs to generate energy, the pressurized air is released by passing it through a burner that gives it that heat extracted in previous phases. Once heated, a set of high- and low-pressure turbines will be used to produce electricity.

So far, there are two plants developed commercially. **Huntorf** (Bremen, Germany) is the first CAES plant commissioned in 1978 and was in operation for more than 40 years. The capacity is 290 MW for 3 hours supporting blackouts to electricity grid (energy efficiency of only around 40%). Last years, this facility gave balance to the grid in contrast with the wind energy. The installation consists of two cavities, 170,000 $m^3$ and 140,000 $m^3$, to store pressurized air (between 48 and 66 bar as operating pressure).

The last commercial CAES facility was developed in Alabama (USA). **McIntosh** began commercial operations in 1991 and is operational for more than 25 years. The storage cavity was mined with a total capacity of 560,000 $m^3$. This facility provides

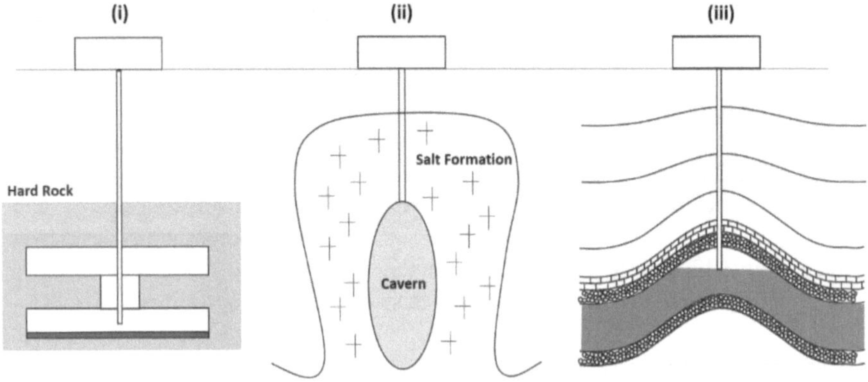


**FIGURE 9.7** (i) Mine cavities, (ii) salt formations, (iii) deep aquifers (Llamas et al., 2017). CAES reaches production capacities from 81 MW to 404 MW (McDowall, 2006) and discharge times from 8 to 26 hours. The useful life is approximately 40 years and has an efficiency of 71%, in case of ACAES (Díaz-González et al., 2012). The main disadvantage is the risk of exploration and infrastructure needed to monitor the underground structure.

a power of 110 MW for 26 hours (energy efficiency of above 50%). The operational pressure in the underground structure is between 45 bar and 74 bar. Based on this design, a new project was proposed in Texas, with a total capacity of 540 MW. The goal of the project is to evaluate the conditions to integrate CAES with wind energy. The underground infrastructure will be an existing brine cavity in salt dome formation.

Finally, one of the largest CAES project was the **Norton** project (Ohio, USA). As a novelty, it was proposed to use an *abandoned limestone mine* as a system to store pressurized air (operative pressure was defined within 55–110 bar). Limestone mine is at a depth of 670 m and with a volume of 9,600 $m^3$. The project was proposed in different stages, with a total capacity of 2,700 MW.

In order to state **ACAES** at a commercial level, it is necessary to define an efficient thermal energy storage (TES): the heat extracted from the pressurized air will be stored (gas–solid exchanger). These subsystems would avoid the use of fossil fuels to heat depressurized air in the electricity generation phase. An energy efficiency improvement of up to 20% has been established. This is a technology still under development, with several projects and different Technology Readiness Levels (TRLs).

Perhaps, the **ADELE project** is the first pilot ACAES (Magdeburg, Germany, 2010). Its storage capacity is 360 MWh for 4 hours. Underground storage is defined in a dome formation (brine cavity). The same infrastructure was used to store natural gas at pressures of 70 bar. The TES is constituted by ceramic material or natural stone reaching temperatures of up to 650°C. This facility is expected to start its operation in 2019 with a capacity of 90 MW.

In parallel, some other researchers focused on the re-definition of small brine cavities to reduce the inherent risk of exploration or decrease the environmental impact during the mining operations (brine cavity digging). Moreover, these smaller cavities are suitable to be connected to renewable resources such as photovoltaic plants or wind farms (Llamas, Laín, Castañeda, & Pous, 2018). Additionally, some other projects evaluate abandoned or obsolete underground facilities to store compressed air: **RICAS 2020** (Ricas2020 webpage, 2019) will develop a large laboratory to test different conditions to store compressed air in a gallery at a depth of 210 m; operational pressure will be around 50 bar, and it will simulate a capacity of 5 MW. The latest project is **ALACAES**. It is defined as a pilot plant to store energy with a capacity of 1 MWh. But the novelty relies on the underground facility: it uses an obsolete tunnel 120 m long and 5 m in diameter.

Table 9.2 summarizes different projects to store energy underground.

### 9.2.2 Túnel-CAES, a Mining Solution to Store Sustainable Energy

In this case, túnel-CAES concept is proposed as an underground infrastructure to store compressed air (Figure 9.8). In this way, exploratory risk and exploitation cost may be reduced considerably. Accordingly, it is possible to operate at higher pressure without greater depths because a self-sustaining and reinforced structure is already available. This remarkable condition increases the energy density, defined as kWh/m$^3$, so it could be possible to increase the capacity to store energy. Insulation system will be based on shotcrete and bolts, and it is one of the technological challenges: the design of a new family of concrete will be necessary. To provide insulation conditions in the tunnel, achieving isolation characteristics against high pressures of gaseous fluid is needed. The design objectives are (1) the identification of underground mines that are already abandoned and (2) consideration of the access galleries that were designed to avoid water flooding (Figure 9.6).

## 9.3 TÚNEL-CAES. CASE OF STUDY

León is a province located in the northwest of Spain, with a long mining tradition. Coal has been a widely exploited substance, although the exploitation of slate as an ornamental rock stands out. According to the regional database, there were up to 146 coal mines, which extracted anthracite and bituminous coal. Most of the produced coal were burned in local power plants, but environmental regulatory due the energy decarbonization had an impact on coal demand in the region. Thus, coal production decreased sharply in the last decade, and it has a direct correlation to the population (Figure 9.9). In accordance with Llamas et al. (2017), another favorable criteria to study the proposed region is the wind farms and solar facilities installed in the region: Castilla y León contributed with 23.1% of total wind generation in Spain (2018) and 10.3% in solar photovoltaic.

For this reason, túnel-CAES concept does not only propose a technical solution to improve the capacity of renewable energy. It is projected for social reason: several regions along Europe have decreased their local economy as a result of the low coal demand. And it could increase the risk of depopulation in several regions in Europe. Túnel-CAES is an economical solution as it is not necessary to construct a tunnel: galleries could be transformed into isolated containers to retain pressurized air. And, local experience would avoid most of the social issues associated to underground energy storage.

**TABLE 9.2**
**Main CAES Facilities in the World**

| | | Underground Energy Storage | Depp (m) and Volume ($m^3$) of the Cavity | Cavity Pressure | Capacity (MW) | Primary Energy | Method | Status |
|---|---|---|---|---|---|---|---|---|
| RICAS 2020, Austria | ACAES | | | | 5 MW | | | |
| ALACAES, Switzerland | ACAES | Tunnel | | | 1 MW | | | |
| Huntorf, Germany | CAES | Brine cavity | 600 m $310 \cdot 10^3 m^3$ | 46 bar–66 bar 40–70 | 290 MW in 3 h | Nuclear | GN | Commercial, 1978 |
| McIntosh, Alabama, EEUU | CAES | Brine cavity | 450 m $560 \times 10^3 m^3$ | 45 bar–76 bar | 110 MW in 26 h | Nuclear | GN + Q recuperator | Commercial, 1991 |
| Norton, Ohio EEUU | CAES | Limestone, abandoned mine | 670 m $120 \times 10^3 m^3$ $9.6 \times 10^6 m^3$ | 103 bar. 55–110 bar | 2,700 MW in 16 h. | | | Ongoing |
| Iowa Stored Energy Park, EEUU | CAES. | Deep saline aquifer | 915 m | | 270 MW total. 2,700 MW | Wind | | Canceled |
| Markham, Texas | | | | | 540 | | | |
| Stassfurt, Germany | ACAES | Brine cavity | | 100 bar. 600°C | 360 MW in 4 hours | Solar and | TES (air through | Ongoing, 2010 |
| ADELE | | | | 70 bar | 90 MW y 360 MWh | wind | brick) 600°C | R&D project |
| Larne, North Ireland. | | Brine cavity | | | 250 MW for 6 h | | | Project, 2020 |

*Source:* Dr. Robert Galler (s.f.), Kim et al. (2012), Song et al. (2012), Evans et al. (2009), RWE (n.d.), Klumpp (2016), Project 1010-CAES Larne, Northern Ireland (n.d.), Klumpp (2016), Díaz-González et al. (2012).

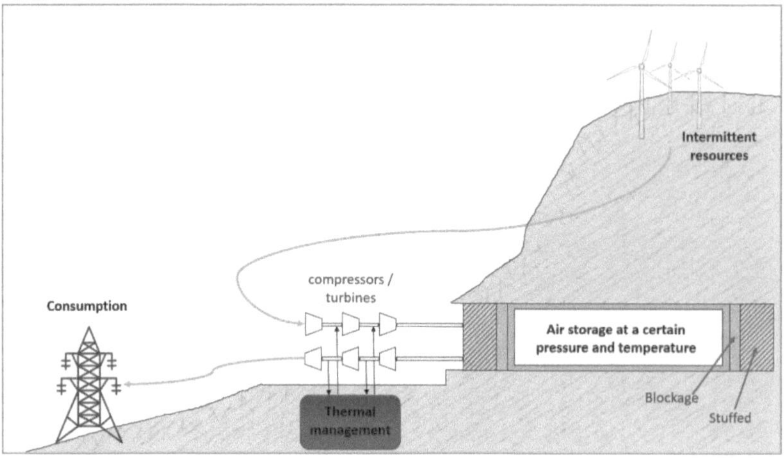


**FIGURE 9.8** Conceptual design of the túnel-CAES, considering abandoned coal mines (UPM).

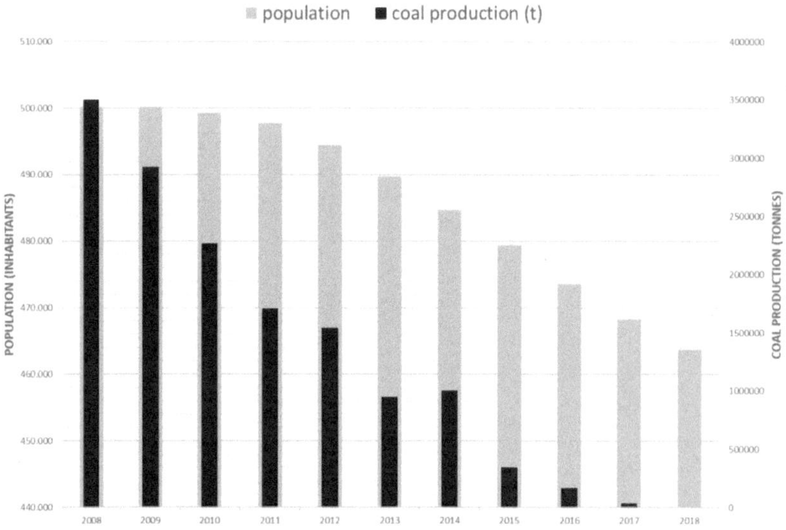


**FIGURE 9.9** Relationship between population and coal production in the region of León (North-West of Spain) (UPM).

The proposed system based on abandoned galleries must consider a series of requirements in order to theoretically assess the energy storage capacity. In this case, of the 146 facilities, 94 underground coal mines have been studied. Thus, coal production will be the key parameter to roughly calculate an available length for each mine; and only coal mines with a production larger than 30,000 t were considered in our study. On the same hand, latest labor plans of several mines suggested that the length could be in a range of 5–10 km (Equation 9.1). Finally, a correction factor (CF) has been applied which is related to the date of operation for each mine; its value will be in a range of 10%–21% in order to provide a sensitive analysis.

$$\text{Volume}\left(m^3\right) = \text{Gallery Section}\left(m^2\right) \cdot L\left[\text{function of coal production}\right](m) \cdot \text{Correction Factor}(\cdot) \tag{9.1}$$

Equation 9.1. Theoretical volume available in abandoned underground mines.

The section of the gallery has been set to 8.54 $m^2$ as the traditional exploitation used to cave the galleries with this section. Length will be estimated according a direct link between coal productions: it will be set to a range between 5 and 10 km. Considering the conditions explained above, 47 abandoned mines were included in this study (Figure 9.10).

Applying Equation 9.1, it is possible to determine the available length for the region of study. As the CF is theoretical, different values have been assigned: 10%, 15%, and 21%, so different scenarios were considered, adding a new economic parameter: minimum length for each site should be at least 1 km long (Table 9.3).

The stored energy capacity in pressurized air should be assessed considering not only the volume but also the pressure gradient. Accordingly, to this last parameter, it would be possible to determine the energy density for different conditions, which will define the requirements to isolate the abandoned galleries (mainly based

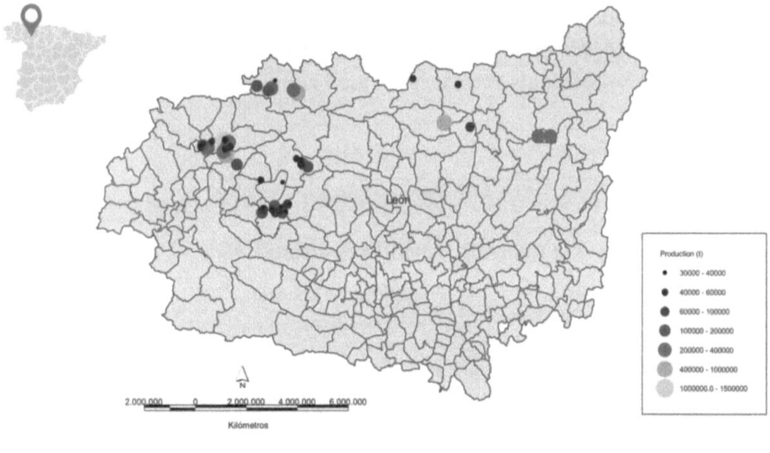


**FIGURE 9.10** Abandoned mines included in the study. Key indicates different range of coal production (UPM).

**TABLE 9.3**
**Theoretical Galleries Available to Store Energy (Túnel-CAES Concept)**

| | Favorable (21% CF) | Expected (15% CF) | Unfavorable (10%CF) |
|---|---|---|---|
| Available length (km) | 54.8 | 5.1 | 1.0 |
| Sites | 47 | 4 | 1 |

Abandoned mines in León region.

on projected concrete). In this case, isochoric containments have been considered (Equation 9.2) (Llamas et al., 2018).

$$B_{\text{isochoric-cavern}} = \left(P_{1,\text{max}} - P_{1,\text{min}}\right)V_1 \frac{T_0}{T_1} \log\left(r_{\text{min}}\right) \tag{9.2}$$

Equation 9.2 Stored exergy, considering pressurized air (Letcher, 2016)

Equation 9.2 determines the recoverable exergy from pressurized air. $P_{1,\text{max}}$ is the maximum allowable value in the cavity, and $P_{1,\text{min}}$ is the pressure when the cavity is empty. Considering brine cavity, $P_{1,\text{min}}$ should be assumed as $1/2{\cdot}P_{1,\text{max}}$, but in túnel-CAES conditions, $P_{1,\text{min}}$ may be defined as atmospheric pressure; finally, $r_{\text{min}}$ represents the pressure gradient (Llamas et al., 2018; Letcher, 2016). Table 9.4 represents the exergy in pressurized air. Isochoric storage may be simplified if a simple throttle to reduce pressure is used. Thanks to this equipment, whatever pressure is present in the cavern at one time is reduced to a constant value suitable for the expansion equipment. The right column of Table 9.4 represents this easy operational process. The left column represents the average deep for conventional brine cavity, and the pressure gradient is showed in the second column on the left.

Finally, it is possible to determine the exergy for different ratio of pressure and volume of cavities (three scenarios were described in Table 9.3). Assuming the two lowest exergy showed in Table 9.4, few energy would be stored in the galleries under consideration (<25 MWh/site). For this reason, those conditions will be ruled out in our study. Table 9.5 represents the exergy which could be stored considering different ratio of pressure.

**TABLE 9.4**
**Quantities of Exergy Recoverable form Pressurized Air**

| CAES (m deep) | Pressure Ratio, r | Exergy (Isochoric) MJ/m³ | Exergy (Isochoric Throttled) MJ/m³ |
|---|---|---|---|
| 110 | 10 | 1.16 | 0.83 |
| 220 | 20 | 3.17 | 2.53 |
| 555 | 50 | 10.71 | 9.11 |
| 780 | 70 | 16.42 | 14.18 |

**TABLE 9.5**
**Total Energy Storage Capacity in Abandoned Mines (León Region as a Case Study)**

| Pressure Ratio, r | Exergy (Isochoric Throttled) MJ/m³ | Favorable (21% CF) MWh | Expected (15% CF) MWh | Unfavorable (10% CF) MWh |
|---|---|---|---|---|
| 50 | 9.11 | 362 (12 sites) | 109 (4 sites) | - |
| 70 | 14.18 | 1,828 (47 sites) | 170 (4 sites) | 33 (1 site) |

## 9.4 CONCLUSIONS

On the way to reducing $CO_2$ emissions in the electricity sector, the use of renewable energy is an unstoppable fact. However, the low load factor offered by intermittent energy resources (solar, wind) makes energy storage necessary in order to decouple energy production and demand. Few current technologies would be available to store significant amount of energy: PHS and CAES could play an important role in the energy management role.

ACAES could increase energy efficiency of the overall system, but it will be necessary to define new approaches to adequate the energy needed to be stored and volume of the cavity. Abandoned mines have been demonstrated as a nice approach to decrease the investment cost of this technology, and its capacity is large enough to be considered as a "management solution."

Coal basins, which are currently in an extremely difficult economy could find a new link between energy and subsurface. Thanks to abandoned mines, it could be possible to state a new economy model with a large capacity to store renewable energy. Moreover, túnel-CAES development would increase not only environmental parameters on the energy sector: it could decrease mine water and gases emissions. Thus, social and environmental points of view are relevant for the development of this technology along Europe and other coal basin worldwide.

According to the gallery isolation and support, minimum pressure of the gallery may be equal to atmospheric pressure, thus low compressor requirements will be needed. So, surface engineering of the túnel-CAES will be less expensive than conventional CAES (brine cavity). Next studies will focus on the confinement of the air underground: mechanical conditions and pressure and temperature gradient will be studied.

Future research will focus on available capacity in abandoned coal galleries as energy storage systems. Flood issues, mechanical properties, and abandoned conditions would be some technical parameters which will be considered in a more accuracy estimations.

## REFERENCES

Andersson, J., & Grönkvist, S. (2019). Large-scale storage of hydrogen. *International Journal of Hydrogen Energy*, *44*, 11901–11919. doi:10.1016/j.ijhydene.2019.03.063.

Beaudin, M., Zareipour, H., Schellenberglabe, A., & Rosehart, W. (2010). Energy storage for mitigating the variability of renewable electricity sources: An updated review. *Energy for Sustainable Development*, *14*(4), 302–314. doi:10.1016/j.esd.2010.09.007.

Caralis, G., Christakopoulos, T., Karellas, S., & Gao, Z. (2019). Analysis of energy storage systems to exploit wind energy curtailment in Crete. *Renewable and Sustainable Energy Reviews*, *103*, 122–139. doi:10.1016/j.rser.2018.12.017.

Díaz-González, F., Sumper, A., Gomis-Bellmunt, O., & Villafáfila-Robles, R. (2012). A review of energy storage technologies for wind power applications. *Renewable and Sustainable Energy Reviews*, *16*(4), 2154–2171. doi:10.1016/j.rser.2012.01.029.

Directiva 2003/87/CE. (2003).

Directiva 2009/28/CE. (2009).

Directiva 2009/29/CE. (2009).

Evans, D., Stephenson, M., & Shaw, R. (2009). The present and future use of "land" below ground. *Land Use Policy*, *26*(SUPPL. 1), 302–316. doi:10.1016/j.landusepol.2009.09.015.

Global storage market to grow from 12 GWh to 158 GWh by 2024 | Wood Mackenzie. (n.d.). Retrieved May 10, 2019, from https://www.woodmac.com/press-releases/global-storage-market-to-grow-from-12-gwh-to-158-gwh-by-2024/

Kapila, S., Oni, A. O., & Kumar, A. (2017). The development of techno-economic models for large-scale energy storage systems. *Energy*, *140*, 656–672. doi:10.1016/j.energy.2017.08.117.

Kim, H. M., Rutqvist, J., Ryu, D. W., Choi, B. H., Sunwoo, C., & Song, W. K. (2012). Exploring the concept of compressed air energy storage (CAES) in lined rock caverns at shallow depth: A modeling study of air tightness and energy balance. *Applied Energy*, *92*, 653–667. doi:10.1016/j.apenergy.2011.07.013.

Klumpp, F. (2016). Comparison of pumped hydro, hydrogen storage and compressed air energy storage for integrating high shares of renewable energies—Potential, cost-comparison and ranking. *Journal of Energy Storage*, *8*, 119–128. doi:10.1016/j.est.2016.09.012.

Llamas, B., de la Cruz Castañeda, M., Laín, C., & Pous, J. (2017). Multi-criteria algorithm-based methodology used to select suitable domes for compressed air energy storage. *International Journal of Energy Research*, *41*(14), 2108–2120. doi:10.1002/er.3771.

Llamas, B., Laín, C., Castañeda, M. C., & Pous, J. (2018). Mini-CAES as a reliable and novel approach to storing renewable energy in salt domes. *Energy*, *144*, 482–489. doi:10.1016/j.energy.2017.12.050.

Letcher, T. (2016). Storing energy. In: S. D. Garvey and A. Pimm (eds.) *Compressed Air Energy Storage*. Oxford: Elsevier; pp. 87–111.

Luo, X., Wang, J., Dooner, M., & Clarke, J. (2015). Overview of current development in electrical energy storage technologies and the application potential in power system operation. *Applied Energy*, *137*, 511–536. doi:10.1016/j.apenergy.2014.09.081.

May, G. J., Davidson, A., & Monahov, B. (2018). Lead batteries for utility energy storage: A review. *Journal of Energy Storage*, *15*, 145–157. doi:10.1016/j.est.2017.11.008.

McDowall, J. (2006). Integrating energy storage with wind power in weak electricity grids. *Journal of Power Sources*, *162*(2 SPEC. ISS.), 959–964. doi:10.1016/j.jpowsour.2005.06.034.

Song, W.-K., Synn, J.-H., Rutqvist, J., Ryu, D.-W., & Kim, H.-M. (2012). Modeling of coupled thermodynamic and geomechanical performance of underground compressed air energy storage in lined rock caverns. *International Journal of Rock Mechanics and Mining Sciences*, *52*, 71–81. doi:10.1016/j.ijrmms.2012.02.010.

Yang, W., Cong, T. N., Ding, Y., Chen, H., Tan, C., & Li, Y. (2009). Progress in electrical energy storage system: A critical review. *Progress in Natural Science*, *19*(3), 291–312. doi:10.1016/j.pnsc.2008.07.014.

## WEB SITES

Dr. Robert Galler, N. W. (n.d.). *RICAS2020*. Retrieved 11 2019, from http://www.ricas2020.eu/

European Comission. (n.d.). *EUROSTAT*. Retrieved 11 2019, from https://ec.europa.eu/eurostat/data/database.

IEA 2019. (2019, 11 18). *The International Energy Agency*. Retrieved from https://www.iea.org/statistics/

National Oceanic & Atmospheric Administration, NOAA Research. https://www.esrl.noaa.gov/gmd/ccgg/trends/, Access on 20th of January, 2020. IEA, 2020. https://www.iea.org/data-and-statistics?country=WORLD&fuel=CO2%20emissions&indicator=CO2%20emissions%20by%20sector, Access on 30th of January, 2020.

*Project 1010-CAES Larne, Northern Ireland*. (n.d.): https://eepublicdownloads.blob.core.windows.net/public-cdn-container/clean-documents/tyndp-documents/TYNDP%202016/projects/P1010.pdf, accessed on 10th of January 2020.

RWE. (n.d.). Adele. Adiabatic CAES for electricity supply. http://www.rwe.com/web/cms/mediablob/en/391748/data/235554/1/rwe-power-ag/press/company/Brochure-ADELE.pdf, accessed on 10th January, 2020.

# 10 Hydrostatically Compensated Energy Storage Technology

*M. Ebrahimi, D. S.-K. Ting, and R. Carriveau*
University of Windsor

*A. McGillis*
Hydrostor Inc.

**CONTENTS**

## 10.1 INTRODUCTION

Renewable and sustainable energy production has recently become an important area of research due to increasing global energy demand [1]. According to various studies, about 80% of the global energy demand is fulfilled by fossil fuels, and this is expected to grow in the next decade [2,3]. Due to the environmental concerns associated with fossil fuel combustion, there is incentive to migrate towards more sustainable energy sources. This, combined with significant reductions in the installation cost, has resulted in a rapid growth in the installed capacity of renewable energy. According to the International Renewable Energy Agency's (IRENA) report [4], the world added 171 gigawatts of renewable energy in 2018 and the global renewable capacity has reached 2,351 GW in the end of 2018. The IRENA report indicates that most of the generated renewable energy is integrated to power networks. Due to the intermittent nature of some of the renewable energy sources, a significant challenge exists in balancing supply and demand over the peak and off-peak time periods. Various solutions have been proposed to address this challenge, some of which have been applied. Energy storage is one such solution and lends promise in furnishing the needed flexibility for the grid. Specifically, the Hydrostatically Compensated Compressed Air Energy Storage (HC-CAES) is an emerging, economically efficient technology. This technology has this potential to rectify this deficiency by managing the received and delivered power to and from the renewable energy sources [5]. This technology not only facilitates the integration of sustainable renewable energy sources with power grid but also helps address the balancing of local and regional generation sources.

This technology takes advantage of the hydrostatic pressure in deep water, applying a series of air compressors and turbines to convert electrical energy into the compressed air and vice versa. This free source of pressure makes HC-CAES technology one of the best fits for storing intermittent offshore renewable energy, such as offshore wave energy, solar energy, and wind energy [6]. Inland and away from the waters, abandoned caverns can be repurposed to serve as air accumulators. Due to the availability of abandoned caverns all over the world, we anticipate HC-CAES technology becoming one of the prominent CAES technologies for the storage of energy. Unlike the majority of the energy storage technologies, in CAES, a high capacity of energy storage for a long delivery time is obtainable. The main equipment used in this technology also has a long operation lifetime which means, after an initial investment for the construction of the plant, the operation and maintenance cost is relatively low. Compared to other types of energy storage technologies, such as battery energy storage, this has a great advantage, making this technology one of the most efficient and sustainable environmentally friendly technologies.

## 10.2 TECHNICAL DESCRIPTION OF THE HYDROSTATICALLY COMPENSATED CAES TECHNOLOGY

HC-CAES technology is an environmentally friendly solution for supporting the power grid for the supply of electricity during peak demand. It is also an efficient offshore energy storage method for handling the challenge of fluctuating and intermittent renewable energy [7]. In this technology, the excess energy generated during the off-peak market is converted into compressed air via air compressors. Compressed air is then stored in air accumulators. Air accumulators can be either air bags installed deep under water or a cavern connected to a water source with a water-filled shaft [8].

The hydrostatic pressure of water keeps the air at a constant pressure, so the potential energy is stored in the air accumulator. When electricity is required, the air is discharged and electricity is generated through the turbine(s). Figure 10.1 depicts the schematic diagram of a cavern-based CAES technology.

In this technology, the main steps to store and dischargee of the electricity energy can be summarized as:

**Compression/Charging**

1. An electric signal is sent to the facility to initiate air compression. An electric motor drives a rotating turbo compressor to compress air.
2. Heat from compression is removed from air and stored in a tank.
3. Compressed air is stored in the air accumulator.

**Expansion/Discharging**

4. Due to the hydrostatic pressure of water, air from the air accumulator is brought back up into the process plant through the same line.
5. Heat from the hot tank is re-introduced to warm the air prior to entering a turbo expander.
6. Air expanding with re-introduced heat drives the rotating turbo expander and electric generator, reproducing electricity.

Figure 10.2 shows the main components of a sample HC-CAES plant with a three-stage compressor and one-stage expander. As can be observed, the technology

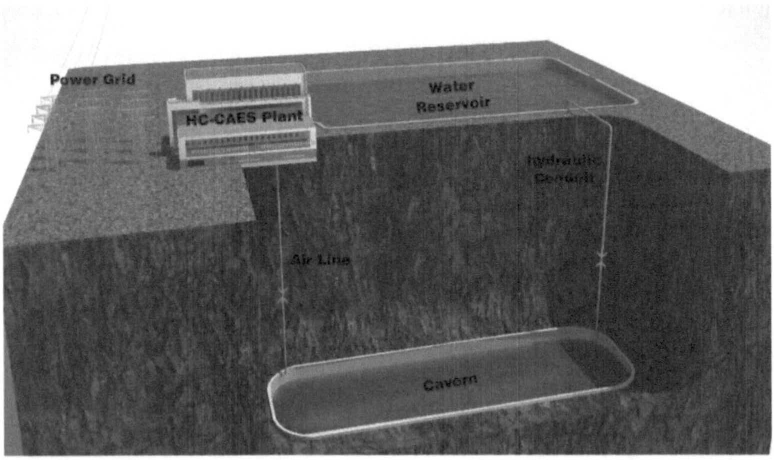


**FIGURE 10.1** The schematic diagram of a sample cavern-based HC-CAES plant.

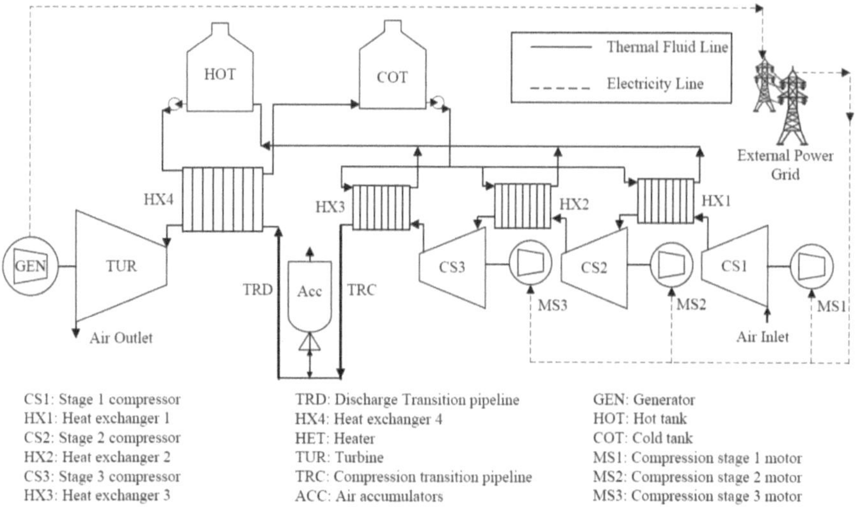


**FIGURE 10.2** The main components of a sample HC-CAES plant.

consists of six main subsystems including the air compression unit, expansion unit, compressed air storage unit, thermal energy storage unit, heat exchanger unit, and transmission pipeline unit.

## 10.3 ENERGY STORAGE CAPACITY OF HC-CAES PLANTS

According to the basis of air storage system design, the maximum energy storage capacity in this technology can be determined for a given stored volume of air and the pressure. Assuming that all generated heat in the compression phase is recovered, the stored energy is then equal to the potential energy of the stored air in the accumulators. It can be calculated by:

$$E_{\max} = (P_h - P_{\text{atm}})V \tag{10.1}$$

where $P_{\text{atm}}$ is the atmospheric pressure and $P_h$ is the system pressure. According to the water column's height, system pressure can be determined simply by:

$$P_h = \rho g h + P_{\text{atm}} \tag{10.2}$$

Assuming that the buoyancy force is much greater than the float weight, the maximum energy storage capacity at the depth of $h$ for a cubic meter of the stored air with adiabatic expansion can be calculated as:

$$E_{\max} = \frac{\rho_{\text{air}} \times R_g \times T_h}{3{,}600} \times \left(1 - \left(\frac{P_h}{P_{\text{atm}}}\right)^{\frac{\gamma-1}{\gamma}}\right) \times \frac{\gamma}{\gamma - 1} \tag{10.3}$$

## 10.4 APPLICATIONS OF HC-CAES TECHNOLOGY IN THE GENERATION OF SUSTAINABLE ENERGY

The main benefits of HC-CAES systems are the ancillary services this technology provides to the power grid. By utilizing HC-CAES, the energy from a variety of intermittent sources such as wind, offshore wave energy, solar energy, and the grid itself can be temporarily stored. The stored energy can be delivered to the grid at the later time, when it is more needed. Unlike most other types of energy storage, this technology is environmental friendly; it has a large life cycle and its discharge duration can be longer than most other energy storage technologies. The primary applications of HC-CAES are explained below.

- Integration with renewable energy plants

  Increasing concern about environmental issues steaming from conventional power sources, their shortages, and rising costs have promoted a growing interest in the integration of renewable energy sources to the power systems. HC-CAES technology can be integrated with renewable power plants to address the instability of generating renewable power generation. In the micro-grid concept, HC-CAES can suppress the micro-grid power fluctuations and improve power supply quality.
- Peak shaving

  Peak shaving means the process of storing electricity during the off-peak periods and discharging stored power during the periods of high demand. This process increases the use of electrical generation facilities during off-peak hours, which decreases the operation and maintenance costs of electricity generators. This application is particularly useful when the electricity is available at low prices during off-peak time periods.
- VAR support from the power grid

  HC-CAES systems cannot only provide electricity but can also be used to generate reactive power. Since HC-CAES systems have a predictable nature (not dependent on wind, sunny sky, etc.), the importance of a reactive power compensator or VAR (voltage-ampere reactive) support from these facilities is more significant. To prevent electricity blackouts, renewable energy sources also need to control energy flux and manage the passive power of the grid. In 2003, the largest blackout in North American history affected more than 50 million people across United States and two Canadian provinces. Since HC-CAES systems can go online relatively fast, they can help the grid to maintain voltage. The effective control of HC-CAES plants could help micro-grids maintain good voltage and frequency stability.

## 10.5 COMPONENTS OF THE HC-CAES TECHNOLOGY

### 10.5.1 Air Accumulators

Large-scale energy storage plants need large volume air accumulators. This makes air accumulators as one of the most sensitive components of HC-CAES technology. They are either an underground cavern or air balloons and access to them is difficult and expensive [9]. They should be sufficiently robust and cost-effective for long-term operation under high-pressure condition and abrasive streams. They also need to fulfill a diverse range of specifications required for industrial applications. Lists of the main requirements for a realistic air are as follows:

- Simplicity with no submerged electrical or controller devices. Therefore, the system can operate for a long time without the need of going underwater and change, repair, or clean the air accumulator's equipment.
- High stability and sustainability for a vast range of operating conditions.
- High flexibility and robustness for not appreciably degrading over its service lifetime.
- Feasibility of installation.
- Being cost-effective to manufacture, install, and operate.

As discussed, air accumulators can be either flexible air balloons or an abandoned cavern. Flexible air accumulators are made of fabric and were first proposed by Laing [10]. In 2015, the first grid-connected HC-CAES system with fixable air accumulator was implemented by Hydrostor Inc. [11]. Figure 10.3 shows Hydrostor's air bags before their installations in Lake Ontario.

Flexible air accumulators can be manufactured in a variety of shapes, though, in order to balance drag forces, it is more efficient for them to be symmetrical on both the vertical and horizontal axes with a single point of anchorage (similar to air balloons) [12,13].

**FIGURE 10.3** Hydrostor's air bags in Lake Ontario.

In this situation, the cable tension $C$, at the equilibrium condition, can be determined by considering the balance of forces acting on the air accumulator.

$$0 = F_b - mg - C \tag{10.4}$$

where $F_b$ is the buoyancy force acting on the air accumulator, $m$ is the mass of bag, and $g$ is gravitational acceleration. Considering buoyancy force for a $V$ volume of air balloon as:

$$F_b = -\rho_h g V \tag{10.5}$$

Cable tension (minimum anchorage strength to hold air accumulators) of an air accumulator then can be determined by rearranging Equation (10.4) as:

$$C = \rho_h g V - mg \tag{10.6}$$

It should be noted that air accumulators are more efficient when connected to the air transfer pipeline from the top. Because the bottom of the balloon-shaped air accumulators is open (there is a direct contact of air and water in the base of air containers), there is a high chance of water leakage to the system. Therefore, if the pipeline is not connected from the top, water can enter to the air transfer pipeline and a much greater back pressure and mass flow rate will be required for the charging phase (which can significantly decrease the performance of the system). On the other side, air leakage from air balloons to water can also happen. Depending on the installation depth, the material of the balloon, and standby duration, air leakage varies. In the balloons of the UWCAES of the Toronto Island plant, the leakage rate was reported about 2.8% per day.

In this technology, the charging and discharging processes happen relatively fast; therefore, it can be assumed that these processes are isothermal at deep-water temperatures. In this condition, the mass required to fill of air accumulators can be calculated by:

$$\text{Required Mass of Air} = \int_{\text{Empty}}^{\text{Full}} dm = \int_{\text{Empty}}^{\text{Full}} \frac{P}{RT} dV = \frac{P}{RT}\left(V_{\text{Full}} - V_{\text{Empty}}\right) \tag{10.7}$$

where $T$ is the water temperature at the depth of air accumulators, and $R$ is the universal gas constant.

Caverns are another type of air accumulators in HC-CAES and salt caverns are a good choice for this technology. Since there are a lot of abandoned salt caverns across the world, this technology has a good potential to grow quickly. The world's first fuel-free cavern-based HC-CAES plant with 2 MW started operations in the Compass Minerals' salt mine in Goderich, Canada [11] in 2020. In any time period, the rate of mass change in the cavern can be described as:

$$\sum \dot{m}_{\text{in}} - \sum \dot{m}_{\text{out}} = \frac{dm}{dt} \tag{10.8}$$

where $t$ is the time, $\dot{m}_{\text{in}}$ and $\dot{m}_{\text{out}}$ are inlet and outlet mass flow rates to/from the cavern, respectively, and $\frac{dm}{dt}$ is the mass accumulation in the cavern. Then air temperature and pressure at time $t$ can be calculated by:

$$T_{c,\,t} = \frac{1}{m_{c,t}}\left(\frac{\gamma - 1}{\gamma}\right)\left(m_t T_{\text{in},t} - m_t T_{\text{out},t}\right) \tag{10.9}$$

$$P_t = X_t h_t + \frac{1}{V_c}\left(n_{c,t} C_{P,t} T_{c,t} - m_t h_{\text{in},t} - m_t h_{\text{out},t}\right) \tag{10.10}$$

In these equations, it is assumed that the cavern is adiabatic. For real problems, the appropriate heat transfer term can be added to these equations based on the cavern wall material.

### 10.5.2 Compressors and Expanders

Typically, pulling or pushing air to or from air accumulators occurs in a semi-isobaric process. This leads compressors and expanders to work at almost the same constant pressure (obviously, mass flow rate is not the same); thus, they can be optimized for a single design point. Considering that energy losses are very small when compressors and expanders are the two main components of the system (because they convert energy between the forms of mechanical work and electricity), it can be concluded that a good efficiency is obtainable from this technology. To determine the actual power consumption and power generation of the compressors and expanders, the final exit temperatures of the compression and expansion processes need to be calculated. Considering these processes are isentropic and assuming air behaves as an ideal gas with constant specific heats, the corresponding final exit temperature can be calculated from:

$$T_f = T_i\left(\frac{P_f}{P_i}\right)^{\frac{\gamma-1}{\gamma}} \tag{10.11}$$

where $i$ and $f$ are the initial and final statuses of the system. The actual power consumption and power generation can then be calculated as:

$$\dot{W}_{\text{comp.}} = \frac{\dot{m}RT_i}{\eta_{\text{comp.}}}\left(\left(\frac{P_f}{P_i}\right)^{\frac{\gamma-1}{\gamma}} - 1\right)\frac{\gamma}{\gamma - 1} \tag{10.12}$$

$$\dot{W}_{\text{exp.}} = \eta_{\text{exp.}}\dot{m}RT_i\left(1 - \left(\frac{P_f}{P_i}\right)^{\frac{\gamma-1}{\gamma}}\right)\frac{\gamma}{\gamma - 1} \tag{10.13}$$

Therefore, the actual final compression and expansion temperatures can easily be determined by:

$$T_{f,\text{ comp., actual}} = T_i\left(1+\frac{1}{\eta_{\text{comp.}}}\left(\left(\frac{P_f}{P_i}\right)^{\frac{\gamma-1}{\gamma}}-1\right)\right) \tag{10.14}$$

$$T_{f,\text{ exp., actual}} = T_i\left(1-\eta_{\text{exp.}}\left(1-\left(\frac{P_f}{P_i}\right)^{\frac{\gamma-1}{\gamma}}\right)\right) \tag{10.15}$$

### 10.5.3 Heat Exchangers and Thermal Energy Storage

HC-CAES systems can become more efficient by using heat exchanger units. In this technology, heat exchangers can be used in both charging and discharging phases. In the charging phase, they extract heat from air not only for efficiency purposes but also to prevent next compression stages from being damaged by air at high temperature. The other main motivation for using thermal storage with an HC-CAES system is to increase the total quantity of exergy that can be stored for a given size of a system [14]. Fiaschi et al. showed that thermal energy storages have the capability to improve the efficiency of cycles by up to 70%–80% [15,16].

In the compression phase, heat exchangers are used for cooling air after the compression stages, and in the expansion phase, they are responsible for heating air before the expansion stages (if the air is not heated enough, its moisture would freeze and block the pipes). The efficiency of heat exchangers is usually evaluated in terms of their effectiveness, which is the ratio of the actual heat transfer rate compared to the theoretical maximum rate:

$$\eta = \frac{\dot{Q}_{\text{actual}}}{\dot{Q}_{\max}} \tag{10.16}$$

By neglecting both the heat storage in the heat exchanger and heat transfer to the environment, the heat transfer rate can be determined as:

$$\dot{Q} = \dot{m}_{\text{hot}}C_{p,\text{hot}}\left(T_{\text{hot, in}} - T_{\text{hot, out}}\right) = \dot{m}_{\text{hot}}C_{p,\text{ hot}}\left(T_{\text{hot, in}} - T_{\text{hot, out}}\right)$$

The pressure drop in heat exchangers also can be calculated based on an empirical equation [17] as:

$$P_{\text{loss}} = \frac{0.0083\eta}{1-\eta}P_{\text{in}} \tag{10.17}$$

## 10.6 LOCATION OF THE HC-CAES POWER PLANTS

Air accumulator type is the main parameter that determines the location of the HC-CAES. When a cavern is used as the air storage, a purposelybuilt water reservoir can be constructed in locations not necessarily located near a body of water (a closed-loop system). The water can be provided from standard municipal, ground, or surface water sources locally. Therefore, any place close to a cavern with proper geotechnical conditions has the potential to be an HC-CAES plant.

When air balloons are used as the air accumulators, the location of the HC-CAES power station is one of the most important design parameters in terms of the cost efficiency of the plant. The location of the plant depends on the distance between the air accumulators and shore, as well as the relative costs of running an underwater transmitting high-pressure air pipeline and a transmitting power line. Therefore, areas with deep water close to the shore are economically efficient for the installation of HC-CAES power plants. In a study by Pimm and Garvey, the two investigated the potential locations for the installation of HC-CAES plants around Europe and North America. They showed that, in this region, there are lots of proper coastal areas with deep water within a reasonable distance of the shore [18].

Overall, when air balloons are used as air accumulators, there are two types of HC-CAES systems based on the location of the power station: land-based and offshore. When air accumulators are installed near shore, it makes sense to locate the power station onshore (Figure 10.4). In this condition, access to electromechanical equipment is easy and the need for an underwater power transmission line is eliminated [19,20]. In such cases, air transmission pipelines run over the seabed (or it is drilled, if the seabed setting is not appropriate) between the power station and the air accumulators. In some cases, access to deep water near the shore is limited. In these cases, it is more cost efficient to locate the power station on a floating offshore platform above the air accumulators (Figure 10.5). In this case, power transmission line runs between the floating power station and the onshore power grid. An air pipeline runs between the floating power station and air accumulators [21,22].

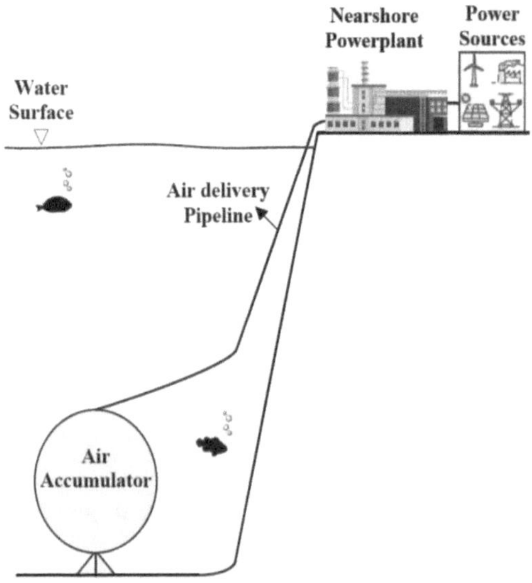


**FIGURE 10.4** A sample land-based offshore energy storage system.

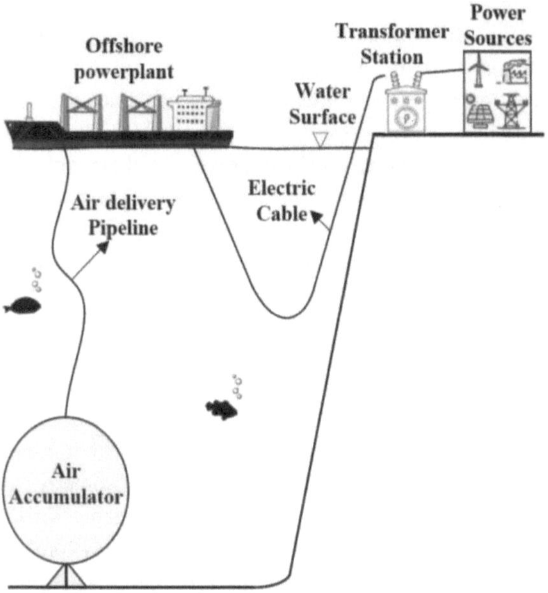


**FIGURE 10.5** A sample floating offshore energy storage system.

The design of the transmission pipeline especially for the floating power plant is a developed discipline and it is beyond the scope of this chapter. Despite the plant type and because of the good thermal conductivity of air accumulators and transmission pipelines, the following terms can be assumed for the air temperature inside the pipeline:

$$T_{\text{air,in}} > T_{\text{air,out}} = T_{\text{acc}} = T_{\text{water}} \tag{10.18}$$

The pressure drop of the air in the transmission pipeline also can be calculated by:

$$P_{\text{loss}} = P_{\text{loss, Vis}} + P_{\text{loss, }k} = \left(\lambda \frac{L}{d} + \sum \varnothing\right)\frac{\rho v^2}{2} \tag{10.19}$$

where $P_{\text{loss, Vis}}$ is the viscous friction pressure loss and $P_{\text{loss, }k}$ is the local pressure loss caused by bends, valves, etc. $\lambda$ and $\varnothing$ are also the friction factor and the loss coefficient.

## 10.7 EXERGY MODELING

Exergy is the portion of energy that can be used to do work and is defined as the difference between the total supplied energy and the portion that is not converted to work. Exergy analysis is a powerful tool that improves the evaluation and analyzation of thermodynamic processes by identifying their imperfections due to irreversibility. Since HC-CAES technology relies on the compression and expansion of air, thermodynamics plays a vital role in the performance of the system. Exergy analysis is one of the most powerful tools that can be applied to develop insights into the thermodynamic processes of an HC-CAES system and aid in performance modeling efforts. It effectively highlights the flow of energy in the system, areas of loss, and potential areas to focus on to improve system performance.

The overall exergy of the system can be expressed by thermomechanical exergy—the maximum amount of useful energy when the system is in mechanical and thermal equilibrium with the surrounding environment. In HC-CAES systems, pressures, temperatures, and mass flux to/from the air accumulator all change continuously. In such cases, exergy analysis, while much more complicated, is extremely useful for understanding the performance of various system components. Such an analysis can quantify a variety of energy transfer mechanisms inside the system, greatly improving the ability to model the performance of the system, especially during transient operation [8,23].

Since the performance curve of the system components is not linear, there are various reasons for the nonlinear behavior of CAES systems. Some reasons include heat transfer between working fluids and various system components (turbomachinery, piping, etc.), redevelopment of a steady-state heat gradient in the system's heat exchangers, and pressure variation of the cavern during charging and discharging phases. In the next sections, the efficiency of this technology is investigated through exergy analysis.

To investigate the performance of this technology, the exergy concept is applied. For processes in which the thermodynamic cycles are well-defined, exergy analysis is a very useful method for quantifying the efficiency of a variety of energy transfer processes inside the system [24]. In an HC-CAES system, the total exergy of any point of the system can be calculated by:

$$\dot{E}_x = \dot{E}_x^{P_h} + \dot{E}_x^{K_n} + \dot{E}_x^{P_t} + \dot{E}_x^{C_h} \tag{10.20}$$

where $P_h$, $K_n$, $P_t$, and $C_h$ superscripts represent physical, kinetical, potential, and chemical exergies. The kinetic energy and the potential energy of fluids inside the HC-CAES system are neglected, and it can be assumed that there is no chemical reaction in HC-CAES systems; therefore, physical exergy is the dominant exergy inside the system so the total exergy at any point of the system can be rewritten by:

$$\dot{E}_x = \dot{E}_x^{P_h} = \dot{m}e_x \tag{10.21}$$

In cavern-based HC-CAES, unlike underwater HC-CAES, there can be some fluctuations in the height of water in the water riser, causing hydrostatic pressure variations in the system. However, in underwater CAES, due to the fixed installation depth of air accumulators, the hydrostatic pressure remains constant. If the pressure variation in the cavern-based CAES is significant, its effect should be considered in the calculation.

Ignoring the potential and kinetic energy, the specific exergy inside the system then can be determined by:

$$e_x = (h - h_0) - T_0(s - s_0) \tag{10.22}$$

where $T_0$ is the environment temperature and $h_0$ and $s_0$ are the specific enthalpy and specific entropy at the ambient environmental condition. To be able to analyze the efficiency of the system components individually, it is necessary to determine how much exergy is destructed inside of each component. This parameter is defined as:

$$\dot{E}_{D,\,k} = \dot{E}_{F,\,k} - \dot{E}_{P,\,k} \tag{10.23}$$

where $\dot{E}_{F,\,k}$ represents exergy of fuel (the input exergy) and $\dot{E}_{P,\,k}$ represents exergy of product (the output exergy). Exergy efficiency of the $k$-th components then can be calculated by:

$$\varepsilon_k = \frac{\dot{E}_{P,\,k}}{\dot{E}_{F,\,k}} \tag{10.24}$$

The total performance of the HC-CAES system at each time step can also be derived by:

$$\varepsilon_{\text{tot}} = \frac{\dot{E}_{P,\,\text{tot}}}{\dot{E}_{F,\,\text{tot}}} \tag{10.25}$$

where $\dot{E}_{F,\,\text{tot}}$ is the total exergy entering into the system and $\dot{E}_{P,\,\text{tot}}$ is the exergy delivered to the grid.

## 10.8 CHALLENGES OF HC-CAES TECHNOLOGY

Besides the benefits this technology provides, it has some deficiencies as well, like any other technology. In this section, the main challenges of CAES systems are addressed.

### 10.8.1 Champagne Effect

The Champagne Effect is a phenomenon with the potential to cause serious problems in the operation of cavern-based HC-CAES plants if not appropriately mitigated. The presence of high-pressure air next to water in the hydrostatically compensated air storage cavern causes air to dissolve in the water [25,26]. The Champagne Effect has only been studied analytically in recent years with the use of computer simulations based on derived equations which have yet to be confirmed empirically [27,28]. In this phenomenon, as air is compressed into the cavern during charging, water from the cavern is forced up the vertical water shaft into the surface-level compensating reservoir. As the water flows up the vertical shaft, the pressure of the water decreases and the dissolved air is released from the water in the form of bubbles, creating a two-phase air–water mixture in the upper part of the shaft. The presence of these air bubbles reduces the density of the water in the shaft, thereby reducing the hydrostatic head of the water column, causing the volume of stored air to expand, thereby accelerating the flow of water up the shaft. In the extreme, this can result in an unstable flow, leading to an unintentional and runaway cavern overcharging event in which the water seal at the lower end of the supply conduit is broken, potentially resulting in catastrophic system failure. Even if an overcharging event does not occur, the loss

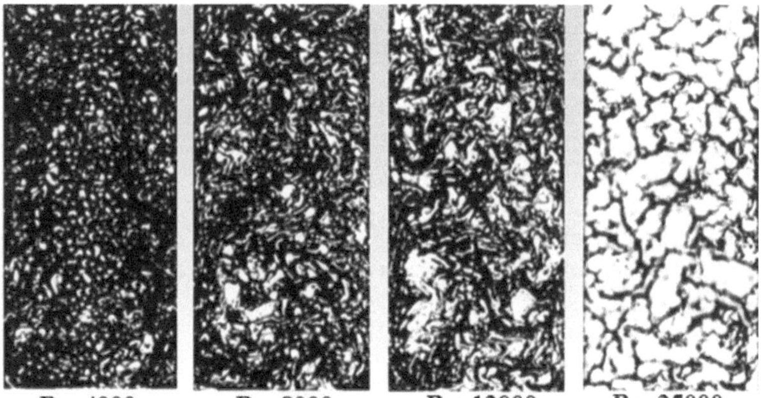


**FIGURE 10.6** Bubble distributions (white areas) in the middle section of the water pipe of an HC-CAES plant at various heights (flow velocity).

of air up the compensation shaft represents a loss of energy and if it is not properly dealt with, could have an impact on the efficiency of the system. As such, it is desirable to minimize the Champagne Effect in an HC-CAES system to both improve system performance and minimize the risk of a cavern overcharging event. To show this phenomenon, the bubbly flow is simulated inside an air–water riser with CFD (Computational Fluid Dynamics). To solve the Reynolds-Averaged Navier–Stokes Equations, a finite volume cell-centered scheme with volume of fluid approach was utilized. An unstructured tetrahedron grid with the maximum face size of 1 mm was also used as the computational grid [29]. In Figure 10.6, the shape and the size of the bubbles in the middle section of a 200mm diameter air riser at various superficial Reynolds number are shown. This figure shows that bubbles' sizes dramatically increase when they move upward in the water pipe.

Fortunately, ready solutions to the Champagne Effect exist. Some of them are summarized in *Compressed Air Energy Storage Symposium Proceedings,* published by the United States Department of Energy, Electric Power Research Institute, and Pacific Northwest Laboratory [30]. This report proposed a number of methods for dealing with this problem such as the use of a restriction to limit velocity and provide damping of the hydraulic compensating system along with a U-shaped water seal. Other suggested methods include oversizing the cavern to avoid blow out, recirculation of water to avoid dissolution of air, and the use of phase separation devices to remove the air bubbles from the water as it rises in the shaft.

### 10.8.2 Response Time of HC-CAES Systems

Response time in energy storage technology is a time that a technology needs to ramp up supply. For energy storage or power production assets, it is desirable for the system to be able to respond quickly to meet the dynamic requirements of the electricity grid. Specifically, to assist with electrical grid stability, it is beneficial for the time it takes the system to begin providing power to the grid after receiving a

**TABLE 10.1**
**Response Time of CAES Technology in Comparison to Other Well-Known Energy Storage Technologies [31]**

| Energy Storage Technology | Power Rating (MW) | Cycling | Efficiency (%) | Response Time Scale |
|---|---|---|---|---|
| Super capacitor | 0.01–1 | 10k–100k | 80–98 | 10–20 ms |
| Flywheels | 0.001–1 | 20k–100k | 70–95 | 10–20 ms |
| Lead-acid battery | 0.001–100 | 6–40 years | 80–90 | <sec |
| Flow battery | 1–100 | 12k–14k | 60–85 | 10–20 ms |
| Hydrogen | 0.01–1,000 | 5–30 years | 25–45 | Seconds–minutes |
| SNG | 50–1,000 | 30 years | 25–50 | Seconds–minutes |
| CAES | 2–1,000 | 20–40 years | 40–75 | Seconds–minutes |
| Li-ion battery | 0.1–100 | 1k–10,000k | 85–98 | 10–20 ms |

dispatch signal—the discharge response time—to be as short as possible. Due to the nature of grid disturbances, it is important that these resources can respond within a matter of seconds to minutes to help stabilize the grid.

In order to fully understand the benefits an HC-CAES brings to an electrical grid, the discharge response time of the system must be known. It is particularly important for large systems (100+ MW) to have an understanding of discharge response times while planning the project in order for the system to be accounted for when planning for impacts to the grid.

The discharge response time of an HC-CAES system is not currently well defined in the literature and will be influenced by a number of factors. These factors include:

- The ability of the turbine and generator to synchronize quickly to the grid speed while having an over speed clutch installed between the turbine and generator.
- The impact of retained heat in the heat exchangers, turbine, and thermal system. Heat retention will be influenced by both standby time and insulation.
- The response time of actuators within the system.

This can cause some challenges when designing an HC-CAES. In Table 10.1, the response time of CAES technology (as the main concept of HC-CAES technology) in comparison to some other well-known energy storage technologies is given. This table indicates that CAES technology in general has a high power rating, long life and also enables long duration storage; however, its response time in comparison to other technologies such as batteries and flywheels is relatively long.

### 10.8.3 Air Leakage

The leakage of high-pressure gas from an underground cavern is a complex two-phase flow phenomenon which is challenging to predict or model with reasonable accuracy. Hard-rock hydrocarbon storage caverns are the only precedent comparable to the design and application of an HC-CAES system [32,33]. There are key differences, however, in the storage pressure and permeability restrictions between these two types of caverns due to the nature of the gas being contained. When hydrocarbon storage caverns are designed, there is no tolerance for leakage because the stored products are toxic. Consequently, these caverns must be located in tight rock formations and the pressure of the stored product must be stored below a "critical pressure" which is sufficiently less than the surrounding groundwater pressure to eliminate the possibility of any product leakage. In contrast, HC-CAES by design have a higher air storage pressure, which will be greater than the surrounding groundwater pressure. Furthermore, since the air contained in the cavern is non-toxic, HC-CAES projects can tolerate some air leakage and thus can be located in caverns with a wider range of permeability than hydrocarbon storage caverns. As there is no precedent to use as a case study to predict leakage from a storage cavern empirically, modeling is required to predict the extent that leakage will occur. Due to the complex nature of the two-phase flow though extremely small pores and fractures in the rock mass, however, a level of uncertainty is inevitable.

All hard-rock hydrocarbon storage caverns operate below their critical pressure, the pressure at which the stored product would escape from the cavern. This is achieved through three primary methods: maintaining a lower pressure for the stored product (i.e., storing less product per unit of volume); implementation of a water curtain with sufficiently greater pressure than the cavern pressure, which increases the groundwater pressure in the rock mass surrounding the storage cavern; and selectively locating in relatively impermeable caverns. HC-CAES projects will almost always operate above the critical pressure when utilizing hard-rock caverns due to the hydrostatic nature of the cavern pressurization; as the cavern pressure is determined by the hydrostatic pressure at cavern depth, only under very rare circumstances could the surrounding groundwater pressure be at a high-enough pressure to eliminate leakage therefore water curtains would be required if the host cavern rock has higher porosities. Groundwater has the ability to impede air flow through porous media; the phenomenon, however, is complex to model and is based on empirical data for various rock types.

### 10.8.4 Cavern Rock Fatigue

Boundaries in underground spaces are subject to high stress due to the pressure difference on either side of the boundary wall. One side of the wall is subjected to very high lithostatic pressure in the rock, while the other is subject to the lower pressure of the cavern. This stress can lead to fracturing of the void boundary, and in a worst case, the void spaces collapse. The likelihood of rock failure leading to these events is further enhanced by cyclic loading of the void space, such as that expected during HC-CAES operation, where the pressure fluctuates between charging and discharging.

In the last decades, several studies have been performed in an attempt to predict the mechanical behavior of gas storage caverns. For example, an empirical model was proposed by Hunsche et al. to describe the steady-state creep deformation of rock [34]. The method was a function of temperature and deviatoric stress in caverns. Aubertin et al. [35] presented a viscoelastic-plastic model from a modified nonlinear

Mohr–Coulomb criterion. Some other studies modeled cavern rock fatigue within the same framework [36].

An HC-CAES cavern, however, is fundamentally different from commercial gas storage caverns in that the pressure cycling is more frequent (daily) and of a lower amplitude (approximately 0.3 MPa). In addition to frequent pressure cycles, the interaction between air and water inside the storage space, creates a dynamic zone of saturation and desaturation in the surrounding rock mass. As such, determining the effects on long-term (20–30 years) rock structural integrity of the thousands of daily pressure and saturation/desaturation cycles associated with HC-CAES caverns is not documented generally in the literature and therefore must be investigated for the local host rock mass in conjunction with a project and it must be adequately addressed for the construction of any HC-CAES plant.

## 10.9 SYSTEM COMPONENTS ANALYSIS

In this section, the main components of the HC-CAES are individually analyzed through some examples. Then the performance and exergy components for a sample plant are investigated.

### 10.9.1 Energy Storage Capacity of HC-CAES Technology

As discussed, in HC-CAES systems, air accumulators can either be stored in underwater balloons or a cavern. In both cases, air pressure is almost the same as the hydrostatic pressure in the surrounding water. Assuming $P_{atm}$ = 101 kPa, and water density is constant and equal to 1,020 kg/m³, the maximum energy storage capacity of HC-CAES systems with air accumulators can be calculated from Equation (10.3). Based on this equation, Table 10.2 lists the maximum energy for a cubic meter of the stored air at various water depths. The energy density in this table is calculated when the expansion occurs in both adiabatic and isothermal processes. It is also assumed that air temperature and humidity are 20°C and 40%, respectively.

From Table 10.2, it can be observed that higher energy density is obtainable with an increase in the air accumulator depth. This owes to the fact that the energy storage

**TABLE 10.2**
**Maximum Energy Density of a Cubic Meter in HC-CAES Systems When Expansion Occurs in Adiabatic or Isothermal Processes**

| Water Depth (m) | Air Density (kg/m³) | Pressure (bar) | Max. Energy Density; Adiabatic Expansion (kW·hr/m³) | Max. Energy Density; Isothermal Expansion (kW·hr/m³) |
|---|---|---|---|---|
| 100 | 13.3 | 11.2 | 1.1 | 0.8 |
| 200 | 25.2 | 21.2 | 2.9 | 1.7 |
| 400 | 48.9 | 41.1 | 7.6 | 4.3 |
| 600 | 72.7 | 61.2 | 13.4 | 7.1 |
| 2,000 | 237.4 | 202.1 | 69.0 | 29.5 |

capacity increases more than linearly with the air accumulator's pressure because of the compressibility of the air. This means that in higher pressure operating conditions, cheaper energy storage costs will be available.

### 10.9.2 Performance Analysis of Compressors in HC-CAES Plants

In this section, a single compressor—one of the main components of HC-CAES system—is thermodynamically analyzed and the exergy components of the compressor are determined. If we assume that the compressor works under adiabatic conditions, according to the exergy definition given in Section 10.7, exergy of fuel and exergy of product can be calculated as:

$$\dot{E}_{F,\,\text{Comp.}} = \dot{W}_{\text{Comp.}} \tag{10.26}$$

$$\dot{E}_{P,\,\text{Comp.}} = \dot{E}_{\text{Air\_out}} - \dot{E}_{\text{Air\_in}} \tag{10.27}$$

To make the calculation easier, we can assume that the reference system is at the ambient temperature and pressure of 22°C and 101 kPa, respectively. Therefore, the input exergy of the compressor $\left(\dot{E}_{\text{Air\_in}}\right)$ is zero. For solving the thermodynamic model, some more input data is required. In Table 10.3, the thermodynamic properties at the inlet and outlet of the understudied are shown.

We know that in HC-CAES systems, the chemical exergy is zero as the chemical composition of the cycle is not changing with the environmental conditions. Therefore, the total exergy of the air at the outlet of compressor can be calculated by:

$$\dot{E}_{\text{Air\_out}} = \dot{m} \times e \tag{10.28}$$

The specific exergy of air can be calculated by:

$$e = (h - h_o) - T_o(s - s_o) \tag{10.29}$$

With the given values in Table 10.3, the total exergy of the system at the outlet of compressor can be determined as:

$$\dot{E}_{\text{Air\_out}} = \dot{E}_{P,\,\text{Comp.}} = \dot{m} \times \left( C_p \times \left( T_C - T_0 - T_0 \times \ln\left(\frac{T_C}{T_0}\right)\right) + T_0 \times R \times \ln\left(\frac{P_C}{P_0}\right)\right)$$

**TABLE 10.3**
**ThermodynamicProperties of Air at the Inlet and Outlet of an HC-CAES Compressor**

| Air Mass Flow Rate (kg/s) | Inlet T (°C) | Out T (°C) | Inlet P (kPa) | Outlet P (kPa) | Isentropic Eff. | $\gamma_{\text{air}}$ | $C_p$ (J/kg·K) |
|---|---|---|---|---|---|---|---|
| 1 | 22 | 120 | 101 | 180 | 0.85 | 1.4 | 1,030 |

$$\dot{E}_{\text{Air\_out}} = 1 \times \left(1{,}030 \times \left(393 - 295 - 295 \times \ln\left(\frac{393}{295}\right)\right) + 295 \times 287 \times \ln\left(\frac{180}{101}\right)\right)$$

$$= 100.7\,\text{kW}$$

According to Equation (10.12), the exergy of fuel also can be calculated as:

$$\dot{E}_{F,\,\text{Comp.}} = \dot{W}_{\text{Comp.}} = \frac{\dot{m}RT_1}{\eta_{\text{comp.}}}\left(\left(\frac{P_2}{P_1}\right)^{\frac{\gamma-1}{\gamma}} - 1\right)\frac{\gamma}{\gamma-1}$$

$$\dot{E}_{F,\,\text{Comp.}} = \frac{1 \times 287 \times 295}{0.85} \times \left(\left(\frac{180}{101}\right)^{\frac{1.4-1}{1.4}} - 1\right) \times \frac{1.4}{1.4-1} = 62.6\,\text{kW}$$

Therefore, the exergy destruction in the compressor with assumed inlet and outlet conditions is:

$$\dot{E}_{D,\,\text{Comp.}} = \dot{E}_{F,\,\text{Comp.}} - \dot{E}_{P,\,k} = 100.7 - 62.6 = 38.1\,\text{kW}$$

The exergy efficiency of the compressor can then be calculated by:

$$\varepsilon_{\text{Comp.}} = \frac{\dot{E}_{P,\,\text{Comp.}}}{\dot{E}_{F,\,\text{Comp.}}} = \frac{\dot{E}_2 - \dot{E}_1}{\dot{E}_{F,\,\text{Comp.}}} = \frac{62.6 - 0}{100.7} = 62.2\%$$

### 10.9.3 Thermodynamic Analysis of Heat Exchangers in HC-CAES Plants

As discussed, HC-CAES plants can be divided into compression and expansion subsystems. These two subsystems are connected via heat exchangers. To analyze the performance of heat exchangers in HC-CAES plants, for instance, a heat exchanger with thermodynamic status as given in Table 10.4 is analyzed.

Considering there is no heat transfer with the environment, the exergy of fuel and exergy of product can be calculated by:

$$\dot{E}_{F,\,\text{HEX}} = \dot{E}_{\text{Air\_in}} - \dot{E}_{\text{Air\_out}} \tag{10.30}$$

$$\dot{E}_{P,\,\text{HEX}} = \dot{E}_{\text{Water\_out}} - \dot{E}_{\text{Water\_in}} \tag{10.31}$$

**TABLE 10.4**
**ThermodynamicProperties of Air at the Inlet and Outlet of an HC-CAES Heat Exchanger**

| Fluid | Mass Flow Rate (kg/s) | Inlet T (°C) | Out T (°C) | Inlet P (kPa) | Outlet P (kPa) | $C_p$ (J/kg·K) |
|---|---|---|---|---|---|---|
| Air | 1.1 | 140 | 75 | 180 | 178 | 1,030 |
| Water | 0.38 | 30 | 77 | 150 | 150 | 4,230 |

To determine the performance of the system and the exergy components, the exergy of the system at all inlets and outlets needs to be calculated first. Assuming that the ambient temperature and pressure are 22°C and 101 kPa, exergy at these points are:

$$\dot{E}_{\text{Air\_in}} = \dot{m} \times \left( C_p \times \left( T_{\text{Air\_in}} - T_0 - T_0 \times \ln\left(\frac{T_{\text{Air\_in}}}{T_0}\right)\right) + T_0 \times R \times \ln\left(\frac{P_{\text{Air\_in}}}{P_0}\right)\right)$$

$$\dot{E}_{\text{Air\_in}} = 1.1 \times \left( 1{,}030 \times \left( 393 - 295 - 295 \times \ln\left(\frac{393}{295}\right)\right) + 295 \times 287 \times \ln\left(\frac{178}{180}\right)\right)$$

$$= 116.5\,\text{kW}$$

Likewise,

$$\dot{E}_{\text{Air\_out}} = 99.4\,\text{kW}$$

For the water stream according to Equation (10.22), the thermodynamic status of system can be determined by:

$$h_3 = C_P \times (T_3 - T_0) = 4{,}230 \times (30 - 22) = 33.9\ \frac{\text{kJ}}{\text{K} \cdot \text{m}^3}$$

$$s_3 = C_P \times \ln\left(\frac{T_3}{T_0}\right) = 0.1\ \frac{\text{kJ}}{\text{K} \cdot \text{m}^3}$$

$$h_4 = C_P \times (T_4 - T_0) = 4{,}230 \times (350 - 22) = 232.6\ \frac{\text{kJ}}{\text{K} \cdot \text{m}^3}$$

$$s_4 = C_P \times \ln\left(\frac{T_4}{T_0}\right) = 0.7\ \frac{\text{kJ}}{\text{K} \cdot \text{m}^3}$$

Then exergy can be calculated as:

$$\dot{E}_{\text{Water\_in}} = \dot{m} \times (h_3 - T_0 \times s_3) = 0.38 \times (33.9 - 295 \times 0.1) = 1.7\,\text{kW}$$

$$\dot{E}_{\text{Water\_out}} = \dot{m} \times (h_3 - T_0 \times s_3) = 0.38 \times (232.6 - 295 \times 0.7) = 9.9\,\text{kW}$$

According to the obtained values, exergy components can be calculated as:

$$\dot{E}_{F,\,\text{HEX}} = \dot{E}_1 - \dot{E}_2 = 116.5 - 99.4 = 16.9\,\text{kW}$$

$$\dot{E}_{P,\,\text{HEX}} = \dot{E}_4 - \dot{E}_3 = 9.9 - 1.7 = 8.2\,\text{kW}$$

Whereas exergy destruction in this heat exchanger is:

$$\dot{E}_{D,\,\text{HEX}} = \dot{E}_{F,\,\text{HEX}} - \dot{E}_{P,\,\text{HEX}} = 16.9 - 8.2 = 8.7\,\text{kW}$$

Exergy efficiency of the heat exchangers also can be calculated by:

$$\varepsilon_{\text{Comp.}} = \frac{\dot{E}_{P,\,\text{Comp.}}}{\dot{E}_{F,\,\text{Comp.}}} = \frac{8.2}{16.9} = 48.5\%$$

## 10.10 EXERGY ANALYSIS OF THE OFFSHORE HC-CAES ENERGY STORAGE TECHNOLOGY

This part carried out exergy analysis for a sample HC-CAES power plant with the capacity of 1 MW. A power plant like the one depicted in Figure 10.2 is considered for the modeling. To avoid judgment about the real performance of the components, unavoidable working conditions are simulated [5,23]. In unavoidable cycles, all processes occurring in the *k*-th component are irreversible while the processes for the remaining components of the system are ideal.

Since here we are not dealing with a real physical plant, for determining temperature *T*, and pressure *P*, some additional assumptions regarding the efficiency and loss in the system must be considered which are shown in Table 10.5 [8].

Applying these assumptions, specific entropy, enthalpy, and exergy can be calculated at any point in the flow diagram. With the presented equations and considered assumptions, one can calculate the thermodynamic properties and mass flow rate for unavoidable and ideal cycles at different state points of the system.

In Table 10.6, the temperature, pressure, and mass flow rate for unavoidable and ideal conditions through thermodynamic modeling of system are compared. Referring to this table and according to Equations (10.23) and (10.24), it can be observed that the total exergy efficiency is about 84.2% and the total exergy destruction ratio

**TABLE 10.5**
**The Assumption of Parameters Considered for Calculation of the Sample System Status for Ideal and Unavoidable Conditions [8]**

| Component | Ideal Condition | Unavoidable Condition |
|---|---|---|
| Compressors | Compresion Ratio = 1.8 | Compresion Ratio = 1.8 |
| | $\eta = 100\%$ | $\eta = 98.5\%$ |
| Heat exchangers | Air pressure loss = 0 | Air pressure loss ≅ 3 kPa |
| | Water pressure loss = 0 | Water pressure loss ≅ 0 |
| | $\eta = 100\%$ | $\eta = 96.0\%$ |
| Transmission pipeline | Air temp. drop = 0 | Air temp. drop = 3 K |
| | $P_{\text{delivered}} = P_{\text{ambient}}$ | $P_{\text{delivered}} = P_{\text{ambient}}$ |
| Air accumulator | Air leak Ratio = 0 | Air leak Ratio = 2.5% |
| TRD | $P_{\text{inlet}} = P_{\text{ambient}}$ | $P_{\text{inlet}} = P_{\text{ambient}}$ |
| | Pressure loss = 0 | Pressure loss = 500 Pa |
| Hot and cold tanks | $P_{\text{outlet}} = P_{\text{ambient}}$ | $P_{\text{outlet}} = P_{\text{ambient}}$ |
| | Water tempe. drop = 0 K | Water temp. drop = 5 K |
| All other efficiencies | $\eta = 100\%$ | $95.0 < \eta < 98.5$ |

**TABLE 10.6**
**Thermodynamic Properties and Mass Flow Rates at Different States Points**

| | Unavoidable Condition | | | Ideal Condition | | |
|---|---|---|---|---|---|---|
| **State Point** | **T (K)** | **P (kPa)** | **MFR (kg/s)** | **T (K)** | **P (kPa)** | **MFR (kg/s)** |
| Inlet | 294.0 | 102.3 | 0.85 | 294.0 | 102.3 | 0.76 |
| CS1 | 371.2 | 188.1 | 0.85 | 369.5 | 189.3 | 0.76 |
| HX1 | 308.1 | 187.3 | 0.85 | 306.2 | 189.3 | 0.76 |
| CS2 | 370.9 | 345.3 | 0.85 | 369.5 | 350.1 | 0.76 |
| HX2 | 307.9 | 344.6 | 0.85 | 306.2 | 350.1 | 0.76 |
| CS3 | 372.6 | 636.3 | 0.85 | 369.5 | 647.7 | 0.76 |
| HX3 | 310.3 | 635.6 | 0.85 | 306.2 | 647.7 | 0.76 |
| ACC | 292.6 | 550.0 | 0.85 | 306.2 | 550.0 | 0.76 |
| HX4 | 469.0 | 543.0 | 0.85 | 475.6 | 550.0 | 0.76 |
| TUR | 279.0 | 102.3 | 0.85 | 279.0 | 102.3 | 0.76 |

is about 15.8% (due to the technical limitation, it cannot be less than this value). Other work [37,38] showed that the efficiency of typical HC-CAES plants under real conditions is about 50%–55%. This gap between real and unavoidable working conditions shows that this technology has a great potential for technology improvement.

## 10.11 CONCLUSION

It is now clear that securing a sustainable future and sustainable development is not possible without access to sustainable and clean energy. The energy industry is the world's largest industry by far and energy services have a profound effect on productivity, health, food, and water security. To cover the increasing demand for energy, the development of renewable energy is inevitable. Most of the renewable energy technologies, however, have intermittent natures. Energy storage is one of the cost-effective and efficient options to rectify this deficiency. This chapter has described the concept of the HC-CAES energy storage technology. It was shown that the configuration of an HC-CAES depends on many factors, including air accumulator type, depth of air accumulators, and profile of power supply and demand. If the HC-CAES is an offshore plant, other parameters, such as the seabed setting and installation distance of the air accumulators from land, also will be important in the design of the plant.

Invoking theoretical data from a sample HC-CAES plant, exergy analysis was applied to investigate the efficiency of this technology. This analysis revealed detailed information regarding the plant performance, identifying the primary sources of energy loss and characterizing the major influences of this brand-new technology on operational efficiency.

Moreover, it was shown that, by applying a proper thermal energy storage unit, the total performance of the understudied HC-CAES system was about 85% under unavoidable conditions. Other studies with real data reported efficiencies of about 50%–55% [37]. This large difference between real and unavoidable operating conditions showed that this technology has great potential for efficiency improvements in future. Finally, this chapter described the outstanding challenges associated with this technology.

## REFERENCES


1. S. K. Mangla, S. Luthra, S. Jakhar, S. Gandhi, K. Muduli and A. Kumar, A step to clean energy - Sustainability in energy system management in an emerging economy context, *Journal of Cleaner Production*, vol. 242, pp. 1–11, 2020.
2. M. Child, O. Koskinen, L. Linnanen and C. Breyer, Sustainability guardrails for energy scenarios of the global energy transition, *Renewable and Sustainable Energy Reviews*, vol. 91, pp. 321–334, 2018.
3. M. Hosenuzzamana, N. A. Rahimab, J. Selvaraja, M. Hasanuzzamana, A. B. M. A. Maleka and A. Nahara, Global prospects, progress, policies, and environmental impact of solar photovoltaic power generation, *Renewable and Sustainable Energy Reviews*, vol. 41, pp. 284–297, 2015.
4. IRENA, Renewable capacity statistics 2019, International Renewable Energy Agency, 2019.
5. R. Carriveau, M. Ebrahimi, D. S.-K. Ting and A. McGillis, Transient thermodynamic modeling of an underwater compressed air energy storage plant: Conventional versus advanced exergy analysis, *Sustainable Energy Technologies and Assessments*, vol. 31, pp. 146–154, 2019.
6. F. Ferrari, G. Besio, F. Cassola and A. Mazzino, Optimized wind and wave energy resource assessment and offshore exploitability in the Mediterranean Sea, *Energy*, vol. 190, pp. 1–15, 2020.
7. C. R. Matos, J. F. Carneiro and P. P. Silva, Overview of large-scale underground energy storage technologies for integration of renewable energies and criteria for reservoir identification, *Journal of Energy Storage*, vol. 21, pp. 241–258, 2019.
8. M. Ebrahimi, R. Carriveau, D. S.-K. Ting and A. McGillis, Conventional and advanced exergy analysis of a grid connected underwater compressed air energy storage facility, *Applied Energy*, vol. 242, pp. 1198–1208, 2019.
9. A. J. Pimm, S. D. Garvey and M. de Jong, Design and testing of Energy Bags for underwater compressed air energy storage, *Energy*, vol. 66, pp. 496–508, 2014.
10. O. Laing and J. Laing, Energy storage for off peak electricity. USA Patent US4873828, 31 March 1989.
11. https://hydrostor.ca/, 10-04-2018. [Online].
12. A. R. Vasel-Be-Hagh, R. Carriveau and D. S.-K. Ting, Flow over submerged energy storage balloons in closely and widely spaced floral configurations, *Ocean Engineering*, vol. 95, pp. 59–77, 2015.
13. A. Pimm and S. D. Garvey, Underwater compressed air energy storage, *Storing Energy*, vol. 2016, pp. 135–154, 2016.
14. T. Davenne, S. D. Garvey, B. Cardenas and M. C. Simpson, The cold store for a pumped thermal energy storage system, *Journal of Energy Storage*, vol. 14, 295–310, 2017.
15. B. Elmegaard and B. Wiebke, *Proceedings of 24th International Conference on Efficiency of Compressed Air Energy Storage*, Novi Sad, Serbia, 2011.
16. D. Fiaschi, G. Manfrida, R. Secchi and D. Tempesti, A versatile system for offshore energy conversion including diversified storage, *Energy*, vol. 48, no. 1, pp. 566–576, 2012.
17. K. Yang, Y. Zhang, X. Li and J. Xu, Theoretical evaluation on the impact of heat exchanger in advanced adiabatic compressed air energy storage system, *Energy Conversion and Management*, vol. 86, pp. 1031–1044, 2014.

18. A. Pimm and S. D. Garvey, Potential locations for underwater compressed air energy storage in Europe and North America. In *Offshore Energy and Storage Symposium*, Edinburgh, Scotland, 2015.
19. Z. Wang, W. Xiong, H. Wang and Z. Wang, A review on underwater compressed air energy storage, *Energy Storage Science and Technology*, vol. 4, no. 6, pp. 585–598, 2015.
20. B. Zanuttigh, E. Angelelli, A. Kortenhaus, K. Koca, Y. Krontira and P. Koundouri, A methodology for multi-criteria design of multi-use offshore platforms for marine renewable energy harvesting, *Renewable Energy*, vol. 85, pp. 1271–1289, 2016.
21. Z. Wang, D. S.-K. Ting, R. Carriveau, W. Xiong and Z. Wang, Design and thermodynamic analysis of a multi-level underwater compressed air energy storage system, *Journal of Energy Storage*, vol. 5, pp. 203–211, 2016.
22. S. Briola, P. Di Marco, R. Gabbrielli and J. Riccardi, Sensitivity analysis for the energy performance assessment of hybrid compressed air energy storage systems, *Applied Energy*, vol. 206, pp. 1552–1563, 2017.
23. M. Ebrahimi, R. Carriveau, D. S.-K. Ting, A. McGillas and D. Young, Transient thermodynamic assessment of the world's first grid connected UWCAES facility by exergy analysis. In *IEEE*, Brest, France, 2019.
24. G. Tsatsaronis, Strengths and limitations of exergy analysis. In: Bejan, A., Mamut, E. (eds.) *Thermodynamic Optimization of Complex Energy Systems*. Springer, Dordrecht, pp. 93–100, 1999.
25. S. Sasaki, K. Hayashi and A. Tomiyama, Effects of liquid height on gas holdup in air–water bubble column, *Experimental Thermal and Fluid Science*, vol. 72, pp. 67–74, 2016.
26. P. Rollbusch, M. Becker, M. Ludwig, A. Bieberle, M. Grünewald, U. Hampel and R. Franke, Experimental investigation of the influence of column scale, gas density and liquid properties on gas holdup in bubble columns, *International Journal of Multiphase Flow*, vol. 75, pp. 88–106, 2015.
27. M. Ebrahimi, A. McGillis, C. Lewis, D. S. Ting and R. Carriveau, A CFD based empirical model for assessing gas holdup in bubble columns, *The Canadian Journal of Chemical Engineering*, vol. 97, pp. 1643–1654, 2018.
28. G. Besagni, A. Di Pasquali, L. Gallazzini, E. Gottardi, L. P. M. Colombo and F. Inzoli, The effect of aspect ratio in counter-current gas-liquid bubble columns: Experimental results and gas holdup correlations, *International Journal of Multiphase Flow*, vol. 94, pp. 53–78, 2017.
29. M. Ebrahimi, A. McGillis, C. Lewis, D. S. Ting and R. Carriveau, A CFD based empirical model for assessing gas holdup in bubble columns, *The Canadian Journal of Chemical Engineering*, vol. 97, no. S1, pp. 1643–1654, 2019.
30. P. Grove, *Compressed Air Energy Storage Symposium Proceedings*, United States Department of Energy, California, 1978.
31. Energy storage monitor: Latest trends in energy storage, World Energy Council, 2019.
32. C. M. Oldenburg and L. Pan, Porous media compressed-air energy storage (PM-CAES): Theory and simulation of the coupled wellbore–reservoir system, *Transport in Porous Media*, vol. 97, no. 2, pp. 201–221, 2013.
33. B. Ye, Z. Cheng, W. Ye and Y. Peng, An analytical solution for analyzing the sealing-efficiency of compressed air energy storage caverns, *KSCE Journal of Civil Engineering*, vol. 23, pp. 2025–2035, 2019.
34. S. Heusermann, O. Rolfs and U. Schmidt, Nonlinear finite-element analysis of solution mined storage caverns in rock salt using the LUBBY2 constitutive model, *Computers and Structures*, vol. 81, pp. 629–638, 2003.
35. M. Aubertin, D. E. Gill and B. Ladanyi, An internal variable model for the creep of rocksalt,*Rock Mechanics and Rock Engineering*, vol. 24, pp. 81–97, 1991.

36. L.-J. Ma, X.-Y. Liu, Q. Fang, H.-F. Xu, H.-M. Xia, E.-B. Li, S.-G. Yang and W.-P. Li, A New elasto-viscoplastic damage model combined with the generalized hoek–brown failure criterion for bedded rock salt and its application, *Rock Mechanics and Rock Engineering Volume*, vol. 46, pp. 53–66, 2013.
37. Z. Wang, W. Xiong, D. S.-K. Ting, R. Carriveau and Z. Wang, Conventional and advanced exergy analyses of an underwater compressed air energy storage system, *Applied Energy*, vol. 180, pp. 810–822, 2016.
38. B. C. Cheung, R. Carriveau and D. S.-K. Ting, Parameters affecting scalable underwater compressed air energy storage, *Applied Energy*, vol. 134, pp. 239–247, 2014.
39. Energy storage monitor: Latest trends in energy storage, World Energy Council, 2019.

# 11 Bioconstruction and Harmonic Complexity of Biomimicry Organisms

*Gabriel Barbeta*
Universitat de Girona

*Yomna Abdallah*
Universitat Internacional de Catalunya

## CONTENTS

## 11.1 THE BASICS OF SUSTAINABLE HABITAT

The construction of our habitat has a huge environmental impact; it almost occupies 3/4 of the energy and resources. We live in energy-inefficient buildings, which account for 45% of the end consumption of energy and 36% of $CO_2$ emissions. The building industry is one of the top consumers of resources, including 40% of all water and 70% of lumber materials (dates World Watch Institute).

As well, the estimate that says that 60% of the buildings needed to house the world population in 2050 have not even been built yet, leads us to consider alternatives technologies. Thus, it is indisputable that we have to take the path of sustainability with ultimate ecology and appropriate technologies. At a time where the modern movement, the post-modern, the high-tech, seemed to mark the future, the line drawn by the environmental balance of our planet redirects the current architecture in a clear way. Thus, architectonic solutions can no longer be internationalized.

Thus in 1993, the International Union of Architects drew up the *International Declaration for Sustainable Architecture* that was ratified in 1996 at the summit meeting in Barcelona. This carefully highlighted the way that prestigious architects design. The analysis of the impact of industrial activity, the construction, and the operation of cities and buildings have matured in such a way that we have realized that it is necessary to clear the emissions of elements such as $CO_2$, CFCs (Chlorofluorocarbons), or VOCs (Volatile Organic Compounds) from air and water, since the continuity of our species is questioned.

The sociocultural, bioclimatic, energetic, and productive parameters of the various parts of the world require approaches that go through the recovery and improvement of the traditional technologies until they are brought to the level of the current safety and durability requirements. It is necessary to think about the buildings of the 21st century in a global way, from the environmental point of view, and locally as a concrete form of action that is adequate to the place and the beings that inhabit them. This can also be derived from a qualitative leap of consciousness from all the technicians and professionals who work in the construction environment. Such an inner change will crystallize in the form of a habitat that provides us with evolution in the process of being humans in real harmony and integration with our time (now) and place (here).

One of the basic needs of the human being is to have a decent home to protect themselves from their surroundings, as well, it is one of the most widely recognized rights. Hence, it is shocking to know that 1.5 billion people do not have access to health services, 2.3 billion to sanitary services, and 1.3 billion do not have access to drinking water. It is essential that the habitat is dignified and healthy. Take into account the cultural, economic, and social conditions, as well as the technological comfort and health, such as ventilation, temperature, humidity, sunlight, lighting, space, noise, drinking water, and wastewater evacuation facilities, which prevent contamination of groundwater levels, and a location that avoids proximity to landfills, cemeteries, swamps, rivers, and streams. Any omission of these factors becomes a point that transmits diseases or affects the proliferation of the neighborhood.

## 11.2 A SUSTAINABLE EQUITABLE DEVELOPMENT FOR THE SOUTH

"We do not believe in the possibility of obtaining a 'sustainable' development that, while respecting the great ecological balances, is done at the expense of the exclusion of a large part of Humanity" (Plataforma São Paulo, for the alliance for a responsible and supportive World. Fundación Charles Leopold Meyer) [1]. As Borges and Díez (1997) questioned the emphasis, the thesis of sustainable development does on intergenerational equity, to the detriment of intragenerational equity. For the resident population in the countries of the south [2], it is very difficult to think about satisfying the needs of future generations when the basic health, education, and social welfare requirements of the present are not covered. According to Professor Asmal, the South African Minister of Water Resources, during the first Meeting of the World Water Society, Stockholm, August 9, 1996, "Growth processes that benefit the most prosperous minorities and maintain or increase disparities between countries, it cannot be considered development. It is exploitation" and he adds

> It would be unfair that emerging economies, which already subsidized enormously, during the colonial era to countries that claim to be 'First World', should now continue to subsidize the entire world by defraying most of the costs of environmental protection [3]

The latter would be caused by inadequate solutions such as limitations on the industrialization of the south, economic measures such as "debt by nature," exports from the north of toxic substances, covered 90% as raw or recyclable materials, the precarious disassembling of merchants, or the transfer of technologies outside the social interests of the recipient countries.

Obviously, the way of consumption of the north cannot be generalized to humanity as a whole, it is necessary to apply the principle of "the polluter pays," thus sanctioning these high levels of consumption, so that there is greater equity. This is why we must generally avoid the use of western technologies in the south. Ninety percent of the construction materials in Africa are imported from the northern countries, such as the cement block or the sheets of galvanized iron. Suddenly, the dignity and sobriety of those traditional houses of mud, logs, and stones have been blurred in a pile of stacked junk wrapped in garbage of synthetic materials, in anti-climatic habitats, and zero habitability. The entire process of technological influences goes against its own cultural roots and economic means and any possibility of self-progress. Missing along the way, a whole technological heritage and popular knowledge have been transmitted from generation to generation. (Figure 11.1 shows an example that follows these ideas.)

It is necessary to develop appropriate technologies [4], in a way of endogenous, environmentally sound, and a fair, balanced international trade [5] and nondiscriminatory that achieves a level of production, although not optimal, but is at least within the capacity of sustainable load. These technologies will arise from local actions designed from globalism based on the following basic criteria:

- Preferably use local materials.
- Minimize technological complexity and industry, to facilitate maintenance and local development.

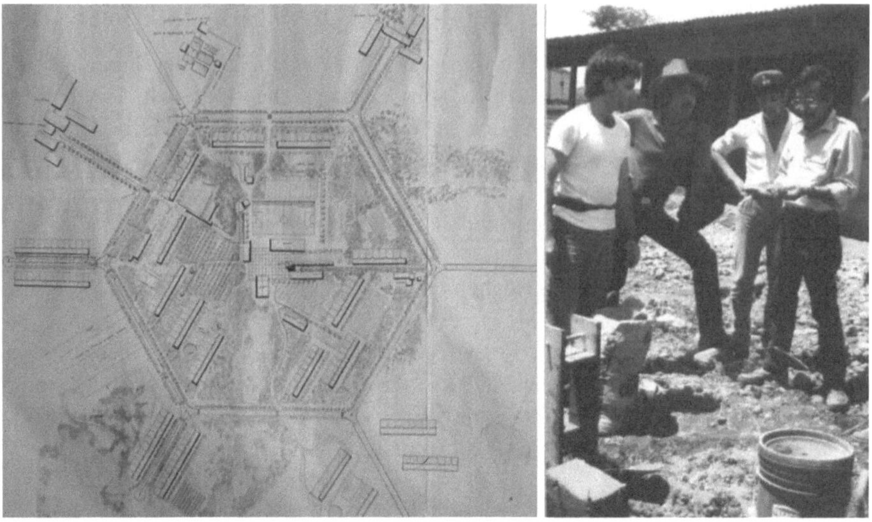

**FIGURE 11.1** G,Barbeta long experience in cooperation, architectural design of Ciudad Segundo Montes, El Salvador 1989–1990. (Photos: Gabriel Barbeta.)

- Have the minimum contribution of imported materials, to save external equipment and technical knowledge, and promote local occupation and training.
- Rejoin traditional knowledge, creativity, and current knowledge to make the solutions appropriate to each situation and appropriated by indigenous cultures.

Architect Antonio Toca, along with other Latin American architects, such as Eladio Dieste with his church of San Pedro in Durazno, or Rodrigo Uribe with his Cali building in Colombia, state that: "...The city is destroyed by modern architecture... By sophisticated technology to artificial control of, lighting, ventilation, and air conditioning; nothing is related to the regions where they are located." "Preserving one's identity is a real struggle." "not just are habitable but also achieved their identification with the characteristics of the people and culture of the place."

## 11.3 THE SUSTAINABILITY OF ARCHITECTURE DEVELOPMENT

In this way, the architects must rethink the way of conceiving the buildings nowadays, intensive in consumption of materials and energy, without being covered in banal and unsustainable esthetic parameters. The law of human attraction says that even back in hunter-gatherer societies, "as villages grow in number of inhabitants, settlements become denser." This discovery has been summarized by the popular scientist Tim de Chant as follows: "Each additional person requires less land than the one before him," means that "a group becomes more efficient in extracting resources from the earth every time the population doubles..." Despite all this, for decades the urban area has been growing at a faster rate than the population in all regions of the world. It requires actions that tend to mitigate the impact of steel and concrete to be more positive globally, applied in the northern industrialized countries especially. As we have seen, construction has an enormous specific weight in human development; therefore, it is indisputable that it must also take the path of sustainability and appropriate technologies. The alternatives are unique and cannot be internationalized; the circumstances of the various places will require very variable approaches from the recovery of local tradition to the incorporation of technologies that contribute to achieving the ideal environmental parameters. Architects and technicians can contribute to the incorporation and dissemination of a greener construction mentality, regarding the environment, international solidarity, renouncing a part of our well-being, and acquiring a more global vision of the problems.

## 11.4 HABITAT'S ENVIRONMENTAL ALTERATIONS

The problem acquires a much more extensive dimension, when other more subtle factors, hitherto omitted, come into the play, leading us to the "sick building syndrome" [6–10], describing the buildings that make their inhabitants sick, when this syndrome reached more than 20% of their occupants. Normally this syndrome occurs in old buildings that have been rehabilitated, according to Enric Aulí Mellado, Director of the Servicio de Sanidad Ambiental de la Generalitat de Catalunya, Barcelona.

The WHO warns that stale air is the main cause of the syndrome, apart from noise pollution. According to NIOSH (National Institute of Security Occupational and Health, U.S.), 62% of buildings lack fresh air, 33% do not have any ventilation, 61% inefficient air filters, of these 18% poorly installed, 58% show dirty ventilation facilities, with the alarming figure of which 22% have contaminated ducts, in conclusion only 25% of the buildings analyzed would be well ventilated.

The World Watch Institute estimates that 30% of new or rehabilitated buildings suffer from the sick building syndrome. Symptoms of such syndrome are dizziness, headaches, pain and irritation in throat and eyes, metallic taste in the mouth, nausea, asthma, fatigue, disorientation, and, in some cases, lipothymy and seizures. This was the case in 1990 affecting 93 of 150 employees of the Commerce Building in Salem, Oregon, or in a similar case in 1976 at an American Legion Convention in Philadelphia with 182 affected by fever, psychic disorders, cough, chest pain, or head, of which 28 were fatalities. Since without treatment, it results in extensive neuropathy, respiratory failure, and death. Caused by a bacterium resident in the air conditioning facilities, it was thereafter referred to as *Legionella pneumophila*, due to the event. According to Professor Jan A. J. Stolwijk, Head of the Epidemiology Department of the University of Yale, a very similar disease is the humidifier fever produced by biogenetic particles, from the textile industry of paper or wood, which nourishes all this usual series of microorganisms (degradation fungi, such as *aspergillus* detected in so many hospital centers, mites and arachnids, viruses, algae, pollen, or insect droppings), causing a kind of flu with work allergies. In the domestic sphere and coinciding with the drop in pollution levels in the street, the introduction of new materials and new activities inside the homes, as well as the new construction techniques and especially the thermal insulation of the building's exterior, have caused a significant increase in indoor pollution. The first serious symptoms that attracted attention came from asbestos disorders in some schools in the United States, the same occurred at Berlaymont EEC headquarters, and the consumer complaints from commercial products that give off formaldehyde, causing respiratory and eye disorders. Currently, these health impacts are much more worrying than those caused by external pollutants.[11]. Among these risk factors, which obviously act synergistically, we must take into account:

**Radioactivity emissions from building materials**. These are stoneware or tiles with thorium salts of metallic tones, granite, alum boards, fly ash, calcium silicate slags of concrete, the phosphor used in drywall and block concrete, phosphate rocks, lightning rods, sanitary facilities, and radon gas of natural origin from the subsoil; the radioactive decay of thorium and radium; and emitting particles $\alpha$, accumulating in basements and ground floors being heavier than air. The synergistic effect of tobacco smoke's compounds (nicotine, nitrosamines, and polycyclic aromatic hydrocarbons) result in the formation of lung cancer or as a mutagenic agent. All these elements are classified as a human carcinogenic element class A, since with only 200 $Bq/m^3$ the risk factor is doubled (R. Caponetto, Dep. Architecture, Catania Univ.) [12]. Thus, this level is adopted by most countries as the maximum limit for new buildings, although in the United States, it is reduced to 150 $Bq/m^3$. The factors that determine the accumulation of these elements depend on the geological structure, the presence of faults, the weather, and the construction materials used (release agents, pozzolans, granites, and lavas, and oils used in concrete "Variation in emission rates highlights need for product-specific data," Healthy Materials, 3, CHH3 SCHL, 1995), and the effect will be directly conditioned by the number of daily air renewals and the type of occupants' lives.

**Emanations of gases and toxic particles** from building materials, furniture, and outside air. These emanations are as asbestos, chrysotile or rock wool, and fiberglass that have already been exposed in their manufacturing process, application, shelf life, or recycling. Also, the emissions from industrial processes such as incineration and dumping of PVC (polyvinyl chloride) can lead to carcinogenic microfibres that affect the respiratory tract (Spanish Senate motion 19/12/95). The pesticides and solvents, involving the formation and emission of toxic, persistent, and bioaccumulative organochlorine substances into the environment (the vapors of vinyl chloride, hexachlorobenzene), inhibit development and affect metabolism. The PCBs-polychlorinated biphenyl, furans, dioxins (WHO classifies dioxin 2, 3, 7, 8-TCDD as a carcinogenic substance in humans) [13], phthalates, and amines that are plasticizers of PVC cause head disorders; nausea; irritations; damage to the nervous, immune, and reproductive systems, liver and kidneys, and are being carcinogenic. Another problems arising from PVC are: high water vapor diffusion resistance; high energy consumption; emission of 19.5 g. of mercury per Tn of chlorine produced; emission of organochlorines when using graphite electrodes; emission, generation, spills, and leaks of toxic substances such as dioxins, dichloroethane, manufacturing charges such as asbestos, stabilizers, and lubricants such as lead and plasticizers such as phthalates and heavy metals such as pigments (antimony, cadmium, lead, chromium, zinc), and blowing agents such as CFC (greenhouse effect); it can only be recycled by 20%; in case of fire, it rapidly releases large amounts of dioxins, cadmium, and hydrochloric acid affecting the respiratory tract, which is why more than 100 European municipalities have banned it in public buildings. Carpets, insulators, and amalgamated wood conglomerate with urea formaldehyde resins and naphthalene in laminates (eye and respiratory tract irritation, asthma, headache, dizziness, nausea, fatigue, allergies, nasal-pharyngeal, and oral cavity cancer). In addition the internal and incomplete combustion processes from heating, cooking, drying ($CO$, $CO_2$, $SO_3$, $NO_3$, and polycyclic hydrocarbons emitted by wood-burning stoves) can lead to nervous and respiratory system disorders. The phenols present in detergents, rubber, gasoline, plastics, tobacco smoke, and synthetic fibers carry the risk of skin and eye irritation, headache, loss of appetite, drowsiness, leukemia, and blood diseases; paints, dyes, lacquers, adhesives, and varnishes (with possible trichloroethylene causing liver cancer, liver, kidney or spleen problems, heart and nerve disorders).

All the relation of toxic substances as aliphatic and aromatic polycyclic hydrocarbons of the law 20/1986 should be avoided. They are attributed a synergistic effect or carcinogenic additive They are present in air fresheners, aerosols, perfumes, soaps, and dry cleaners.

In the last decades, more than 100,000 toxic chemical compounds have been released to water, air, and used without knowing how long they would remain inside us or what consequences they could have on our health. These are POPs (persistent organic pollutants), designed to resist the passage of time, that accumulate in our body for decades, as indicated by Miquel Porta, Professor of Medicine and Preventive Health. These are associated with a wide variety of effects: immunosuppression, cytotoxicity, tumor promotion, endocrine disruption, neurotoxicity, epigenetic, neurodegenerative effects, infertility, and cancer, and there are people whose body reacts negatively much more exaggeratedly and eventually develops chemical sensitivity multiple (SQM), a disease characterized by absolute intolerance against odors or chemicals.

**Electromagnetic fields**: Harmful electromagnetic fields can come from the power grid [14, 15] (High-voltage network, transformers, electrified train tracks) and the interiors (appliances, TV and computer screens, electrical ring installation, undersized conductors, lack or inefficiency of the ground, radio alarm clocks). The threshold of harmful electric field would be [16] 6 V/m and dangerous from 6 to 14 V/m.

For electromagnetic field, it would be 65 nT (nanoteslas) as a harmful threshold and 65/250 nT hazardous exposure zone. A carcinogenic effect (risk factor × 1.7), producer of chronic myeloid leukemia (childhood risk factor at 300 ηT × 3.8), synergistic action in the production of brain tumors (×2.15), has been demonstrated [17–19]. According to Professor Karl E. Lotz, of the Hochschule für Bauwesen in Biberach an der Riss, the minimum distances to a house should be 250 m, 450 m if it is to the south, and 112 m if it is to the north. See also the writings of José Luis Bardasano, about bioelectromagnetism and pineal gland/non-ionizing radiation in Biology and Medicine, published by the Instituto de Bioelectromagnetism Alonso de Santa Cruz, Universidad de Alcalá de Henares. Michael Paul Nelson, at Oregon State University, found the parasitic relationship between EMF and white blood cells, which absorb magnetite from the environment and can give false positives when trying to relate it to leukemia, as published on *Nature*, 1995. However, Russel Reiter, Professor of Neuroendocrinology, at the University of Texas at San Antonio, establishes the regulation of the pineal gland based on the principle that the hormone produced, called melatonin, has an anti-oncogenic effect and therefore protective against cancer but is suppressed by EMF. A study by Dennis Henshaw, a researcher at the University of Bristol, published in the *International Journal of Biological Radiation* states that electromagnetic forces release the existing radon in the walls of buildings, an element that is assimilated by the body at serious risk to those people with unchained cancers. Nancy Wertheimer and Ed Leeper (University of Colorado), studying cases of childhood leukemia in Denver, discovered a relationship between the observed cases and the proximity to high-voltage power lines or large transformers (first epidemiological study on the subject). Savitz (University of North Carolina) conducted studies for seven years of cancer cases in children under 15 years taking into account the electromagnetic fields produced by appliances.

Their results reported that children exposed to elevated fields run from 1.3 to 1.6 more risks of getting cancer than those not exposed; and in the case of leukemia, the risks are doubled. Michael Repacholi, WHO Electromagnetic Fields Project Director, 1998: Research aimed at establishing the relationship between the onset of cancer and diseases of the central nervous system and cell phones, power lines, and radar. The Department of Energy of Orlando, USA, verified (in 1992) the synergistic effect between noise, chemical contaminants, and electromagnetic radiation with serious consequences in the early stages of pregnancy.

In this line, to avoid the effects of reverberation of the fields in structures and metal meshes, at the University of Girona, the author of these pages works on a line of research for the execution of concrete reinforced with plastic and organic fibers (bamboo, hemp, chestnut, rice husk) [20].

**Alterations of the natural terrestrial radiation field** [21]. These are associated with sleep disorders (reduction of serotonin), headaches, synergistic effect, or carcinogenic additive (reduction of thymus activity, hyperactivity of the adrenal gland, and alteration in brain waves, electrolyte alterations in the heart and in the polarity of the cells). Professr Karl E. Lotz (Hochschule für Bauwesen in Biberach an der Riss), in his book *Ecobioconstruction*, states that natural terrestrial radiation originates from the decay of radioactive particles from the earth's crust, giving $\alpha$, $\beta$, and $\gamma$ rays that in turn destroy beryllium nuclei, emitting thermal neutron beams towards the surface of 25 meV, being diverted and concentrated by the movement of groundwater (formation of mV currents/cells work at 60–90 mV) or geological faults. These high-frequency bands cause disturbing effects on the biological regulation. Geo-pathogenic zones cause galvanotaxis, marrow necrosis, bifurcations, galas, and helical growth in trees. According to Dr. Ernst Hartmann (Heidelberg University, Heidelberg), the existence of the north-south geomagnetic network of $2 \times 2.5$ m verifies the existence of medical pathologies associated with its alteration due to geological faults and groundwater. A research of the Animal Physiology Unit of the Faculty of Biology conducted by Valencia, Dr. Javier Núñez and Manuel Núñez demonstrated the different immunological response of mice to geomagnetic variations.

According to J. Kulczycki in *Basis of Electromagnetic Hygiene*, the alterations of terrestrial magnetism in nano tesla per meter (nT/m) are established. The harmful threshold would be 2,000 nT/m and a disturbed area of 3,000/10,000 nT/m.

The nuclear physicist Nobel Prize Ernest Rutherford, and the astrophysicist Mc Lennan, Goeckel and Kohluster, demonstrated the existence of ultra-penetrating cosmic waves. Professor Piccardi (University of Florence) showed that the water molecules were resonators of these waves, modifying the EMF, as demonstrated by G. Lakhorsky.

**The ionization of the air produced by the described radiations** and by the release of electrostatic charges from synthetic materials such as carpets, plastics, metals, fibers, and synthetic textiles of PVC, polyamide, nylon, and aramid. It can also come from large metal masses such as cars; hence, it is important moving the garage away from rest areas, apart from all the toxic volatile compounds that come off the hot engine (cadmium, lead, hydrocarbons...) that can penetrate the house.

**Noise pollution**. The 60 dB marks the maximum threshold to avoid possible damage. It affects irritating the central and vegetative nervous system causing permanent stress, an increase in respiratory and heart rate, decrease in peripheral circulation and gastric motility, and an increase in blood pressure (Barcelona is the second European capital with more noise pollution).

## 11.5 THE WASTE ISSUE

There is a need to initiate actions prior to recycling, which are reuse and reduction. This implies that at the time of choosing a product, the reuse possibilities that it has; the recovery of the container; the volume of waste it generates; the pollution caused in its production, use, and discharge; and its degree of biodegradability should be assessed. This gives rise to the following basic design principles:

- Incorporate in the project phase, adequate functional and constructive criteria, to significantly minimize waste originating from the construction.
- Re-incorporate the excavation waste in the same work to reduce them to the minimum. The Spanish regulations for concrete include the use of up to 20% recycled aggregate in concrete (according to UdG Germans Xirgu research line for manufacturing recycled concrete).
- Use materials and construction solutions that cause little waste, both in the construction phase and in the life of the building.
- Those products that contain hazardous materials should not be used, since they will be considered special industrial waste. These are materials such as welding products; putties for joints based on bitumen and asbestos; anti-oxidants, termiticides and creosotes; paints and varnishes; dirty accessories and materials without drying used when painting; various chemicals such as anti-corrosive, drying, fungicides, insecticides, solvents, diluents, acids, bases, abrasives, detergents; asbestos; asbestos cement; and muds originating in perforations as well as metals such as zinc, lead, tin, chromium, nickel; laminated-glued or pressure-treated wood; and hydrocarbon products obtained from soot, hydrocarbons.

At the project stage of the building, all the recycling and reuse possibilities must be programmed when entering the demolition stage.

- These have a certain degree of reuse: beams, trusses, precast concrete, doors, windows, cladding, tiles, skylights, veneers, boards, sandwich plates, prefabricated slabs, screens, movable partitions and railings.

The degrees of recyclability we can count on are as follows:

- The stone demolition debris, which represents 95% of the total weight of the building, is recyclable as a sub-base material for road works and the granules to make new concrete. In the case of reinforced concrete, it is necessary to separate the reinforcement, which is a complex and expensive operation. The presence of polymers in the mass of fiberglass waste can cause durability problems when recycled into new pastes or concrete, recycling of glass fiber or asbestos and cement compounds is not recommended (Les composites à matrice cimentaire, CSTB, Paris, 1994). The Danish experience of annealing waste from the demolition of construction walls, which allows easy separation of lime or cement mortar, is remarkable. These are usable again, and the bricks are recycled as granules. The granulate from ceramic waste, if you want to use it for concrete, cannot exceed 10% and the characteristic resistance is less than 125 kg/cm$^2$. Besides, they cannot be used in aggressive environments, and they cannot contain plaster of the coatings, which is a serious problem for recycling. In Viladecans (Barcelona), there is a manufacturing plant for concrete blocks with recycled aggregates, separating the plaster by an expensive screening and washing process. The granulate from cement waste, if to be used for concrete, cannot exceed a characteristic resistance of 250 kg/cm$^2$. Apart from that, they can only be used in environments I and II.
- Wood can be made of beams and trusses, wood/cement panels when shredded and mineralized, or laminated to make parquet flooring, as long as it is not pressurized with toxic products, and as a last option for energy recovery by burning it.
- Neoprene, rubber, asphalt, and bituminous materials can be re-incorporated in mass to make road pavements.
- Metals have always been recycled, since their environmental impact and manufacturing cost and very high, their oxidation does not prevent recycling. The most common are: lead, copper, iron, steel, cast iron, zinc, aluminum, and alloys.
- Thermoplastics are the most easily recyclable: polystyrene (allows granules, cassettes, or panels), polyethylene, and polypropylene (loads low-quality wastes due to thermal or light degradation, this does not allow recycling for the same use of pipe). Although PVC is 20% recyclable to make pipes, floors, and cables, it has an environmental impact so high that it must be separated, as in the case of thermosetting it is very difficult to recycle plastics with polyester resins, phenolic, or epoxy, polycarbonate, and polyurethanes.
- Plaster in plasterboard or cardboard, while gypsum is recyclable as it can be re-incorporated as raw material, but in the case of coating adhered to other surfaces, it is complex and expensive.

- Insulations based on mineral fibers (slag wool and rock wool), as more homogenous, are more easily recyclable by fusion, while it is not advised in case of sheets that have aluminum bonded or other materials such as cardboard, as they cause health problems due to the fibers and vapors of urea-formaldehyde resins that evaporate when melted.
- The glass is recyclable as raw material, although it can be used for other uses such as street furniture in Catalonia. It is significantly difficult to recycle the mineral residues that give the color to the glass, the glass ceramic, and the composite glass.
- The slate must always be recycled, since its extraction has a high environmental impact by representing that 90% of the material excavated in the quarry is waste. The possible applications, some of them still under study, are: as artificial blackboards agglomerated with resins or cement; as fillers in bituminous paints and coatings; mixed with slag from blast furnaces to make mineral wool; manufacturing of cellular concrete parts; in pre-kneaded concrete or as light granules.

## 11.6 CONSCIOUS ECO-ARCHITECTURE

> Originality is to return to the origins, so original is that which with new means returns to the simplicity of the first solutions.
>
> *Antoni Gaudí*

Sensitivity, sustainability, rationality, and esteem throughout nature serve to forge more conscious projects with the current human landscape, as shown on Figure 11.2. This can be called **conscious eco-architecture**. In this new human cycle that begins, a new architecture is born where the architect becomes only an **integrating channel** of all the environmental, technical, functional, and psychological conditions of the occupant, and stems from calm, serenity, observation, esteem, and understanding, leaving aside his most egocentric part, forgetting abstract genius and neurotic mental wanderings. When we are able to open our heart, and the heart is followed by the mind, we realize that the current planetary emergency leads us to a new consciousness, something beyond ideologies and thoughts. An architecture based on obtaining low-cost living spaces in an equitable and conscious manner; an architecture that does not neglect the millions of people who do not have housing; an architecture that makes us see that 15% of the world's population live with 80% of the world's resources and that this forces us to minimize energy consumption, in the use of the building, in the execution process, and in the consumption of raw materials. An architecture that tells us that we are obliged to recover and improve indigenous technologies in search of local sustainable solutions; an architecture that proposes the principle of thinking "globally and acting locally" to achieve true sustainability. An architecture, in short that proposes us to apply the technical principles of bioclimatic, of bioconstruction, of geo-biology, of permaculture, without losing consciousness, in this task, of our condition of human beings. It also takes us to the integration of the landscape and

**FIGURE 11.2** G. Barbeta, energy rehabilitation, Sta. Cristina d'Aro. (Photo: V. de la Dehesa.)

the topography of the land and leads us to the design made from the sun and the light, so that it reaches all areas of the house.

Sun's rays, especially ultraviolet, purify the air, kill microbes, and regulate the excess humidity. At the same time, this natural light is antidepressant, it favors the segregation of serotonin, the good ocular functioning, and it nourishes us with prana. Likewise, it guides us to bioclimatic design for maximum adaptation to the local climate and vegetation. It is interesting to use geothermal energy and thermal inertia generated by burial of part of the house, greenhouses, trombe walls, interior courtyards, porches, eaves, landscaped roofs, buried ventilation pipes, and solar chimneys. Eco-architecture takes us to the consideration of the influence of electromagnetic and geomagnetic fields, which can help us reduce certain physical pathologies and allow us to heal with radiant walls with solar water preheating. This new constructive awareness raises the collection of rainwater and gray water (bath, kitchen, washbasins, and shower) for reuse. In the same way, it invites us to the use of renewable energies (solar panels, collectors, windmills).

Ecological construction makes us consider bioconstructive materials and technologies. It is necessary to look for appropriate technologies to our climate and to the native resources that involve little energy consumption, recyclable, have good performance of breathability and insulation, and do not contain toxicity of any kind.

The use of binders such as clay, plaster, lime, and natural cements should be sought. It is necessary to use insulation with crushed natural cork, wood fiber, hemp, or recycled paper; the wood must be recycled or from native trees; the covers must be vaulted or dome-shaped; and the armed ones can be made with native bamboo fibers; paints, natural, and without toxic solvents; and water pipes, with recyclable and healthier materials. The presence of metals and materials with high-energy consumption in its manufacture should be minimized. By doing a methodical study of the life cycle of the material, we deduce that the solution is to use stabilized earth, mud walls, concrete blocks, or compressed bricks with hydraulic press (CEB). At this time, the CEB-Bioterre compressed with hydraulic system can be found on the market and is one of the construction materials with the lowest energy consumption and environmental impact. Figures 11.3 and 11.4 show the use of Bioterre in housing construction in Vall Llemena, Girona.

**FIGURE 11.3** Construction details of walls. (Photos: Gabriel Barbeta.)

**FIGURE 11.4** Gabriel Barbeta, housing construction with Bioterre, Vall Llemena, Girona. (Photo: Gabriel Barbeta.)

## 11.7 THE EASE OF BUILDING WITH COMPRESSED EARTH BLOCK

We must demonstrate to other technicians, insurance companies, organizations, and professional associations that earth as a building material, a millennium material, is safe and easy to control in comparison with the other construction technologies. For that, it is essential to ensure comparable compressive strength, durability against water and salt, predominantly, and control of the architectural benefits obtained.

Based on the wide experimentation and the methods outlined in the doctoral thesis of Gabriel Barbeta (http://www.tdx.cat/TDX-1105102161519/index_cs.html), we can say that it is now feasible to have quality-standardized material made of earth. CEB is the technology that obtains these objectives most easily. The reasons for this are:

1. The technology is comparable to traditional ceramic brick, in terms of both the type of equipment needed and the way to work with it.
2. Workers do not need to be highly qualified; any builder can carry out work with this material.
3. It is easily industrializable, which implies a more homogeneous and controllable material.
4. In turn, it allows the use of hydraulic compression machines, which give greater compactness and strength.
5. As the pieces are small, problems caused by shrinkage due to drying or the expansiveness of clay are not an issue and do not affect the proper functioning of the walls.
6. Transportation within a 70 km radius of the Bioterre manufacturing plant was affordable. That meant only a 10% energy impact increase in the cost of the material, making it better than using the expansive earth taken from the same place.
   - The Bioterre CEB is a block type 5, standardized by the Spanish UNE 41410 structural material with great flexibility: it can be one of the best anti-seismic and allows it to withstand traction (ININVI regulations, Peru), and overcome resistance to the collapse of reinforced concrete structures. For cement soil, according to Carlos de Souza Pint, that comes from Brasileira Cement Portland Association, $E$ = 45.000–90.000 kp/cm$^2$ at 7 days and 90.000–175.000 kp/cm$^2$ to 90 days. The dynamic modulus of elasticity is 30%–25% higher than $E$ compression. Young's modulus = 25410 kp/cm$^2$
   - Compressive strength in N/mm$^2$, $\sigma c$ = 7–10, exceeding 5 N/mm$^2$ block marks the CTE F. Traction 10%. Shear $\tau = 1/8$

     Compression $\leq 3.442 \pm 0.057$ N/mm$^2$.

     Bending $\sigma f = 3.57\ (\sqrt{\sigma c'}/7)$. 0.88.
   - The compressive strength when wet goes down by almost half is largely dependent on the type and degree of stabilization employed.
   - The breakdown in the press is never instantaneous; it breaks gradually rearranging its molecules, compressing. Even when it is stabilized, it retains its plastic state.

- The Poisson's ratio (according to FELT) is 0.2–0.3, at compression of 0.08–0.24.
- Fire resistance >240 minutes TIPE M0.
- Thermal insulation $e$ = 30 cms. Standards UNE-92-001-90/91 UNE 92-201-89 UNE-92-202-89 ISO/DIS 8990.

$$\lambda = 0.415\ \text{W/m} \cdot °\text{C} \quad a\ 0°\text{C} \quad K = 1.06\ \text{W/m}^2 \cdot °\text{C}.$$

$$\lambda = 0.546\ \text{W/m} \cdot °\text{C} \quad a\ 35°\text{C} \quad K = 1.30\ \text{W/m}^2 \cdot °\text{C}.$$

$$\boxed{\mu(\text{damping factor}) = \exp\left(-e \cdot V\pi c\delta / \lambda T\right) 0.82}$$

$$\varphi(\text{thermal lag}) = T/2 \cdot e/\lambda \cdot V\lambda c\delta/\pi \cdot T = 7.99\ \text{hours}.$$

- Acoustic insulation $e$ = 30 cms F·500 Hz.
  Rg en db (A) = 54.0

### 11.7.1 Other Outstanding Properties of CEB

- Great thermal inertia, heat build-up in summer and/or during the day for winter and/or night. Specific heat: 0.27 kcal/kg.
- No radioactive or toxic particles, other than the very nature of the place.
- Resistance to climate change. The coefficient of thermal expansion is 0.012 mm/m°C.
- Volumetric stability, from the stabilization of the clays that compose it. Coefficient of linear shrinkage of 3 mm/m (unstabilized).
- Permeability and regulation of water vapor. Water absorption coefficient of 5%–8% (dry weight).
- Adherence to wood and organic materials.

### 11.7.2 Generic Parameters for the Construction of a CEB Wall

- Mortars: Faster natural cement or mortar M-2.5.
- With the addition of 50% prepared earth maximum aggregate size <5 mm.
- Joints: 15 mm.
- Rig: Perforated bricks are valid and solid bricks with labyrinth seal. Chases max: 25 mm in segments <1.5 m.
- Expansion joints: <40 m and <20 m in morphology LoU Hollow: Width >2.1 m.
- Between holes, hole-encounter walls >0.5 m.

## 11.8 SUSTAINABLE ECOBIOCONSTRUCTIVE DESIGN

In the following diagram, you can see the main factors that the architect must consider in order to have the most global and conscious vision of the architectural project. The architect acts as an integrating channel for various technical solutions, mainly around the global housing problem, minimizing energy consumption and $CO_2$ emissions, to project a healthier architecture that is respectful of the environment and human beings. The technological and industrial development of this last century has represented a new contact with new kinds of energy such as electricity or nuclear materials such as asbestos or the almost 50,000 chemical substances present in our homes. Right now, we begin to know how they affect our health and that of the earth, which is also the responsibility of architects and builders.

This global vision is what makes it possible to analyze, before making a decision, for example, if an element can be made of wood or not. To know this, we must know its origin, if its extraction does not exceed its capacity for self-regeneration, the environmental impact that occurs throughout its life cycle (LCA), if there has been an excessive consumption of energy in its transport and distribution, or it is an appropriate technology for a certain culture and climate. It is evident that this leads us not to use tropical timber in our country or that it is not totally sustainable to import American hardwoods due to the energy expenditure in their transportation.

That is why in sustainability, the resolution of projects based on the "native" stands out, recovering and improving many traditional technologies. However, we must not forget that the new technologies displaced the traditional ones with a certain reason for durability, including adobe. Thus, we must be realistic; a true sustainable constructive alternative must be in accordance with the degree achieved by the technique and current quality of life. It is absurd to pretend that people continue to revoke their adobe walls annually after the rains or in some cases cannot eliminate insects and parasites. The consequence has been the massive use in the southern countries of inappropriate and unsustainable technologies typical of the northern countries, relegating the materials of the area and their popular knowledge.

The task is not easy because it involves questioning which technology, material, or architectural solution in each case, for each person, culture, place, or socioeconomic situation. This implies a scientific cognitive basis on new technologies and sustainable ecobioconstructive materials, such as earth material, gypsum, lime, natural and pozzolanic cements, wood, and natural fibers, to improve them and really turn them into quality alternatives to serious global climate and habitat problem.

It is of utmost importance and urgency to develop research in this field, to give new impetus to the evolution of architecture and construction.

This analysis base will serve at the same time to understand the ecobioconstructive framework of application of this technology in more organic and environmentally correct architectures.

Apart from the search for appropriate technologies to solve the global housing problem, in this type of sustainable globalizing analysis, we must anticipate the enormous specific weight of the energy issue and how it directly affects greenhouse gas emissions, producers of acid rain, and ozone destroyers. Therefore, in these design bases, there is a need to make an energy analysis of the life cycle (ACV) of all building materials and systems using renewable energy and design with the help of bioclimatics. An example of a bioclimatic building is exhibited in Figure 11.5, designed by Gabriel Barbeta for housing in Montebatres, Madrid.

**FIGURE 11.5** Gabriel Barbeta, bioclimatic housing, Montebatres, Madrid. (Photos: Gabriel Barbeta.)

To reduce $CO_2$ emissions by half, the insulation of buildings should be improved and greater energy self-sufficiency of buildings (NZEB, near zero emission building) should be achieved based on the application of bioclimatic principles and current certification of **Passive House** [22]. It's essential to replace the electric energy coming from thermal power plants, which are the main emitters, by renewable and non-polluting energy. Even positive balances should be reached, to use surpluses in mobility, and in turn be a regenerating building for the environment, a living building, which is encouraged in the American certification of the Living Building Challenge. Thus, it would be summoned what passive bioclimatic principles that can be integrated into a project consist of:

- Solar path and location. Evaluate angles of incidence and orientation openings.
- Design consideration of the shape and orientation, the various forms of heat transmission by convection, conduction, direct, diffuse and reflected radiation, and natural convective ventilation phenomena to obtain thermal comfort (Olgyay or Givoni graphic).
- Optimization of the insulation on the outer face for curing unnecessary heat loss in winter.
- Spaces. Use of spaces with lower comfort needs, such as toilets, warehouses, garages, storage rooms, and stairs in poorly oriented areas (north-west).
- Thermal storage: thermal inertia. Increases the mass and considers the climatic effect of the soil and walls to dampen the day–night, sequential, and seasonal thermal jumps. It highlights the thermal properties of the earth for this purpose, as well as its greater heat capacity and thermal inertia than concrete or ceramics.

- Passive heating: trombe walls, the use of overhangs to prevent the entry of solar radiation, columns of light, proper orientation of the house with respect to the sun, and the prevailing winds.
- Solar walls: use of the "thermosyphon air panel," a type of insulated collector with selective surface and simple glass, placed in the south walls to generate hot air towards the north side.
- Solar preheating systems, using pipes under cover.
- Greenhouses, galleries, and transparent insulators.
- Passive cooling: low-emissivity glass and thermo glass; radiation absorption measures, reduction of albedo due to the effect of the integration of vegetation, which can affect 85%–90%; in the same way, there is an absorption of pollutants, highlighting species such as *pine halepensis* or olive.
- Use of hybrid cooling systems, taking advantage of night cooling (it is estimated that in dry and hot climates, a cooling potential of 45–220 $Wh/m^2$ can be achieved), aluminum night air cooling collectors (performance 16–190 $Wh/m^2$ night, depending on the weather), thermal mass, floors or cooling panels with water at 15°C–20°C, use of evaporation cooling systems, or precooling by the use of river water.
- Use of sun protection systems in skylights and windows with Venetian or Mediterranean blinds or shading atriums.
- Reduce solar absorption through high albedo materials and colors.
- Forced ventilation through "solar chimneys," "exhaust lantern," and "turbine cap" curved static vacuum cleaners and fins to prevent dust and atmospheric pollution.
- Control and calculation of solar radiation using the mask method or simulators.

## 11.9 THE HARMONIC COMPLEXITY OF ORGANICITY

Nature seeks at all times to develop optimal ways to save resources and energy, while generating an optimal environment for the development of life. We do not find hard, cubic, or pointed shapes. Organicity gives life, resistance, and harmony.

There is also a law and tendency of the universe towards entropy: complexity of essentiality and the beauty of complexity.

Getting to drink from the source of the infinite wisdom of nature is not to recreate ourselves in projecting and building new copies. Maybe it's not just mimicking either. It's as green painting our habitats already granted them the ecological condition, and it really is not so and is much more now that the term Eco is sold so easily.

The walls behave like a vibratory continuum, so the incidence of energy fields generates their transmission, concentration, distortion, reverberation, and amplification. A part can be re-emitted from these and reverberated to the air again, it can even generate harmonics. The shape of the wall, the dimension, the proportion, the crystalline, and molecular characteristics of the material determine it. In the design process, we can assess it based on mindfulness, sensitive crystallization, biotic quality, or musical vibration.

That is why we felt it is important to recover a new feeling and a more humble and tranquil design from the proportion. The divine or golden proportion is one of them that we have forgotten in time and that can help us refine our buildings to facilitate the resounding of the melodies of our own souls. In the solstice mandalas that we use, recovered from the regulatory layouts of cathedrals by engineer Raymond Montercy, other derived proportions also appear. The diagonals of the different families of fractalized squares or their projections usually give the main roots used in sacred geometry as $\sqrt{2}$ $\sqrt{3}$ $\sqrt{5}$. Among them, we highlight the minor diagonal, which comes out of the middle of the side of the square, which gives us the value of $(1 + \sqrt{5})/2$. This expression has a value of 1.6180 and was baptized by the Greeks as the number of gold or golden ratio and that the letter of the Greek alphabet is associated $\phi$. It is also the mathematical relationship between the diagonals and the sides of the pentagon, and it is in the relationship of all sides of the pentalfa star that comes out of its diagonals. It is also present within platonic solids: dodecahedron area = $15\ \phi/\sqrt{3} - \phi$; dodecahedron volume = $5\ \phi^2/6 - 2\phi$; and icosahedron volume = $5\ \phi^2/\ 6$. It is the result of a second degree equation of a parabola: $\phi^2 - \phi - 1 = 0$. Therefore, it is an irrational algebraic number. From this come equivalences such as $\phi - 1 = 1/\phi$ or $\phi = \sqrt{1} + \sqrt{1} + \sqrt{1} + \ldots$

It has always been present in architecture, and we know that it was used by the Egyptians in the pyramids, the Greeks in temples such as the Parthenon, in the Romanesque, Gothic, Renaissance (the divine proportion of Leonardo da Vinci, [23] or of Palladio), in the same modernity with renowned architects such as Le Corbusier in the Villa Savoye, or on the facade of the Unité d'habitation de Marseille, and in painters such as Salvador Dalí. Vitruvius devised a mathematical calculation system of pictorial division to divide the spaces in equal parts and thus achieve a better composition. It was based on the general principle of contemplating a rectangular space divided, in broad strokes, into thirds, both vertically and horizontally. It could also be done by making the bisector of a square and using the diagonal of one of its halves as a radius to extend the dimensions of the square until it becomes a "golden rectangle." By placing the main design elements in one of these lines, awareness is taken of the balance created between these elements and the rest of the design. In 2007, the author of these lines had the opportunity to meet casually on a visit to Albert Sastre, a rigger of Miravet Castle, in Southern Catalonia, who had done his graphic survey and had verified the use of the phi number in the plant of the church (17.6 m length/ 11 width). On the Ebro river stands this austere, but imposing medieval castle, one of the best examples of the architecture of the Order of the Templars in Europe, by master Pere de Rovira, who rebuilt and became one of the main centers of Christian power of the Iberian Peninsula around 1153. Serenity, peace, and connection are present when you have the opportunity to be alone inside. It gives the sensation as if something resonates within us, a fact totally related to the same human proportion, as seen in the image of the Vitruvian Man. The total height would be $\phi$ multiplied by the height of the navel.

In 1871, the mathematician L.A. Quetelet ratified it with statistical studies on body measurements in European men. Also the MITT electrotechnical engineer Dan Winter has studied the presence of the golden ratio in the DNA spiral and its relationship in the implosion of life. On the one hand, it relates the measures of height (34 Å) and diameter of the double spiral (21 Å), but we cannot ignore that the structure of the nucleotides are regular pentagons and hexagons.

In Nature we also find $\phi$ at a more macroscopic level: as in the proportions of flowers, many are pentagonal following Ludwig's botanical law; in the type of growth of many plants and animals; in the relationship between the nerves of the cut of a leaf; in the relationship between the number of males and females of bee hives; in the arrangement of the leaves of many plants at an angle of 137° 30′ 28″, forming an upward spiral, which allows them to capture solar energy without creating shadows between them, since phi is the irrational number that takes longer to converge. (As an architectural example of all these, see Figure 11.6.)

One of the most widespread examples that illustrates it is the structure of the nautilus shell. This expresses the golden spiral, following the Fibonacci series. Each radius of the circle encompasses the sum of the previous two. Following the formula $n = n - 1 + n - 2$, the fundamental series is 1, 1, 2, 3, 5, 8, 13, 21, 34, as the numbers get larger, the quotient between two consecutive tends to approach the number $\phi$. A similar fact occurs in the spiral of Archimedes.

A gold spiral can be constructed, from a succession of gold rectangles and squares, drawing quarters of circumference within each square and tangents to it. This spiral approaches the golden spiral, a logarithmic spiral of center at the intersection of the two diagonals indicated in Figure 11.7, a point called the Eye of God.

Acoustics studies at the University of Oviedo [24–26] reveal to us how the rooms that use the golden ratio solve many problems of reverberation and have less modal coincidences in the low-frequency range. It is necessary to reduce the degree of symmetry, and that the measures of the orthohedron are different in addition to being irrational numbers. The golden design of the listening rooms, in addition to providing an esthetic aspect, significantly improves its timber, the character of the sound dependent on the proportion of harmonics or overtones. It is necessary to observe with conscience how the golden ratio acts. When we use it, it usually gives us the feeling that things have a correct and balanced size. It is as if we identify with universal information, with the unified quantum field, information that is also within us.

**FIGURE 11.6** Gabriel Barbeta, pentagonal housing, Montseny. (Photos: Gabriel Barbeta.) Published in Gabriel Barbeta, Esteve Navarrete, "A pentagonal block home," 31–36. In C. Mileto, F. Vegas, L. García Soriano, V. Cristini (eds.), *Earthen Architecture: Past, Present and Future*. Taylor & Francis Group, London, 2014.

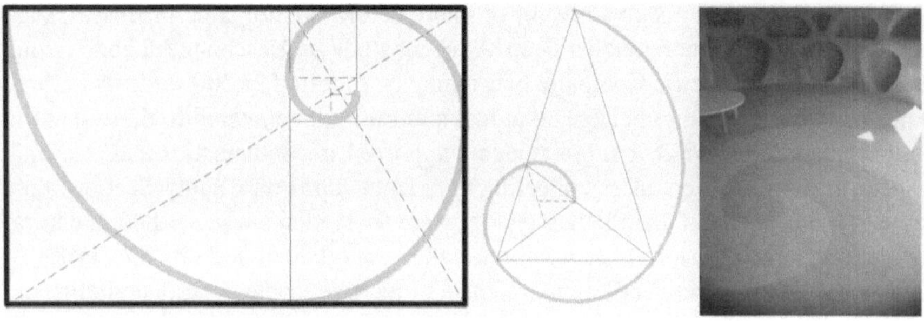

**FIGURE 11.7** Gabriel Barbeta, use of the golden spiral in the layout of a linoleum pavement for a school. (Photos: Gabriel Barbeta.)

Without realizing it, we are using it daily in the sizes of credit cards, publications, doors, and an endless number of everyday examples.

## 11.10 THE SCHOOL PROJECTS

The school projects that we do are buildings of pure awareness, intention, and love in keeping with the current situation in Gaia and with the mental and spiritual health of the children. The future is in their hands, so if we want to divert our current course, we must educate, live, and inhabit in different ways. We are talking about growing up in living, bioconstructive, resonant, and harmonizing schools, where everything is pure resentment of the inner potential of each human being. It is not just an architecture to be seen but really to feel, breathe, play, and vibrate. For this reason, throughout the process of materialization of the idea, the way of working, agreeing, and dialog has been taken care of allowing human love to remain permeated in its walls. This is where it becomes essential for parents, teachers, and children themselves to generate engagement processes. We have realized that the process is as or more important than the building itself. We cannot rest on those Machiavellian bases that the end justifies the means. This has led our society to not know how to differentiate when we are doing a school, a prison, or a pig farm, as a group of teachers claimed in the project of its scholar center in Sant Llorenç dels Morunys. They only dominate execution parameters in record time, zero maintenance, and almost police security, within unmovable standards and administrative methodologies. The result is not far from making unsustainable, soulless, humane shoeboxes. On the contrary, we are advocating for schools like tuned Stradivarius, hand-made and with the wisdom of the teacher, making spaces that resonate and enhance the best within us. However, the school must be a place of growth and evolution, avoiding many of the psychosomatic and physical ailments caused by lack and effective burdens, or blurring the patterns and forms of thought that we carry over. This makes us feel the need to make the walls of a school healthy for children at all levels. The general forms that emerge are based on criteria of biomimetics, golden proportion and geometries based on solar motion, in particular angles and solstice proportions, as exhibited in Figure 11.9, with criteria recovered from the ancient European masters of cathedrals. Sunlight is an activator of many natural and organic processes, such as the pineal gland that regulates sleep. A recent study by Heschong Mahone Group for Pacific Gas & Electric Company (www.hmg.com) tested 21,000 students in three states and found that those classes that had natural light compared to those that did not obtained 20% and 26% more students approved in Mathematics and Reading, respectively. In addition, after collecting data from Johnston County Schools, they came to the conclusion that the best grades were for students who studied in natural light and that their performance was higher than the other students by 5%–14%.

Natural lighting and ventilation are ensured by practicable central skylights and good orientation of the openings, as exhibited in Figure 11.8.

Other benefits of moderate sun radiation are of sun exposure that can improve the signs and symptoms associated with different skin conditions, such as acne, psoriasis, or jaundice. As well as favoring it, elimination of toxins from the dermis strengthens the immune system, increasing the number of white blood cells, which are the first barrier in our body to prevent and fight infections; contributes to the body's synthesis of vitamin D, while helping to regulate cholesterol levels and to have strong and healthy bones; UV rays increase the production of neurotransmitters such as endorphins, which are related to well-being, and serotonin, which is involved in regulating sleep and body temperature and in a natural antidepressant state. Regarding energy fields, the use of healthy walls and underfloor heating and solar preheating of water combined with the use of renewable energy the influence of electromagnetic and geomagnetic fields should be considered, which could help us reduce certain physical ailments. In a technician world and invaded by wave-based systems, we need daily spaces that facilitate our own cellular regeneration and external protection. Children and people with a weak immune system are the most sensitive and most in need. This is achieved by generating distance to antennas, high or medium voltage lines, Wi-Fi, and inviting the cell to close. At the same time, screening can be done based on the density of the earth and the effect of the tree, and in severe cases, it can be done by a good grounding of metal mesh or the graphene paints themselves.

**FIGURE 11.8** Gabriel Barbeta, rehabilitation Viladrau School and Yoga Center, using linoleum, natural bark cork, and the golden spiral. (Photos: Gabriel Barbeta.)

Really, schools should be white areas, without EMF (as in France) or in the Euskadi school ordinances. The EEA-supported Bioinitiative Report (2007) claims to prevail over cable by removing Wi-Fi from schools, kindergartens, libraries, and leisure (Salzburg Department of Public Health in Austria, 2005). In addition, the European Parliament (2008 and 2009) is committed to minimizing exposure by weighing the "Available and effective technology options," a line already recommended by the German cable government over Wi-Fi. Switzerland gives tips to minimize that too.

## 11.11 EARTH SCHOOL, A CASE STUDY

The nursery project revolves around a central courtyard of elliptical shape, suggesting the protection of the womb, which defines the volume and geometry of the building in its most subtle expression. As if it were a sacred space in the center of the courtyard, there is an oak tree and symbols around, which gives a source of life organizing the school.

The intention is to work in terms of not only building efficiency and in terms of technical aspect in construction and energy savings but also achieving the aspects of health and well-being of users in the widest sense. Since the beginning of the creative process, the project has been integrating natural materials, working with a great environmental awareness and minimizing the presence of metals and steel structures. From this awareness derives the solution of sustainable building elements for covering by working directly in compression. One way is to design with common sense, which for millennia, man has used and the nature manifested in many forms organic, effective, and efficient.

The design process includes concepts of bioclimatic architecture such as the trombe walls, made of salt ionizing bricks and green pergolas for sun protection in the classrooms of the south facade, the roof garden with native species, recovery systems for rain and waist waters, cross-ventilation systems, biomass boiler, and solar panels for underfloor heating. As shown in Figure 11.9, the school entrance, located in a corner, is framed by a large arched wooden girder. The building's low height with reference to the access road makes the roof garden full of autochthonous species visible and classrooms decorated with mosaics of broken ceramic tile, called "trencadís" in Catalan, a distinguishing characteristic of the architectural movement called Modernism. The classrooms have their own colos, both inside and outside, to help children recognize and differentiate each one. On the inside, the school is organized around a central oval courtyard, inspired in the shape of a uterus and providing a microclimate and a better view of the rooms.

The courtyard is more public than the entryway and the multi-purpose room. In its north-east corner, it opens out onto the only bit of nature in the area: a park, a creek, and views of the massif of the Montseny Natural Park in the background. Wooden beams radiate from the center of the courtyard, which is presided by a fountain and a noble oak tree, as if it were a sacred place, as shown in Figure 11.10.

Beyond the entryway and the hall, the building is divided into two parts: to the north, an administration and service area, and to the south, an arc formed by five domed hexagonal classrooms, facing the green area, and by another one further down the hallway at the eastern end, made intentionally in the shape of a heart.

FIGURE 11.9 Gabriel Barbeta, school, Santa Eulalia of Rončana, exterior view. (Photos: Gabriel Barbeta.)

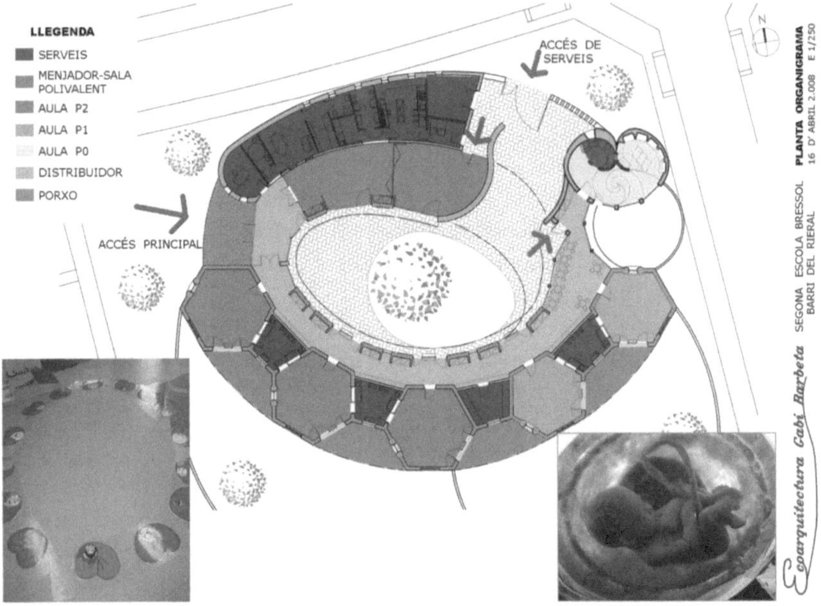


FIGURE 11.10 Gabriel Barbeta, school, Santa Eulalia of Rončana, general functional floor plan. (Photos: Gabriel Barbeta and Esteve Navarrete.)

The classrooms are arranged so that older children are near the entrance and the lunch room, and the smallest are at the end, where they have the peace of mind they need. The shape of the classrooms promotes work in a circle, typical of group dynamics.

The area, radius, and height of each one changes subtly, depending on the age of the children, with areas ranging from 39 to 31 $m^2$. The hexagonal-shaped perimeter CEB wall is 30 cm thick and 208 cm tall. Over it is a 25 cm concrete hexagonal perimeter hoop and Nubian domes between 1.5 and 1.65 m height, depending on the section and based on the golden ratio. The radius of the circular path of the domes is 3.3 m, and the heart shape is formed by working with three radius: 3.69, 2.25, and 1.38 m. Lavatories and changing rooms are found between the classrooms to give teachers greater versatility and more support.

The access corridor is especially wide to make room for side cabinets, to open up the classrooms, and to promote complementary activities such as snacking and game playing.

The northern part contains the services (bathrooms, kitchen, and administrative offices), and the building is separated from the busy street by a CEB wall 45 cm thick with small controlled apertures to minimize heat loss. The multi-purpose/dining room is open to and perfectly illuminated by the central courtyard, and when subdivided by a folding wall, it is independent of the nursery, thanks to the northern access. A curved rammed earth wall reusing clay soil from the excavation of the foundations provides the backdrop for the room (Figure 11.11).

### 11.11.1 Bioconstruction Technologies

Since the beginning of the creative process, the project has been integrating natural materials such as the CEB (block compressed earth), earth walls (rammed earth), finished with lime mortar and ground insulation, cork, natural fibers, wood, eco-friendly paints, flooring cork-based linoleum, and tree bark façade finishes.

The weight-bearing walls of the school were built with double walls of Bioterre CEB, most with a thickness of 30 cm. They stand on a reinforced concrete foundation of hexagonal paving stone interconnected by braces because the ground was at risk of settling and expanding. The wall alternates thicknesses of 15 cm indoor

**FIGURE 11.11** Gabriel Barbeta, school, Santa Eulalia of Ronçana, exterior view of the entrance and courtyard fountain. (Photos: Gabriel Barbeta.)

**FIGURE 11.12** Gabriel Barbeta, bioconstruction materials in the classroom. (Photos: Gabriel Barbeta.)

and outdoor 30 cm expanding to 45 cm on the north side that borders to the bustling main street. Inside view shows a wall with earthy color CEB transparent water-repellent treatment on the basis of latex, silicone resin, or potassium silicate. It has also been used with stabilized earth walls (rammed earth); adobe clays identify the place and finished with lime mortar and soil cover that is mechanically projected (Figure 11.12).

## 11.12 RAMMED EARTH

Despite the ease of construction with CEB, and even greater ease for curved walls (Figure 11.5), it was decided to also use earth from the site as an ecological gesture. Using earth from the excavation site to build walls is certainly one way to minimize the impact of construction. Also, reviving the rammed earth technique seemed like a good option given the local traditions and the type of wall that was to be built. The walls have been rammed using a manual tamper. The rammed earth is composed of several layers of earth roughly 15 cm thick, poured into a curved formwork (wooden), and rammed with a rammer (manual or pneumatic). After compaction, each layer is 8–10 cm thick. This procedure is repeated until completion of the wall, as exhibited in Figure 11.13.

Thousands of rammed earth buildings constructed centuries ago are still giving excellent service in Spain, France, Algeria, Morocco, Central America, China, and the Middle East. In Australia, there are many rammed earth buildings over 100 years old still in use; therefore, earth construction could be considered as a durable kind

**FIGURE 11.13** Gabriel Barbeta, rammed earth with reserve for a hole. (Photos: Gabriel Barbeta.)

of construction. Durability in the context of earth construction means the ability of the structure and all its elements to withstand the destructive action of weathering and other actions without degradation to the expected service life. Rain and frost are the most destructive natural actions causing erosion and deterioration of the earthen elements. Accidental abrasion is also a significant agent of deterioration.

Rammed earth has relatively good strength in compression but generally poor strength in shear and tension, especially when moist. The mechanical strength of a soil is very much dependent on the voids ratio of the soil after ramming, cohesive strength of fines content, aggregate strength, and moisture condition during testing. Density of the soil is a very important factor for the strength of the soil. To carry out this task, it was essential to check the properties of the earth and of the stabilization system used. Although linear shrinkage is a major handicap, it also provides water protection. Therefore, modern rammed earths appeared in which binders were added, such as cement, hydraulic, calcium lime, or chemical admixtures. They are called "stabilized rammed earth." The main advantage of stabilizing the rammed earth is to increase its durability (with respect to water attack), and mechanical performance (compressive strength) reduces shrink and swell and provides waterproofing qualities (see example of Figure 11.14).

The results obtained in the specimens were: retraction of 1 cm on the longitudinal test 60×7×4 cm (1.67% of the total). The fiber dosage compensated the content of expansive clays to obtain an excellent compressive strength of 192.11 kp/cm$^2$.

A linear shrinkage test with different dosages

a. 10% lime.
b. 12% cement, 7% perlite, and 24% brown fibers.
c. 12% cement and 7% perlite.

FIGURE 11.14 Retraction tests with the stabilized local soil. (Photos: Gabriel Barbeta.)

## 11.13 CTB DOMES

Vaults are structures used to horizontally distribute vertical loads. They are built using materials that primarily work via compression. Vault-roofed spaces have been with us even longer than we might think (Guillaud, 2016). The vault and the dome are the elements with the longest history in construction. At first, they may have been caverns or natural shelters, but over the years, they have gradually evolved in terms of materials, techniques, and uses, shifting from woven plants, adobe, bricks, stones, Roman concrete or even wood, to modern concrete, steel and high-tech wood. These domes also offer us many advantages such as providing a greater volume compared to flat ceilings, enhancing the sensation and perception of space, allowing a better climate pattern during summer, thanks to the possibility to allow hot air to go up to be evacuated through the practicable skylight, minimize the area covered affected by solar radiation, thanks to the geometry of the dome, compared to a flat roof, and finally offer a natural finish with multi-tonal Bioterre seen without the need for any lining.

Their use became as commonplace as reticular reinforced concrete structures are today, and they were used for roofs that covered spaces used for a wide range of purposes. Of all the techniques that can generate vault-roofed spaces, we have focused on vaults built without form work. There are two kinds of vaults built without form work:

- Leaning vaults, in which the bricks are joined to each other on their largest side, called the table in Spanish, which is delimited by the length and width and is placed such that each arch-shaped row of bricks rests upon the previous one, as the Mexican vault. Mexican vaults are more closely associated with spheroidal shapes, with the peculiarity that they spring from polygonal shapes in most cases. They have been built for over 200 years in the center of the Republic of Mexico (Alfonso Ramirez Ponce, Ramón Aguirre, 2016) using a brick with a specific size: $20 \times 10 \times 5$ cm. There are many variations, but the most common one is the kind that springs from the corners with leaning arches that rest on the previous one, forming pendentives that build upon each other, enclosing the space to be covered until only one piece can fit.
- Timber vaults, [27–33] which look like a warped partition (Bassegoda). In them, the bricks are joined on their smallest side, which calls for the use of quick-drying mortar and a more skilled hand trained at laying them. These are highly sustainable compared to conventional roofs; they consume 1/3 of the energy and produce only 1/4 of the $CO_2$, in addition to being economic and competitive in terms of completion time. In addition, it is perfectly competitive in price, the technical solution of a flat wood frame is 13% more expensive than domes CEB solution (dome cost: 79.15 €/m$^2$). However, it would be essential to design the dome with a shape that does not require an additional layer of compression. The options are provided by vaulted systems with an alternative that not only is competitive in terms of price and time but can also be made with compressed earth blocks, thereby doing without almost any kind of form work. Likewise, because the bricks are joined on their largest surface, the technique to add a piece to the vault is simpler and does not require a quick-drying agglomerate to join them perfectly, which ultimately makes it easier for a less-skilled person to build a vault.

The construction system of the elliptical Nubian domes as exhibited in Figure 11.15 is based on the path from the fixed center and radius variation in each of the different courses. The technique is based on site knowledge and improvement of work developed by Hassan Fathy in Egypt and in India in the 1960s.To do this, we used something new: a rotary telescopic metal ruler, functioning as a compass.

This technique results in a stepped, irregular, and absorbent surface when block is left in the exposed, significantly improving the acoustic performance. The blocks have a stagger of 17° from the radius to allow for greater grip and facility of implementation, although one of the best natural cements in the country was used, the same that has been used for the vaults of the Sagrada Familia.

The dome is 15 cm of Bioterre complete with a cover of 7 cm of concrete reinforced with polypropylene fibers and the isolation of two layers of natural cork bound with cement glue. The CEB, made with raw earth, has excellent absorption and plasticity

**FIGURE 11.15** Gabriel Barbeta, building of Nubia's domes without form work. (Photos: Gabriel Barbeta and Xavier Puig.)

characteristics, due to the clays that bind it, not to mention its high thermal inertia and low environmental impact as construction material. The design and calculation of the domes were made from a radius of 3.3 m and a point or height of 1.75 m, although among them were small dimensional variations. A waterproofing layer of flexible polyolefin was placed over it. The exterior finish is made of mosaic, called in Catalan "trencadís," a characteristic element of the Art Nouveau architectural movement. In the various domes, different colors were used, a range for each classroom, made from recycled ceramic tiles. Lighting and bioclimatic natural ventilation are guaranteed by deposing central skylight with a highly reflective tube and deflectors to transmit sunlight to the center of the classroom.

The loads considered were of 2.92 $kN/m^2$ for blocks of land, 0.08 for cork isolation thickness of 6 cm, and 1.3 $kN/m^2$ for the final layer of mortar with waterproofing and multi-colored ceramic recycled tiles. At the same time, we considered the variables of wind loads, earthquakes, and snow and follow the mandatory requirements established by the Technical Code for Construction in Spain (CTE-SE-DB-AE). The design values were 0.4 $kN/m^2$ for wind and 0.5 $kN/m^2$ for snow because the altitude is below the 300 m above the sea level. Transiting use loads were not taken into consideration because the shape of the domes slopes and the finishing materials were considered not passable. The calculation of the seismic force resulting from the own loads in relation to the height was made based on Spanish rule NCSE-02.

To study the stability of the section we used the static method by funicular polygon graphic computer drawing. We analyzed the average worst section of the domes, using the following assumptions:

- The dome is similar to an infinite continuum of circular sections. We analyzed a circular area of the dome's horizontal sliced section.
- The scale of the graph lines is proportional to the real thrusts.
- The balance of the dome is achieved by containing the different thrust lines obtained within the central third of the cross section of the different segments. The horizontal thrust must be counterbalanced at its base by a perimeter concrete ring beam.

- At the top of the dome, the reaction there is a horizontal force equal and opposite to the thrust at the base. We place it situated in a third of the most external section.
- The axis lines of the reactions and the total charge of the section run through the center of gravity of the section, which is directly determined by the command "mass properties" in the AutoCAD drawing, having previously assigned "volume" to the Auto CAD object, if the traced polygonal axis lines of the reaction fall outside the middle third center.
- The axis-force line must be relocated inside the middle third center changing the inclination and tracing back the force polygon.
- The resulting polygon conditioned us to increase the section in the lower area of the dome, filling it with concrete to prevent traction and the use of steel reinforcement.

The linear thrust obtained was 634 kp to be added to the seismic thrust in order to determine the amount of steel reinforcement in the hexagonal perimeter hoop. During the construction of the domes, due to the great flatness of the last rows of CEB and, despite using the Nubian construction system block offset to the radius of the track, it was decided to use a fast natural cement brand named "Marfil" from the local company Cementos Collet.

### 11.13.1 Characteristics of Natural Cement

One of the main properties of natural cement is its speed setting, and good resistance to sulfates combinatorial, mostly in the mine of origin, which are of higher quality and more consistent having uniform properties, and containing more than 20% of tricalcium aluminates, C3A. This allows them to combine with gypsum up to 7% above the Portland cement to form calcium sulfoaluminate expansive, while we can have resistance to the 30 N/mm$^2$. This makes it an excellent binder for the execution of domes and vaults and so was used to ensure durability in the work of Antoni Gaudí's Sagrada Familia church.

Another property of natural cement is its resistance to moisture and its ability to harden in the presence of this; it even gains more strength with time. It has a low-energy cost of manufacturing 2.6 MJ/kg, a saving of 65% compared to other types of cement.

### 11.13.2 Regulations

Currently, the use of vaults is not part of the technical standards applied in construction. For example, they are not mentioned in the Spanish Technical Building Code.

It is worth pointing out the CEBs referred to here do comply with the CTE-SE-F Spanish Building Code, with a resistance equivalent to a solid ceramic brick. They do appear in the UNE 41410:2008, "Compressed earth blocks for walls and partitions"; although this standard does talk about the proposed material, it logically does not mention vaults specifically. Therefore, their use remains within the discretion of the person advocating it.

In Mexico, the NMX-508-ONNCCE-2015 and NMX-513-ONNCE-2015 standards were passed last year, approving the stabilization of CEBs and their use in vaulted roofs (Aguirre, 2016).

### 11.13.3 Comparative Costs of Execution

It presents the actual costs and real time for the execution; they were used for the official certification of the building construction site, to assess the feasibility of the technology today, comparing the existing solutions (flat roofs and CEB domes) analyzing the supporting layer plus the insulation, ignoring the finish coat and waterproofing, and obviously the economic reduction by the construction company:

| Domes | Price |
|---|---|
| CEB dome | 79.15 €/m |
| Compression layer of polypropylene fiber concrete | 41.96 €/m |
| Natural cork 8 cm thickness bonded insulation | 17.14 €/m |
| Overall d | 138.25 € |

| Flat Roof | Price |
|---|---|
| Structure of wooden beams | 61.98 €/m$^2$ |
| Double tongue and groove timber 22 with 3 cm of rock wool intermediate | 65.76 €/m$^2$ |
| Compression layer 5 cm thick with steel mesh and a wooden frame connector | 12.49 €/m$^2$ |
| Natural cork 8 cm thickness unattached insulation | 15.90 €/m$^2$ |
| Overall wood roof | 156.13 € |

Analyzing the above tables, we see that considering the supporting layer over the insulation covering the technical solution of flat wood frame is 13% more expensive than domes CEB solution, demonstrating that they are perfectly competitive in price.

The time spent for the construction of the 6.6 m diameter dome was eight days of work by a crew of two in an 8-hour day, although in the first two domes, two more days were necessary due to lack of training, site organization, and excessive time spent cleaning the mortar joints of the seen CEB. Nonetheless, the speed with which a Mexican vault can be built is at a pace of 2 hours/m$^2$ for an experienced builder.

This would be more economical in the case of on-site block manufacturing, at the present time, the price of Bioterre in our country is high compared to ceramics bricks, because this is a small-scale new technology. There is one differential factor, namely, the former requires more labor and less material.

| × m² | Weight (kg) | Emissions (kg) | Energy (MJ) |
|---|---|---|---|
| Vault | 161.00 | 05.82 | 161.3 |
| Ring beam | 2.16 | 32.80 | 338.98 |
| Slab | 434.00 | 103.73 | 1,032.45 |
| Concrete | 325.00 | 32.04 | 177.00 |
| Steel | 12.00 | 44.00 | 550.00 |
| Wood | – | 02.24 | 33.56 |
| Lost formwork | 102.00 | 25.45 | 271.89 |

Per slab surface. Information on the roof and the perimeter concrete beam are based on the BEDEC Bank of the Institute of Building Technology of Catalonia.

### 11.13.4 Environmental Sustainability

This kind of vault requires less material to cover the same space, and none of these materials is concrete or steel, which means that its environmental impact in terms of energy expenditures and $CO_2$ emissions will be much lower.

If we examine the comparative table, we can see that the vault weighs less than half of what a concrete roof weighs, that its $CO_2$ emissions are one-third lower, and that it needs less than half the energy to process the materials involved in it. If we bear in mind the calculation of the perimeter concrete beam, if the latter is not needed, the energy needed to build it is nine times lower than a reticular roof.

## 11.14 CONCLUSIONS

Finally, once exposed to the eco-architecture design values, we can conclude that they are competitive and perfectly acceptable in the construction of contemporary buildings in highly developed areas, although prevalent social concepts are intrinsic to the technology of the self-construction. It is an open path that offers options and hope to an essential sustainable development in this planet. If we become aware of our own nature, it shows us shapes and future solutions at the big problems actually. It will be easier to integrate into the landscape and discover a new regenerative architecture.

The technological aspect of solidarity in the use of low tech is to find a common path to all humanity, should not be forgotten, and must be present in our consciousness as technicians.

In turn, the duality between economy and time should not condition the increase of the quality of our buildings, respecting the environment using materials and technologies with low environmental impact and to think in the maximum comfort, convenience, and health of the users.

We have been implementing raw material-based solutions to improve the quality of our buildings, to respect nature, and to use environmentally low-impact materials, considering at the same time the maximum comfort, welfare, and health of users. Working with this type of construction is also an opportunity to use earth, wood, straw, and bamboo as structural materials. These materials were chosen for its qualities and is perfectly suitable. In addition to their many advantages, these technologies and materials also influence the perceptions of users. According to the teachers, feelings inside the building are very positive. This is reflected in the opinion of the educators and parents who work in this kind of schools; their feelings inside the building are very positive, characterized by a general feeling of well-being, calm, and loving place, an architecture to be felt and accessed, not to be seen from afar, designed in harmony with nature and for the benefit of people who use it.

## REFERENCES


1. Plataforma São Paulo, The alliance for a responsible and supportive World. Fundación Charles Leopold Meyer).
2. T. Borges, C. Díez, "CITMA, Cuba: política ambiental a tono con los nuevos tiempos. Proyecto de Ley de Medio Ambiente", *Temas*, 9, 13–19, 1997.
3. K. Asmal, *1st Meeting of the World Water Society*, Stockholm, August 9, 1996. (Declaration of the International Seminar of Cocoyoc). A. Ángel, Towards an environmental society, El Labrador / Fundación Medio Ambiente y Desarrollo Alternativo, Bogotá, 1989.
4. C. González Lobos, *"Nueva arquitectura en América Latina", Por una arquitectura apropiada y apropiable*, Gustavo Gili, Barcelona, 1990.
5. X. Arnella, *Converses Transversals. Aproximación a la investigación, a la economía y a la política bioambiental*, Beta, Barcelona, 1997.
6. E. Bardana, "The sick building syndrome", *50th Congress of the American Association of Allergy and Immunology*, Chicago, 1992. Term adopted by WHO as SEE.
7. K. Harada, K. Hara, C. N. Wei, S. Ohmori, O. Matsushita, U. Ueda, "Study VOCs indoor air before and after remodelation, and sick building syndrome", *International Journal of Immunopathology and Pharmacology*, 20 (2), 69–74, 2007.
8. A. Seki, T. Takigawa, R. Kishi, K. Sakabe, S. Torii, M. Tanaka, T. Yoshimura, K. Morimoto, T. Katoh, S. Kira, Y. Aizawa, "Review of sick house syndrome", *Nihon eiseigaku zasshi. Japanese Journal of Hygiene* , 62 (4), 939–948, 2007.
9. A. Kanazawa, Y. Saijo, M. Tanaka, T. Yoshimura, H. Chikara, T. Takigawa, K. Morimoto, K. Nakayama, E. Shibata, R. Kishi, "Nationwide study of sick house syndrome: comparison of indoor environment of newly built dwellings between Sapporo city and Southern areas", *Nihon eiseigaku zasshi. Japanese Journal of Hygiene*, 65 (3), 447–458, 2010.
10. J. Singh, "Biological contaminants in the built environment and their health implications", *Hutton+Rostron Invest. Gomshall / Building Research and Information*, 21 (4), 1993.
11. J. Fernández Solá, S. Nogué Xarau, "Sensibilidad química y ambiental múltiple", *Jano*, 1662, 14–20, 2007.
12. R. Caponetto, S. M. Cascone, L. Marletta, "Environmental issues of building materials", Environmental friendly cities, *Proceedings of Plea'98*, Lisboa, 1998, 529–530, James & James Science Publishers Ltd.
13. WHO classifies dioxin 2, 3, 7, 8-TCDD as a carcinogenic substance in humans.
14. N. Wertheimer, E. Leeper, "Electrical wiring configuration and childhood cancer", *American Journal of Epidemiology*, 109, 273–284, 1979.
15. S. Martín, "Allergies, cancer and autoimmune diseases have a lot to do with environmental quality". January 1999.

16. J. Kulczycki, *Fréquences radioélectriques et hyperfréquences. Critères d'hygiène de l'environement*, WHO, Geneva, 1981.
17. M. Feychting, A. Ahlbom, *Magnetic Fields and Cancer in People Residing Near Swedish High Voltage Power Lines*, Institutet för Miljömedicin (IMM), Karolinska Institutet, Stockholm, 1992.
18. Occupational exposure to EMF, 1992, Department of Neuromedicine, National Institute of Occupational Health, Solna.
19. *American Journal of Epidemiology*, 134 (9), 1991.
20. L. Torres, "Estudio numérico y experimental del comportamiento en servicio de estructuras de hormigón armadas con barras de composites de matriz polímero (FRP)".
21. I. Santiago, G. Barbeta, "Sinestesia Geométrica". *GEA*, 104–107, 2019.
22. M. Wassouf, *De la casa pasiva al estándar Passivhaus La arquitectura pasiva en climas cálidos*, Gustavo Gili, Barcelona, 2014.
23. F. Corbalán, *La Proporción áurea*, RBA, Barcelona, 2010: proportion study of the face of the Gioconda.
24. J. M. Álvarez Gómez, J. Suárez González, *La importancia del estudio de la sección áurea en el diseño de espacios sonoros estereofónicos*, Ingegraf, Santander, 2002.
25. G. Barbeta, "La volta catalana, una solució sostenible", lecture, COATG, Girona, 2011.
26. G. Barbeta, J. Caminero, E. Navarrete, "Earth domes in a school building in Santa Eulàlia de Ronçana, Catalonia". In VV.AA., *Domes in the world proceedings*, Nardini editore, Florence, 2012.
27. E. Colin, R. Ulrich, "View through a window may influence recovery from surgery", *Science*, 224 (4647), 224–225, 1984.
28. J. García Sanz, M. Fortea, A. M. Reyes, "Comparative analysis on environmental sustainability between masonry vaults and concrete structures", *Revista Ingeniería de Construcción*, 27 (1), 05–22, 2011.
29. G. Marc, I. Hannebicque, *Plantes depoullantes pour la maison*, Eugen Ulmer, Stuttgart, 2008.
30. K. Frances, W. Sullivan, "Environment and crime in the inner city: Does vegetation reduce crime?", *Environment and Behaviour*, 33 (3), 343–367, 2001.
31. M. Gernot, *Manual de construcción en Tierra*, Ecohabitar, Artieda, 2010.
32. P. A. Roger, *P.F.C. Estudio de la resistencia a cortante y a compresión de una fàbrica de BTC Bioterre*, Universitat de Girona, Girona, 2010.
33. D. Velarde Ma, G. Fry, M. Tveit, "Health effects of viewing landscapes-landscape types in environmental", *Urban Forestry & Urban Greening*, 6 (4), 199–212, 2007.

# 12 Back to the Basics
## *Return to the Origin, Gaudí and Nature*

*Alberto T. Estévez and Judith Urbano*
Universitat Internacional de Catalunya

**CONTENTS**



## 12.1 INTRODUCTION: BACKGROUND AND DRIVING FORCES

The Amazon rainforest is in flames, an entire continent like Australia is burning, the ice of another continent, Antarctica, is melting, numerous islands are evacuated due to rising sea levels, etc. Whether people deny climate change or not, there is no doubt the Earth is undergoing apocalyptic deforestation and extinctions of entire species, not to mention water and food crises, wars, and social riots that are uprising everywhere, even in developed countries. Sustainability does affect not only ecology but also sociology, in an inseparable manner. However, immense fortunes are spent looking the other way: the colonization of Mars, holidays in a space station, a cottage on the Moon, etc. The idea of a millionaire businessman who wants to colonize Mars with a million people by 2050 is a lucrative business. This outrageous budget could be used to create a million self-sufficient oases in a million deserts on our planet, which is what we need right now. The world's most powerful people have it easy to make history today, not only for one million people but for billions.

Although this may seem obvious, for a sustainable future, we humans need to be sustainable in our actions, processes, and infrastructures and in the energy required to support those actions, processes, and infrastructures. Therefore, at last, especially in recent times, sustainability has become a pressing priority for the entire planet. It has become urgent, a real emergency, especially now that we are facing the threat of nearing a point of no return. Luckily, although slowly, all these issues are increasingly seeping into public consciousness. This includes those people who run large multinationals. By taking small steps to reduce the harm brought on by their companies, they have the unique opportunity to generate an enormous effect and make change happen. For example, in a relatively short period of time, the media have echoed the studies on the disastrous effect of plastic bags, or something as specific as plastic drinking straws, on the environment. After raising public awareness, clamors, and complaints, there has been a wave of plastic straw bans by companies worldwide, and those problematic items are gradually being withdrawn from the market. All this was generated by a simple decision of corporate leaders, and not a single client was lost in the process. It is clear that the surprising speed at which this has happened has to do with social network empowerment, enabling citizens to get involved in corporate issues and take a stance, triggering an immediate cascade. However, much remains to be done to overcome the complexity and contradiction observed in human behavior, as can be seen in Figure 12.1. On one side, some big companies offer plastic artificial grass to cover landscapes, and, on the other side, other large companies are replacing plastic cutlery for metal cutlery, which is a very commendable decision taking into account what it represents in terms of costs. Our planet, however, can no longer bear those economic profits, because the need for a sustainable planet today is priceless. On one side, some shops care about selling in-season fruits and vegetables, and, on the other side, the shop next door to our home sells spring fruits, which come from the antipodes by polluting airplanes, in winter. We are all responsible, and we must refuse to maintain unsustainable behavior. For example, something as simple as avoiding to buy fruits and vegetables out of the natural season they are harvested in helps, because those foodstuffs come from very far away and this generates contamination. Consuming those products is a whim, not a vital necessity. Maybe we can afford this expense, but the damage that we do to our planet today cannot be undone tomorrow.

We have reached this point precisely because of the technological advances achieved. Those breakthroughs, especially the ones related to artificial intelligence, have increasingly significant implications in all areas of human activity. Therefore, open and specific debates on these issues should be considered more necessary than ever. It is clear that humanity can obtain an unquestionable benefit from technological progress, although it depends on the ethical use we make of these new possibilities available. In parallel to the immense technological progress, an appropriate development of responsibility and values is required. From north to south, from children to adults, extensive training in those values is essential. It is much more urgent for children to know all the principles of sustainability that should govern our daily lives

**FIGURE 12.1** Complexity and contradiction observed in human behavior. Some big companies offer plastic artificial grass to cover landscapes (a), while other large companies are changing plastic cutlery for metal cutlery (b). Some shops care about selling seasonable fruits and vegetables (c), and, the shop next door to our home sells spring fruits, which come from the antipodes transported by polluting airplanes, in winter. (Photos: Alberto T. Estévez.)

than other useless information that fills their brains and that will be forgotten after they have used it in an exam.

Each one of us has to do more than strictly our part. For this reason, since the 1980s, the architect who signs this chapter wanted to learn and he started putting into practice the principles of passive solar architecture (Estévez, 2019). In 1996, he founded the first architecture school in the world (ESARQ - UIC Barcelona) that included compulsory undergraduate subjects on sustainability (with contents on combining ecology and architecture) and international cooperation (with contents on merging sociology and architecture). He also promoted two pioneer master's degrees in sustainability and international cooperation, the latter successfully founded in 2004 thanks to Amanda Schachter and Alex Levi's valuable help.

This is the **background** we have, the present landscape, the challenge! And each and every one of us, without exception, is the **driving force** to take on that challenge.

## 12.2 NATURE, WHAT ELSE?

And now, let us look around. We do not need to go far. It is right there, before our very eyes, in our own cells, bones, and physiological systems. It is present in all plants and animals, large and small, that surround us, even if we do not see it. This is where to find the model, the top model for sustainability: **Nature, what else?** (Allow us to talk about "Nature" as the human convention provided by its definition, as an abstract entity, as a human figuration of the set of elements that form our environment).

Nature's "laboratory" has been improving living beings for millions of years, with the main objective of just "living." It has developed from the simplest to the most complex. It would therefore be folly not to go to the source of that enormous experience and wisdom in order to adjust our main objective today, which is also "living," and the survival of our planet. Our goals, those of Nature and those of humanity, converge. Nature, however, has a huge advantage over us as far as its "science" and applications are concerned. Therefore, we have to focus on learning from it, through biolearning (Estévez, 2014).

Each field of human knowledge must find its own biolearning path. In our case, we will try to undertake biolearning to apply it in the fields of engineering, architecture, and design. This is where a great champion, a great pioneer, Antoni Gaudí enters the scene. He was the first to transcend the immediate inspiration Nature offers beyond the purely formal application to architecture and design, which is what architects and designers sometimes do.

Most people know his unique works which attract millions of tourists to Barcelona every year. Thanks to the Internet, it is easy to find out about some aspects of his biography, so we are not going to cover this in the present chapter (or check Estévez, 2002a, 2010). However, only a few know what is behind the forms of his architecture: real and deep biolearning of Nature. When someone asked Gaudí where he had learned his architecture, he pointed to a tree near his workshop saying: This is my teacher! (Bergós, 1974). Indeed, when he was a child, because of his delicate health, he spent long periods of time in the countryside, surrounded by mountains and stones, trees and flowers, birds and insects, transforming his weakness into his strength. "Little Gaudí collected the purest and most pleasing images of Nature, that

Nature he always called 'my teacher', and he remembered them with pleasure even in his old age" (Bergós, 1974). At the same time of being "blessed with a remarkable gift of observation, he developed a personal architecture based on applying what he learnt from Nature to his work" (Bonet, 2000). That is also the lesson of Gaudí for each one of us: to learn from Nature from the point of view ("with glasses on") of our work, which, in the case of Gaudí, was architecture and design.

In fact, it is easy to marvel at the forms of Nature seen from the outside, but it is not enough. As soon as we pay closer attention, its processes and functionalities become fascinating. Sunflowers, for instance, rotate their heads during the day to track the sun's movement across the sky. Gaudí paid tribute to sunflowers at the beginning of his professional career, as can be seen in the central photo of Figure 12.2. He integrates them into his first two works in ceramics, especially in *El Capricho* (1883–1885), but also in a small part of *Casa Vicens* (1878–1888). It is as if he felt from the very beginning not only that he was going to stick to a formal inspiration of Nature but that he would be the first architect in history to transcend to a deeper level of learning from Nature, of biolearning. Undoubtedly, his special sensitivity for Nature provided him with it. When Gaudí was working on his first house, *Casa Vicens*, he felt sorry for the little yellow flowers on the construction site that were doomed to disappear, so he decided to preserve them forever in the decorative ceramic tiles, which covered the whole building, as can be seen in Figure 12.2b. The palmetto leaves in the iron fence were included for the same reason. Antoni Gaudí was a magnificent example of someone who respected the environment, long before anybody talked of ecology (Estévez, 2002a).

In that sense, the fact of going beyond the mere exterior form of Nature, until—as he said—discovering the secret laws of the universe,

> Gaudí's work is an advance from traditional architecture to new architectural structures based on the mechanics and experiments such as those of the catenaries, but, at the same time (...) it truly penetrates in the world of natural morphology, which does not copy but transfigures and integrates into an architectural or structural-ornamental factor.
>
> **(Cirlot, 2001)**

**FIGURE 12.2** (a) Sunflower field. (b) Sunflower façade decoration (Gaudí, *El Capricho*, Comillas, 1883–1885). (c) Little flower façade decoration and palmetto leaves in the iron fence (Gaudí, *Casa Vicens*, Barcelona, 1878–1888). (Photos: Alberto T. Estévez.)

Gaudí did not adopt the different arches defining each architectural style (Roman, Visigoth, Romanesque, Gothic, Renaissance, Mannerist, Baroque, Mudejar-Mozarabic, Hindu, etc.), as his contemporaries used to do in the architecture of the late 19th and early 20th centuries. He would have his own, the catenary arch. That is, his own constructive system, which cannot be identified with any historicist style.

This functioning of the catenary arch can be understood, by looking at Figure 12.3, where a free hanging chain (catenary) is transferred to parabolic-catenary arches at the *Teresian College* designed by Gaudí (Barcelona, 1887–1888). On the right of Figure 12.3, Gaudí's peculiar architectural design methodology is shown. He created string models, hanging from the ceiling tying bags of sand to them to calculate the weight proportional to that of the real construction. When turning the photo of the model upside down, the structure of the designed building emerged in the most structural and efficient way.

Only engineers using mathematics and physical science in the construction of bridges, railway stations and iron markets, and Catalan masons applying centuries-old tradition and popular experience have approached the catenary arch, as it is the most consistent with the natural function of structures. Gaudí went further than those engineering infrastructures and anonymous popular constructions and, by using his artistic rationality, he was the first to take architecture to a whole new level. He was scientifically aware of its structural effectiveness. This fact confirms Nature was his definitive source of inspiration and learning. He did find inspiration not only in its "flesh," formal and external, but also in its "bones," in its structures and functions. Gaudi's works are never dressed in arbitrary ways, but they respond to the same criteria of tectonic functionality as those of Nature. He managed to provide maximum structural efficiency to the traditional compression architecture by following those criteria, closing that chapter to start modern traction architecture. Thus, Gaudí offered the alternative of the unbeatable catenary arch in architecture when compared to the poor static performance of arches of historical styles (semi-circular, pointed,

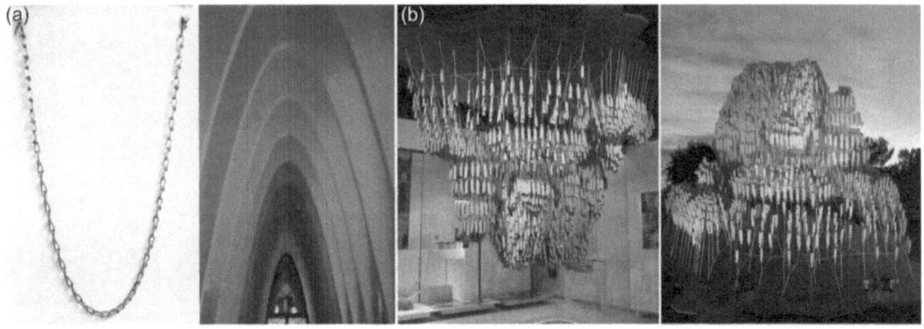


**FIGURE 12.3** (a) Catenary (free hanging chain) and parabolic-catenary arches (Gaudí, *Teresian College*, Barcelona, 1887–1888). (b)Right: image of the sString model, hanging from the ceiling with bags of sand to calculate the weight proportional to that of the real construction, following Gaudí's peculiar architectural design methodology. When turning the photo of the model upside down, the structure of the designed building emerged in the most structural efficient way. (Photos: Alberto T. Estévez.)

horseshoe, etc.), which presented structural deficiencies in roofs, vaults, walls, and vertical pillars as well. The catenary arch solves everything better, converted into a surface via successive repetition (which forms the characteristic diaphragm arches in Gaudinian spaces), repeated through space along a curve that is perpendicular to its own and of similar geometry. These catenary arches, and their similarity to the parabola, and their corresponding vaults of ruled geometry and resulting inclined pillars follow the natural line of thrust of the loading: they achieve the ideal structural work in the buildings with the minimum necessary material and with the minimum energy cost, as Nature does. Gaudí innovated the history of architecture taking compression construction to its highest point. Furthermore, we should recognize that these resulting sinuous curves are those of Nature, in whose natural evolution the most appropriate structural geometries have been selected, so that plants and bones are—apart from cost-effective—maximally strong. Trunks, branches, stems, and all the apparatus of living beings are made up of this geometry, of conoids, ellipsoids, hyperboloids, paraboloids, and helical growth patterns (Estévez, 2002b).

As we can see in Figure 12.4, paraboloids, hyperboloids, and helical growth patterns govern the shaping of trees. And Gaudí, learning from Nature, designed with paraboloids, hyperboloids, ellipsoids, and helicoids the whole interior of the *Sagrada Familia Church*, where the pillars of the Basilica spread out like tree branches (see Figure 12.5), following the natural lines of thrust from the structure above.

Nature never stops teaching us, patient, from the origin of life on this planet, as if it were waiting for us to be sufficiently prepared and mature. We only need open and curious minds and sensitive and attentive spirits in order to learn its lessons. For example, as seen in Figure 12.6, plant leaves achieve the most efficient balance when using the minimum material (less energy needed in the internal processes of growth), the minimum thickness, and covering the largest possible area, to absorb

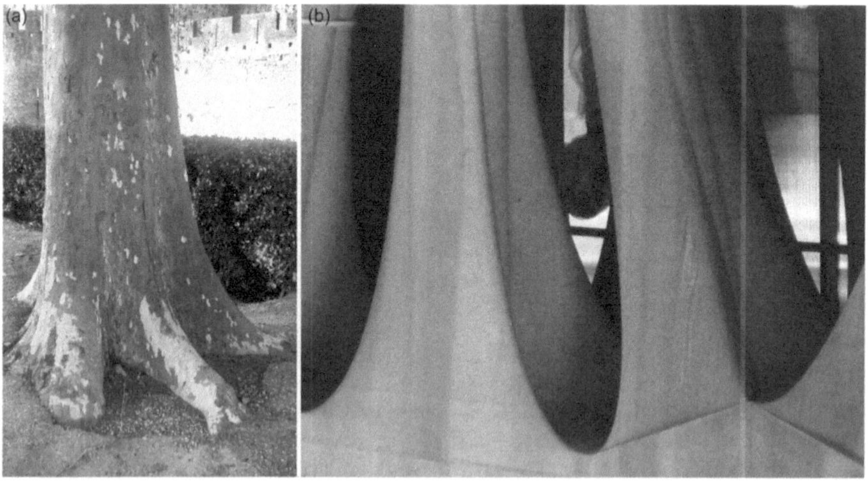


**FIGURE 12.4** Paraboloids and hyperboloids of a tree (a). Paraboloids, hyperboloids, ellipsoids, and helicoids inside the *Sagrada Familia Church* (b).

**FIGURE 12.5** (Below) The pillars spread out like tree branches, following the natural lines of thrust from the structure above. (Photos: Alberto T. Estévez.)

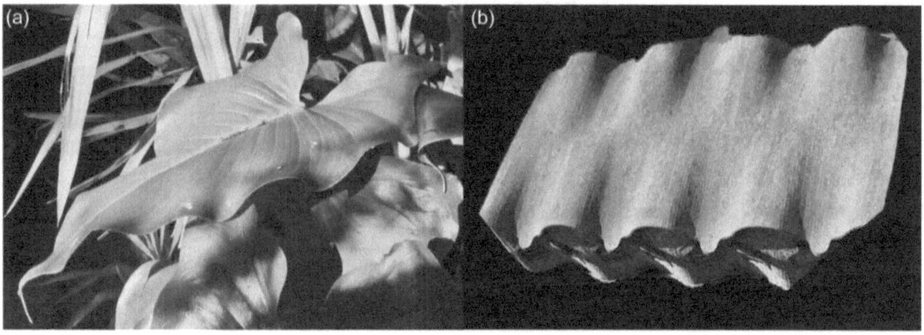


**FIGURE 12.6** (a) Plant leaves achieve the most efficient balance when using the minimum material, the minimum thickness, and covering the largest possible area, to absorb the most light and to allow water to flow, thus avoiding overloads in cases of rain. To do this, their surface curls, giving greater rigidity to the whole. (b) Gaudí, *Sagrada Familia Schools*, 1908–1909, design of the roof, following the same concept of efficiency using twisted ruled surfaces (conoids). (Photos: Alberto T. Estévez.)

the most light and to allow water to flow, thus avoiding overloads in case of rain. To do this, their surface curls, giving greater rigidity to the whole, more than if the surface is flat, which is the case of conventional human buildings, furniture, and numerous objects. Gaudí knew this and designed the roof of the *Sagrada Familia Schools* (1908–1909), following the same concept of efficiency using twisted ruled surfaces and conoids in this case.

Similarly, as shown in Figure 12.7, bones in Nature achieve the most efficient balance when using the minimum material (less energy needed in the internal processes of growth) and their resistance, following hyperbolic forms. They are more resistant

**FIGURE 12.7** (a) Hyperboloids in an image of bones, (b) balustrade (Gaudí, *Casa Batlló*, Barcelona, 1904–1906), and (c) pillars of a gallery (Gaudí, *Sagrada Familia Church*, Barcelona, 1882–present). (Photos: Alberto T. Estévez.)

than the shape of a cylinder with the same amount of material, as those found in conventional architectural columns. Again, Gaudí knew this, and in *Casa Batlló* or in the *Sagrada Familia Church*, he designed the vertical elements with hyperboloids, following the same efficiency principle.

Many details can be learned from Nature, as long as we keep our eyes wide open. As Figure 12.8 shows, the tiles of Gaudi's *Casa Batlló* (1904–1906) follow the principle of fish scales; besides, they are provided with hydro-dynamicity, because they need to let the water flow in the best possible way.

Gaudí only needed Nature and its guiding principles to develop all his work as an architect and designer. He also managed to synthesize a basic point of Nature, which ended up configuring a value, a value of intelligence, an indicator of quality: continuity. And he said: "the continuous forms are the perfect ones. (...) Polyhedral forms and those mistakenly called geometric forms do not abound in Nature. Even those that man makes flat (doors, tables, boards) become warped over time" (Puig-Boada, 2004). In Figure 12.9, we can see the continuity of forms as an essential characteristic in Nature, where each part and each function is solved with continuity on the whole. This is undoubtedly because continuity provides better structural and growth conditions. Gaudi's architecture offers even greater advantages by designing his works in this same continuity as Nature. This can be seen in the exterior of *Casa Milà* (Barcelona, 1906–1911) or in the interior of *Casa Batlló* (Barcelona,

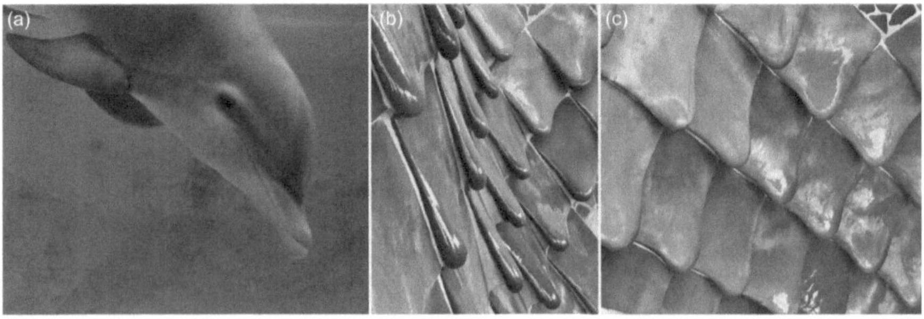


**FIGURE 12.8** Dolphin and roof tiles like fish scales of Gaudí's *Casa Batlló*, Barcelona, 1904–1906. (Photos: Alberto T. Estévez (a) and Judith Urbano (b and c).)

**FIGURE 12.9** (a) Continuity in the forms of Nature, lily and little snail "surfing" on it. (b) Continuity in the forms of Gaudí's works, exterior of *Casa Milà* (Barcelona, 1906–1911), interior of *Casa Batlló* (Barcelona, 1904–1906). (Photos: Alberto T. Estévez (the two images on (a) and Judith Urbano (the two images on (b).)

1904–1906). Exterior continuity helps the flow of air and water. Interior continuity improves thermal and acoustic characteristics. This is form and function in perfect convergence and harmony.

It is clear that Nature, as an abstract entity, has no awareness of beauty. Nature does not look for it as artists or architects do, promoting or discarding living beings in relation to their beauty, but only in relation to their functional efficiency. Despite this, as if it were subjective, we find Nature beautiful in its continuity and flow. This occurs in a kind of "automatic" human agreement, beyond cultures and traditions. It is thus Gaudí himself who recognizes that there is an "objective beauty," which is precisely what Nature shows us through its achievements of functional efficiency. And this is always accompanied by color.

Gaudí tells us:

> Nature does not present any object to us in a monochrome way, completely uniform in terms of color, neither in vegetation, nor in geology, nor in topography, nor in the animal kingdom. The contrast of color is always more or less vivid, hence we must use color in a part or in an entire architectural element.
>
> **(Ràfols, 1999)**

This is said by an architect who used Nature as a model precisely in order to learn the architecture that must be performed consciously from it.

Once it is understood that we have to give color to what we do, a second point appears: color in Nature has gradations and blurring, different shades, and chromatic nuances. This is something, as can be seen in Figure 12.10, Gaudí also applied in his works: either by introducing these gradations of colors in the material itself, as in the glazed ceramic tiles at *Casa Batlló*, or by introducing them into the light that tinges the spaces, thanks to installing colored stained glass, as the reflection we can see in the structures of the *Sagrada Familia Church.*

A third consideration would be to observe how color appears optically "vibrant" in Nature, pixelated as we would say today by analogy with the digital world. In Figure 12.11, the mentioned optical vibration of every little point can be seen in Nature, from a microscopic size to a larger scale, as for example by looking at flowers

**FIGURE 12.10** Color gradations of Nature (a), ceramic roof tiles at *Casa Batlló* (b), and the light reflected on the structures of the *Sagrada Familia Church* (c). (Photos: Alberto T. Estévez (a and c) and Judith Urbano (b).)

**FIGURE 12.11** From left to right, the optical vibration of every little point that can be seen in these microscopic images of a butterfly wing (4x), a butterfly eye (8x), and in a tree with its little flowers and leaves. Next, the ceiling of *Park Güell* in Barcelona (1900–1914), built with small ceramic tiles, all different, and of different white tones. Although each element is similar to another, the small differences in size, shape, position, color, and light together produce a special and mysterious attraction to the human eye, more so than artificial elements, which are all always absolutely identical in their manufacture and homogeneous color: this gives connotations of coldness, artificiality, less humanity, less attraction, and less visual comfort to the whole. (Photos: Alberto T. Estévez.)

and leaves. Gaudí also introduced it in his works, in his typical colorful "trencadís" made with small ceramic tiles. As in Nature, the little pieces of his building surfacing are all different in color, even the white tones found in *Park Güell* (Barcelona, 1900–1914), for example. Although each element in Nature is similar to another, the small differences in size, shape, position, color, and light together produce a special and mysterious attraction to the human eye, more so than artificial elements, which are always absolutely identical in their manufacture and homogeneous color: this gives connotations of coldness, artificiality, less humanity, less attraction, and less visual comfort to the whole. Gaudí also followed Nature with respect to color application.

Clearly, for such an effective inspiration, we need sincere fascination, which is only possible by looking at Nature in a loving and sensitive manner, behind which there is respect for the environment, to such an extent that Gaudí can be qualified as the first environmental and ecological architect, at a time the words ecology and sustainability did not even exist (Estévez, 2010).

The emergence of modernity in architecture and design implied a rejection and demolition of quality values that were traditional in history. Thus, in contemporary times, the safest quality value was consolidated around the level of originality that a work could have. Given the empty desire for originality that architects and designers sometimes had, Gaudí said one of his most famous sentences: "Originality means returning to the origin" (Martinell, 1967), and the origin is Nature. This is a sentence that brilliantly channels the superficiality of just looking for name and fame towards the deepest root where the authentic quality of a work lies, the learning that any work could take from Nature. This is because the authentic quality is reflected in the intelligence we use to do something. We know that sensitivity is also intelligence, together with ingenuity. And the intelligence that is seen in the conformation of living beings, in Nature, is still very far from being surpassed by the human being, if he ever succeeds. Therefore, it would be good for scientists to humbly recognize it, instead of appearing in the media exuding self-assurance and complacency, for every tiny achievement they accomplish by joining a few molecules. In contrast, Gaudí's humility, the sublime synthesis of his work and thought, made him exclaim:

> My ideas are of an indisputable logic, but I just do not understand how they have not been applied before, and I am to be the first to open fire. This is the only circumstance that sometimes makes me hesitate.
>
> **(Bassegoda, 1989).**

Gaudí had no doubts because even his reading of history was made under the principle of "return to the origin." For example, he said "we do not have to stick with what our ancestors did. They came to Nature to learn" (Matamala, 1999). Nowadays, in the greatest and most serious challenge that humanity faces, paradoxically, to go forward we must look "backwards," towards the origin of everything, "return to the origin, back to the basics." We must learn from Nature and from all its potential to teach us the most perfect understanding of sustainability we can apply. That is the origin, the basics, and the most effective source we can turn to for solving the sustainability of the entire planet. In fact, Nature had already ensured such sustainability for millions of years, before the human being disrupted it in just a few decades. Nature solves the planetary sustainability with life. Human beings must follow the same path. Something shouted out loud for many years: from the origin to today's evolution in architecture and design, taking us to concepts such as Genetic Architecture (Estévez 2003), Biodigital Architecture (Estévez, 2015), and Metabolic Architecture (Dollens, 2017).

Sometimes we are able to introduce solutions from Nature, with living elements, in a very simple and traditional way, as presented in Figure 12.12: living plant elements, which are sustainable, renewable, recyclable, applied to architecture; façades, which block the direct incidence of the sun with its shadow on the walls, prevent heating in summer, emit oxygen, and absorb $CO_2$ and dust. Today, any gesture in favor of introducing life into architecture and design should not only be well received but is necessary.

**FIGURE 12.12** Images of (sustainable, renewable, recyclable) living plant elements applied to architecture; façades which block the direct incidence of the sun with its shadow on the walls, prevent heating in summer, emit oxygen, and absorb $CO_2$ and dust. (Photos: Alberto T. Estévez.)

However, virtually all academics (with few exceptions) of current architecture consider including life in architectural projects suspicious, as they are still governed by the principles of an obsolete architecture, devised in the 1920s and 1930s of the last century, very far away from the understandings of nowadays avant-garde architecture, of biodigital architecture: which includes key words as biology and digital, concepts as natural intelligence and artificial intelligence, biolearning and machine-learning, organic forms and digital tools, and bio-manufacturing and digital-manufacturing. If Gaudi currently lived, would he fit into this?

We must pay attention to the signs of the times, to the *Zeitgeist*, or we will look ridiculous when our history is written, as today, in addition to the great development experienced by biological technology, we can count on the immense help of digital technology. To the power of the media that we have in our hands today, we have to add intelligence and willingness to face the challenge of our time. Sensitivity, for adjusting the objectives and research methodologies and the uses we give to the results, is also required.

## 12.3 CONCLUSION

We can see the true meaning of "back to the basics, and return to the origin," and let biolearning give us the clues to solve the urgent needs our global society faces. The closer the proposals are to Nature, to life, the better the results we will obtain. Nature means efficiency and sustainability and, therefore, *Green Energy and Infrastructure: Securing a Sustainable Future.*

For this reason, we must help to increase research, which is essential to find better sustainability solutions. Let us involve politicians, who are the ones who must direct their citizens towards the real common good, and this can only be done with a global vision of international solidarity. We must also take advantage of all the digital technology that gives us enormous advantages for such research and we have to implement its solutions. Biological and digital techniques combined with biodigital architecture offer us new and advanced possibilities.

In conclusion, if technological progress causes increasing inequalities, it cannot be considered real progress. If the so-called technological progress of humanity becomes an enemy of the common good, it would lead to an unfortunate regression, to a form of barbarism dictated by the law of the strongest. Therefore, our work must engage in an effort of civilization, which will also be measured by the objective of reducing economic, educational, technological, social, and cultural inequalities. Obviously, we all have to act from our field of expertise, as when there are no more areas or people that are not affected by (or that do not affect) global sustainability in a positive or negative manner. In our case, it is the field of engineering, architecture, and design.

It could be claimed that the responsibility requested is really fulfilled when one begins to live in harmony with oneself. This means to live in harmony with Nature and with human beings at the same time. There is an intercommunication between the entire cosmos, where they are neither exclusive nor excluded. The project for everyone to live a full life can only be created in this way. Such an understanding of life is characterized by the connectivity and harmony of relationships between water, territory and Nature, community life, and culture. Do we want to live well? We can only do so if we do things well, making responsible use of all the goods that are available to us in this world. We need to adopt responsible habits, protect the land, and change our culture of excessive consumption, solid waste production, and stimulate reusing and recycling. We need to reduce our dependence on fossil fuels and the use of plastics. Changing our eating habits (excess of meat, fish, and shellfish) for more austere lifestyles is necessary and so is being actively engaged in planting trees, in looking for sustainable alternatives in agriculture, energy, and mobility respecting the rights of Nature. We need to promote the education of integral ecology at all levels and look for new economic models and initiatives for a sustainable quality of life, for an integral and self-sustainable development. We urgently need the development of new clean energies and policies that drastically reduce carbon dioxide ($CO_2$) emissions and other polluting gases. All companies should establish monitoring systems for their supply chains in order to ensure that the products they buy, create, or sell are produced in a socially and environmentally sustainable way. The responsibility is personal, collective, corporate, and political.

Luckily, a better world is possible precisely thanks to technological progress. However, it is only possible if it is accompanied by an ethic based on the vision of the common good, an ethic of freedom, responsibility, and fraternity, capable of promoting the full development of people in relation to others and to the environment. We need all of it, in the only world that we have under our feet in the short and medium term.

## REFERENCES


Bassegoda, Juan, *El gran Gaudí*, Ausa, Sabadell, 1989, p. 27.

Bergós, Juan, *Gaudí, el hombre y la obra*, Universidad Politécnica de Barcelona, Barcelona, 1974, p. 18 and 104.

Bonet, Jordi, *L'últim Gaudí / The Essential Gaudí*, ECSA / Pòrtic, Barcelona, 2000, p. 12.

Cirlot, Juan-Eduardo, *Gaudí: una introducción a su arquitectura*, Triangle postals, Barcelona, 2001 (1966), p. 14.

Dollens, Dennis, *Metabolic Architectures: Turing, Sullivan, Autopoiesis & AI*, ESARQ (UIC), Barcelona, 2017.

Estévez, Alberto T., *Gaudí*, Susaeta, Madrid, 2002a.
Estévez, Alberto T., "Antoni Gaudí: Arquitectura cromática". In Judith Urbano (ed.), *Gaudí a París l'any 1910*, Sites Books / ESARQ (UIC), Santa Fe / Barcelona, 2002b, pp. 7–10.
Estévez, Alberto T., *Gaudí (Enciclopedia del Arte)*, Tikal, Madrid, 2010.
Estévez, Alberto T., "Genetic architectures / Arquitecturas Genéticas". In VV.AA., *Genetic Architectures / Arquitecturas Genéticas*, Sites Books / ESARQ (UIC), Santa Fe / Barcelona, 2003, pp. 4–17.
Estévez, Alberto T., "Learning from nature: Architecture and design in the first biodigital age". In Alberto T. Estévez (ed.), *2nd International Conference of Biodigital Architecture & Genetics*, ESARQ (UIC), Barcelona, 2014, pp. 8–23.
Estévez, Alberto T., *Biodigital Architecture & Genetics: writings / escritos*, ESARQ (UIC), Barcelona, 2015.
Estévez, Alberto T., "Sustainable living? Biodigital future!" In Jacqueline A. Stagner, David S-K. Ting (eds.), *Sustaining Resources for Tomorrow*, Springer International Publishing / Springer Nature, Berlin, 2019, pp. 137–162.
Martinell, César, *Gaudí: su vida, su teoría, su obra*, Colegio de Arquitectos de Cataluña y Baleares, Barcelona, 1967, p. 141.
Matamala, Juan, *Antoni Gaudí. Mi itinerario con el arquitecto*, Claret, Barcelona, 1999, p. 342.
Puig-Boada, Isidre, *El pensament de Gaudí*, Dux, Barcelona, 2004 (1981), p. 165.
Ràfols, Josep Francesc, *Gaudí 1852–1926*, Claret, Barcelona, 1999 (1928), p. 95.

# 13 Triple Bottom Line Analysis, Methodology, and Its Implementation

*Amit R. Patel*
Maharaja Sayajirao University of Baroda

*Prabir Sarkar, Himanshu Tyagi, and Harpreet Singh*
Indian Institute of Technology Ropar

**CONTENTS**

## 13.1 INTRODUCTION TO TRIPLE BOTTOM LINE ANALYSIS

"Triple bottom line" analysis is the assessment of a process or a product considering economic, environmental, and social impacts. The concept is slowly but steadily being accepted into the corporate world and the management structures. The expression of triple bottom line analysis was first coined by John Elkington; he denoted the importance of triple bottom line analysis with the help of a pictorial representation of *Cannibals with Forks* on the front cover of his book, representing cannibal as a business entity and triple bottom line analysis as a tool, shown with the help of the three prongs of fork (Elkington 1999). They stated that although a company's bottom line is what the company earns financially, however, the company should also consider the environmental impact as well as the social implication of their product into account.

The notion of triple bottom line analysis is wider than this. The concept of the triple bottom line analysis is believed to be a comprehensive measurement of an impact or the change that is proposed or being carried out in the system. Triple bottom line analysis is also referred by some authors as simply another term for corporate social responsibility and is a step leading towards a broader concept of corporate sustainability (Hohnen and Potts 2007). Under triple bottom line analysis, a process requires to satisfy the analysis along with all the three components simultaneously (Brova 2008; Swarr et al. 2011), these three bottom lines are linked and dependent on each other (Murakami 1999). Literature suggests adhering to the triple bottom line analysis is a cornerstone of corporate social responsibility and all the three pillars are linked to each other, for instance, if the environment situation of a society may deteriorate, then it may influence economic condition as well and this, in turn, will influence the societal change all in a cyclic manner. Therefore, all three must be given equal importance in the company's balance sheet (Jacobsen 2006).

Although the triple bottom line analysis is seen as a constraint to the business from the corporate world, in end, it is largely agreed that the triple bottom line analysis is a long-term approach and is receiving a positive response (Jacobsen 2006). It has also been increasingly followed in management, consulting, investing, and even in non-governmental organization circles (Norman and MacDonald 2004). Janeiro (2011) in his thesis referred to a very popular equation, which is originally proposed by Kloepffer (2008), that states that life cycle sustainability assessment (LCSA) of the products or a process can be seen as a total effect of life cycle assessment (LCA), life cycle costing (LCC), and social LCA (sLCA).

As stated, LCSA consists of three components, and given the fact that all the three components are differing fundamentally from one another, the results are also fundamentally different. The issue of interpreting results having three different dimensions is not only difficult but also subjective in nature. It is argued by Janeiro (2011) that in the absence of a specific objective and measurable conceptual framework of sustainability, deciding the required complexity level of the sustainability study is important, any inadequacy will possess a risk of missing the core objective of study and make entire process a pointless exercise.

Another issue in dealing with LCSA is choosing and interpreting the results in terms of indicators. It is proposed to integrate (or aggregate) the results obtained from the three analysis under certain framework like Coordination Action for innovation in Life Cycle Analysis for Sustainability—CALCAS (Janeiro 2011). However, at times, the results obtained from the three pillars are not merged, instead, it is left to the user to weigh the outcome and interpret it. This is because, certainly, the results in all three dimensions are dependent, but even so, they are unequal in nature and do not pose uniformity due to highly qualitative nature. Although each pillar has a separate benchmark or qualifying standard, it is also important to note that the results from the three pillars are not replaceable by one another. Judging the importance of trade-off between the three pillars is an essential part of doing triple bottom line analysis. For instance, higher air pollution emission from a process cannot justify lower score on the social front or higher profit cannot justify higher air pollution emission (which is the case of many processes involving emission).

### 13.1.1 Environmental Sustainability

LCA is a systematic, analytical, and comprehensive method to identify, evaluate, and minimize the environmental impacts of a specific process (Morrison and Sinclair 1998). It considers all the stages of a product's life cycle, such as extraction of raw material, processing the material, manufacturing the parts of the product, use or operation, and end of life (Williams 2009).

According to ISO standard 14040 and 14044, LCA technique has four basic steps, shown in Figure 13.1, which also describes the framework for the LCA analysis (ISO 2006), (ILCD Handbook 2010). The brief scope of the four steps are as follows:

a. Definition and scope, that determines the need for study, data specificity, decides appropriate collection method and data presentation,
b. Life cycle inventory, carrying out exact process diagrams, executing data collection, and evaluation of the data,
c. Life cycle impact assessment, in which the impact categories are defined, and their weights are decided, and subsequent results are discussed, and
d. The final report, which includes significant data pertaining to the study, data evaluation and its interpretation, final conclusions, and recommendations (Williams 2009).

The outcome of an LCA study is a quantitative determination of environmental impact and opportunities for improvement by quantitatively justifying a change in the process for environmental impacts (Williams 2009). The results of the study are conveyed through instruments such as eco-labeling, eco-design, carbon footprint, and green public procurement (ILCD Handbook 2010). In other words, LCA can

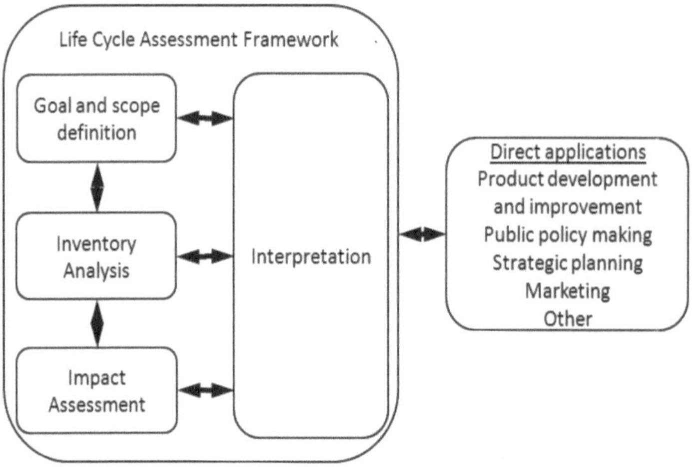


**FIGURE 13.1** Life cycle assessment framework as per ISO 14040 and 14044, 2006. (Adapted from O'Connell 2005.)

help in determining hotspot(s) in the process chain so that alternatives may be evaluated with an aim to reduce overall environmental impact (Hassan, Jaramillo, and Griffin 2011).

In one of the relevant studies, conducted by Roberts et al. (2010), LCA was applied to evaluate the use of biomass under pyrolysis option. The study compared the effect of adapting biochar route for the straw residue obtained from the pyrolysis, against the conventional utilization of biomass thorough combustion. The study considered 1,000 kg of dry biomass as a functional unit in the pyrolysis process. Emission factor (EF) in both cases was determined in terms of kg of $CO_2$/1,000 kg of biomass. Results observed a negative EF for biochar route compared to the direct combustion option, for which EF was positive (Roberts et al. 2010). Singh et al. (2010) applied LCA to estimate energy consumed and air pollution emissions released from a bioenergy alternative against a conventional means of generation of energy. Later, Pant et al. (2011) applied LCA as a method to support decision-making for new technology alternatives from bioenergy.

### 13.1.2 Social Sustainability

The objective of social sustainability is to try and ensure that future generation should have the same or greater access to social resources as the current generation by implementing required processes, adopting suitable systems, organizing necessary structures, and building healthy relationships; thus, its measurements are healthy and livable communities (McKenzie 2004). Although a social dimension to sustainability is widely accepted, exactly what this means has not been very clearly defined and agreed (Bramley et al. 2006).

Practitioners of LCA have recently started giving attention to social issues. This gives way to a new but interesting sector of sLCA (Andrews et al. 2009).

sLCA vouch for social and to some extent socioeconomic aspects of product and their potential positive as well as negative impacts along the life cycle. Although, sLCA provides an element of thoughts and fails to offer any basis for the conclusive decision, its importance as a third pillar in the triple bottom line analysis is significant. Interpretation of subcategories is expressed in light of the impact category. Social impacts are interpreted as the outcome of the complex relationship to reflect the set of categories considered for the study.

Guideline suggested by United Nations Environment Programme (UNEP) for conducting sLCA is believed to be the most comprehensive and widely used (Benoît et al. 2009; Andrews et al. 2009). The guidelines offer a skeleton and help in defining key elements to include in the analysis. It also provides information for defining the goal and scope, inventory, impact assessment, and interpretation phases of sLCA and offers the basis for the development of databases. The stakeholder category and subcategory are used to consider the effect on the impact category (Benoît et al. 2009), these impacts are measured qualitatively or quantitatively.

### 13.1.3 Economic Sustainability

Economic sustainability using LCC methodology is also known as cost–benefit analysis. Under LCC, the present value of the different cost component is obtained. These cost components are capital investment cost, the replacement cost of equipment, operation and maintenance cost, fuel cost during operation, cost of disposal at the end of life of the equipment, etc. The present value of all these cost component is summed up to obtain LCC of a process (Ravindranath et al. 2006). The advantage in applying discounted cash flow method of obtaining LCC helps in estimating and comparing installing and operating cost for the different types of energy generation processes. In other words, it helps in bringing different types of project with different cost structure on a single scale for judicious decision-making.

One of the disadvantages of LCC method is the lack of the provision to consider the risk into the analysis. Sensitivity analysis, used along with LCC results, is a fairly good tool to resolve the issue. Sensitivity analysis will help quantify and, therefore, determine the risk associated in accepting different input data, such as discounting rates and escalation rates, considered for the study. Kneifel (2010) applied LCC method to determine if investment in the capital cost burden that arises due to the proposed modification or required additional capacity increment is justified or not. Under this, a "base case" scenario is obtained and the LCC cost of this scenario is compared with the other alternate option(s) that is proposed for the purpose of screening (Kneifel 2010).

Literature recommends that LCA and LCC should not have the same system definition, method of allocation, treatment of time, aggregation, etc. for a given problem, instead LCA and LCC are two distinct, and certainly non-inclusive, ways of extracting indicators from exactly the same system (Heijungs, Huppes, and Guinée 2009).

Literature also highlights the possible discrepancies in determining the cost into the study. The beneficiaries may influence the scope of the cost input for the study, and therefore, the result may alter significantly. If the cost of damage done to the environment is also considered in the LCC study by the stakeholder, the results would be drastically varying in many cases. For instance, the cost of building a hydropower plant may significantly differ if the government includes the cost of the compensation required to pay to the displaced people. However, the construction cost of the dam site is the only criteria for a construction company. Additionally, the external cost due to inclusion of such externalities is not only subjective but are also different for different stockholders, for instance, customers and service providers often have conflicting views about what is an adequate level of service that should be provided or offered to the customers. This varies the LCC from service to service and from actor to actor throughout the supply chain system for a product or for a service (Finkbeiner et al. 2010).

## 13.2 A CASE STUDY

The present section will concentrate on the application of triple bottom line giving all the details about a practical example, while the following sections will deal with the further analysis. The sections will provide the reader with an in-depth understanding of Triple Bottom Line (TBL) process.

### 13.2.1 Problem Description

Open-field burning of wheat straw in the fields is a grave concern for the policy developers in many countries, due to its inherent nature of generating a large amount of air pollution emission that is taking place in the rural area (Garg 2008). Air pollution emission is a cause of concern, the way agricultural residues are currently burned in the open fields of Punjab, India. Combined harvesting technologies, which have become common in rice–wheat cropping system in India, leave a portion of the stem in the field after the harvest leaving behind large quantities of straw in the field. To remove the straw and prepare the field ready for the next harvest, farmers follow widespread burning of the residue.

Air pollution emission due to this open-field burning of agricultural residue is a significant component, comprised of approximately 61% in Asia and 39% for the rest of the world (Garg 2008). Statistics suggest that currently less than a quarter of rice straw in India, around half in Thailand, and almost all in the Philippines, are being burned in open fields (Adam 2013). The burning of 1,000 kg of straw will generate approximately 3 kg of particulate matter (0.17%), 60 kg of CO (3.5%), 1,460 kg of $CO_2$ (85%), 199 kg of ash (12%), and 2 kg of $SO_2$ (0.12%) (Gupta et al. 2004). Burning of straw also leads to adverse health effects (Kumar and Kumar 2010).

Researchers are equally concerned about the effect of the addition of pollutants into the air due to the consumption of diesel in the rural area for the generation of off-grid electricity. Both issues lead to large air pollution emission generation and waste of available resource and is a cause of concern. If the use of the large agricultural waste, the straw that is being burned in the fields, can be directed to generate electricity by use of any process, in the rural area, the twin problem of air pollution emission reduction and availability of electricity in the rural area can be effectively addressed.

To address both of these air pollution emission generation issues simultaneously, researchers believe pyrolysis can be an appropriate technology option (Gupta et al. 2004; Kumar and Kumar 2010). Pyrolysis provides an alternate; to open-field burning, by avoiding the burning of straw, and to reduce the use of diesel, by using biooil and gases in dual fuel mode to generate electricity. A sustainable power generation alternative using straw against the existing widespread open-field burning not just will pass on the positive environmental benefit but also would offset the air pollution emission due to the use of diesel and other fossil fuels.

### 13.2.2 Pyrolysis Process

To identify a possible solution for the above, we discuss the brief background of the pyrolysis process. Pyrolysis is a thermochemical conversion process in which feedstock is heated in an oxygen-free environment to obtain vapor and char (R. J. Evans and Milne 1987). Vapor fractions are condensed in a condenser, while char can be utilized for thermal applications or as a fertilizer in the field (Woolf et al. 2010; Major 2009). Upon condensation of vapor fraction, viscous dark liquid can be obtained called biooil (Bridgwater 2012), and some of the non-condensed gases will be left over.

Pyrolysis can be of three types. Fast pyrolysis in which biomass is subjected to high heat transfer rates, typically 6.8 W/cm$^2$ (Bridgwater and Peacocke 2000), and low residence time, typically two to three seconds, to maximize biooil generation (Bridgwater 2012). Slow pyrolysis in which biomass is subjected to slow heating rates and longer residence time, in terms of hours to maximize the char production (Bridgwater 2012). Intermediate pyrolysis is characterized by low heat transfer rates, typically less than 0.01 W/cm$^2$ and short residence time for vapors, typically two to three seconds and longer residence time for solids, typically 20–30 minutes (A. R. Patel 2018).

Intermediate pyrolysis is a recent version of pyrolysis that could address the limitation of both fast and slow pyrolysis, it helps generate an almost equal amount of biooil, biochar, and gases (A. R. Patel 2018). The intermediate pyrolysis process has been developed by the European Bioenergy Research Institute (EBRI) at Aston University, Birmingham, UK, and has been tested with wheat straw for results at Indian Institute of Technology (IIT) Ropar, Punjab, India (Hornung 2011).

The biooil generated during intermediate pyrolysis process can be mixed with diesel or biodiesel in various ratios depending on the application, for instance, in the ratio of 1:3, as fuel for off-grid power generation. The gas of low calorific value produced during intermediate pyrolysis is a by-product and could be utilized for low-grade energy heat-generation applications, such as boiler feed water heating. Non-condensable gas fractions generated at the pyrolysis will be flared in the present study. However, the non-condensable gas fraction can also be used for converting heat to any thermal applications or it can be fed to the engine under dual fuel mode and can augment the power generation using biooil blend.

Biochar is another by-product of intermediate pyrolysis. Biochar can be used as a fertilizer in the farm (Lehmann and Joseph 2009; Novak et al. 2009; Roberts et al. 2010; A. Zhang et al. 2010; Lehmann, Gaunt, and Rondon 2006). The molecular framework of biomass is rearranged during the process to form an aromatic structure with many dangling bonds. This offers biochar a high porosity, high surface area, and highly reactive nature. Aromatic structure of biochar is recalcitrant in nature and gives the effect of long-term carbon sequestration in soil. The structure can enhance the availability of macro-nutrients such as N and P (Atkinson, Fitzgerald, and Hipps 2010). Its addition increases the pH of humus and gives growth of microbial community (Antal and Grønli 2003). The heterogeneous surface of the biochars' hydrophilic, hydrophobic, and acidic and basic properties makes them react with soil substances (Atkinson, Fitzgerald, and Hipps 2010). Biochar in soil may alter its structure, texture, porosity, and particle size distribution, and density can provide refuge for micro-organisms like mycorrhizae and bacteria. Use of biochar improves soil–water permeability, increases pH and the availability of minerals, free-living N-fixing soil bacteria, increase in soil–water content or water-holding capacity, and increase in amonified ($NH_4^+$) nitrogen in the soil (Atkinson, Fitzgerald, and Hipps 2010). Addition of biochar may improve the plant productivity in the range of 20%–220% (Atkinson, Fitzgerald, and Hipps 2010).

The intermediate pyrolysis technology works with a reactor system called "Pyroformer" which is specially designed to separate ash-rich residues from fuel (Hornung 2011). The Pyroformer is a reactor with a two coaxial counter-rotating screw system to which the pellets made from the different biomass agriculture residue are fed. Pellets are conveyed through the inner screw. The feed is then passed from the feed inlet to the char outlet and vapors are collected and condensed in a condenser. The char formed at the end of the reactor is partly transported back into the reactor via the outer screw and is mixed with the fresh feed charged to the reactor. The mixing of hot char ensures better heat transfer to the fresh feed and offers a catalytic effect on the vapor products formed. The char can be collected separately in a container. The speed of the screw can be adjusted to achieve a required residence time of the pellets. The reactor is monitored and controlled with respect to the reaction temperature obtained using electrical heaters. Schematic diagram of the Pyroformer is as shown in Figure 13.2 and pictorial view of Pyroformer in practice is shown in Figure 13.3.

### 13.2.3 Process of Intermediate Pyrolysis of Wheat Straw

Energy generation using intermediate pyrolysis begins from the field where the wheat straw residue is left unused after harvest. Straw is collected and fed to the chopper, and the small chopped pieces are collected and transported to laboratory unit or plant site. Chopped pieces are supplied to a pelletizer which will produce pellets. The pellets are then fed to a pyrolyzer unit. Pyrolyzer will heat the pellets to generate vapors. The vapor is transferred to a separate container where it is condensed in a direct mixing jet-type condenser in presence of biodiesel or diesel as a quenching medium. The vapor condensed in this fashion produces biooil. Biodiesel is used as a quenching medium at the condenser to obtain a well-mixed blend of fuel. Biooil blend (a mixture of bio-petroleum and biodiesel in 30:70 vol.%) is prepared during pyrolysis at the condenser and is furnished to a Lister engine of 10 HP having power

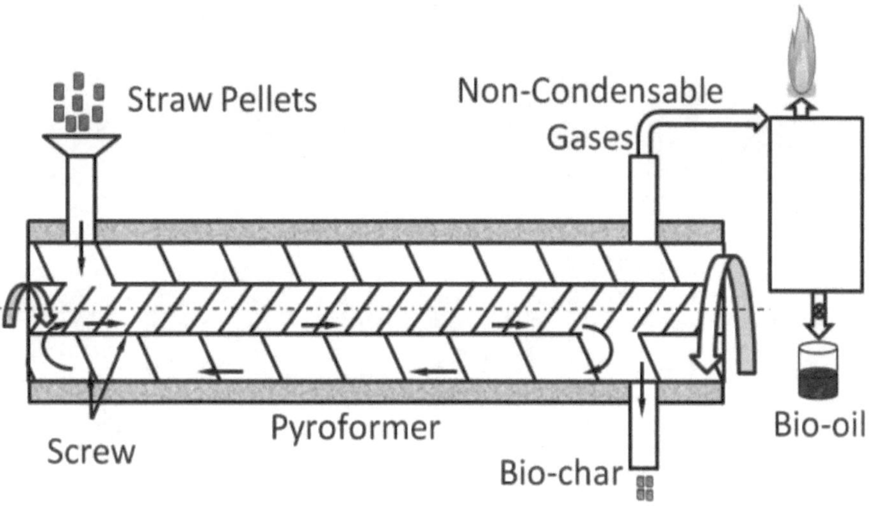


**FIGURE 13.2** Schematic of Pyroformer arrangement. (Adapted from Hornung 2011.)

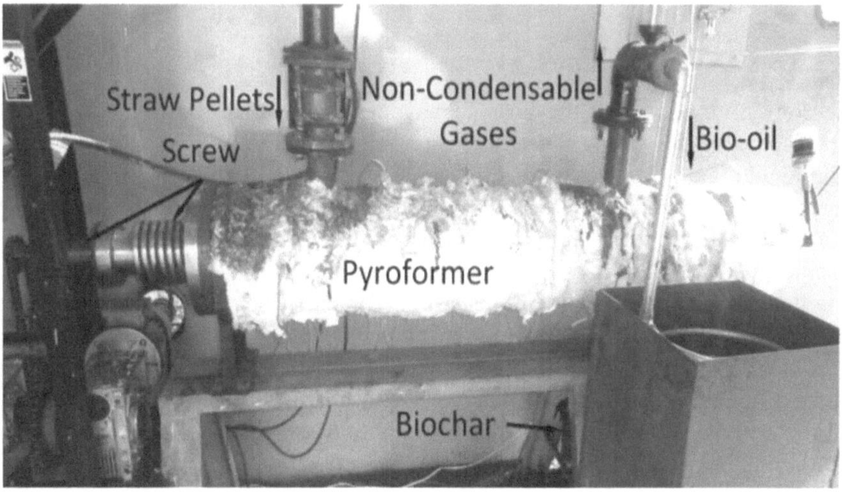


**FIGURE 13.3** Pictorial view of Pyroformer in practice (A. R. Patel 2018; A. Patel et al. 2016).

generation capacity of 7 kW with the assistance of a diesel generating set. Biochar obtained from the pyrolysis process can be utilized for the combustion. Instead, it can also be utilized as a fertilizer in the farm field and get the added advantage of carbon sequestration (Ouadi et al. 2012). Figure 13.4 shows the flow diagram of the intermediate pyrolysis process.

**FIGURE 13.4** Wheat straw pyrolysis process from farm to its utilization (A. Patel et al. 2016).

### 13.2.4 Scenarios Considered for the Study

The conventional way of the farm management in Punjab, India, involves two important pathways, using wheat straw residue in the field by way of open-field burning and the use of diesel for the farm activity and for the off-grid electricity generation. The former option is adopted because of lack of labor and shortage of time to clear the field for the next harvest, while the latter option is practiced due to short supply of electricity in the remote locations. Both options are responsible for a huge air pollution emission generation, which is the cause of concern these days. This present practice of farming is considered as the reference scenario and is termed as "business-as-usual" (BaU) scenario. As an alternative to this, the option of intermediate pyrolysis technology to utilize residue straw is proposed, and the corresponding scenario is referred herein as "pyrolysis-assisted system" (PAS) scenario.

In order to analyze the improvement in the existing practice, it is essential to compare both the scenarios. In the existing scenario, two streams of air pollution emissions are talked about: air pollution emission from open-field burning of straw and air pollution emission resulting due to the use of diesel for the power generation.

To give equal weightage to both the type of the emissions, it is essential to bring them to equal scale. For this, it is assumed that both the types of air pollution emission streams have a consumption of 500 kg of straw or equivalent (in terms of energy content). This means that 500 kg of straw are considered in the analysis that is being burned in the open fields for the purpose of generating air pollution emission, while 192 kg of diesel (which is equivalent to 500 kg of straw in terms of energy content) is considered in the analysis that is being consumed for generating power for the purpose of assessment of air pollution emission. The combined air pollution emission effect of both the streams is considered for analyzing an existing scenario. On the other hand, 1,000 kg of straw is considered for the pyrolysis-assisted scenario. Since both the streams now have the same energy input in terms of straw or diesel, depending on the stream, this makes an easy comparison of the change in the output and for the triple bottom line analysis. Although, the BaU and PAS are appearing to be two fundamentally different scenarios and choice of 500 kg of straw and 192 kg of diesel in BaU scenario seems to be arbitrary. The combined effect of these units makes a direct comparison with intermediate pyrolysis process, which offset the emission from open-field burning of straw, generates the power by use of biooil, and avoids the use of diesel, an easy and straightforward task. Figure 13.5 describes BaU and PAS scenarios.

The PAS will have three stages, which are pellet making from straw, pyrolysis of straw pellets, and power generation using biooil blend, and will generate two useful by-products, namely, biooil and biochar. The biooil can be used as a fuel for the generation of electricity, and biochar can be used as a fertilizer in the field. Figure 13.6 gives a brief description of the system boundary and different steps considered in the process under both the scenarios.

### 13.2.5 Intermediate Pyrolysis Process

The intermediate pyrolysis process is performed and tested in the laboratory, be it on the small scale, the process chain will consist of a laboratory-scale pelletizer, a pyrolyzer, and a small capacity engine, herein referred as laboratory-assisted intermediate pyrolysis (LAIP) process chain. Laboratory-scale units typically have low production rates, inefficient use of manpower, and low net electricity production. To overcome these shortcomings and improve the production rate and to make an effective utilization of the resources in the process an industrial-scale intermediate pyrolysis (ISIP) chain is proposed in the present study. The capacity of units proposed in ISIP is of higher size. These capacities are arrived at based on the available

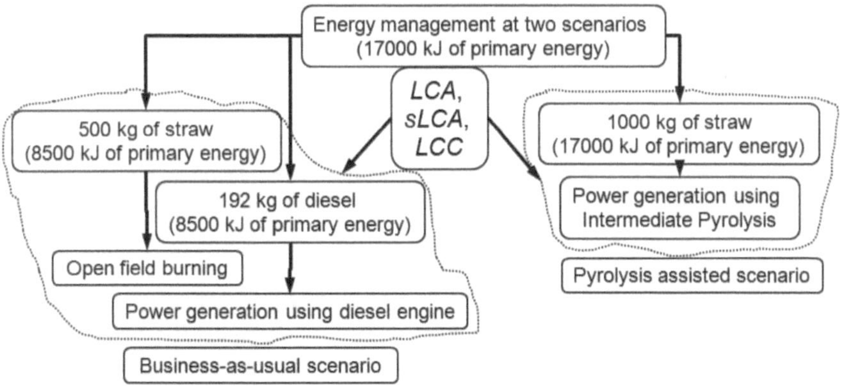


**FIGURE 13.5** Schematic of the BaU scenario.

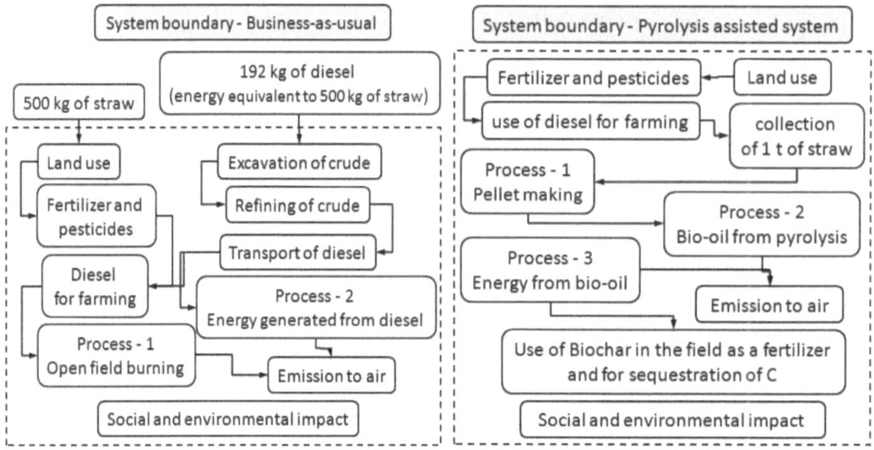


**FIGURE 13.6** System boundary and different stages for BaU and PAS considered for the study.

capacity machine in the market and its successful use for the similar applications in industry and the practice of operation in the region. For instance, 1,000 kg/h pellet-making machine is very common in the region and such machines are used to produce pellets for cattle feed. Also, a large diesel generator set is very common in the industry located in the rural area. These sets are mainly procured to act as a standby source of electricity generation, as the reliability of the supply in the rural area is poor. Considering both of these, an ISIP unit is possible for the operation. Figure 13.7 shows both the process chains, showing clearly the capacity of each unit. The dotted line represents LAIP process chain and dashed line represents ISIP process chain, while a solid line shows the process in both the cases. Figure 13.8 shows the capacity of the different unit (shown with the solid line box) and energy and mass balance for the LAIP process (shown with the dotted line box). Options I and II in the figure refer to the power generation using diesel and dual-fuel mode, respectively. The plus and the minus signs for the energy show that the energy is either produced or it is consumed during the process, respectively.

To effectively evaluate the stated scenarios, two additional end-use routes, concerning to the way carbon is associated in the system, are considered in the study. System boundary, represented as *A*, deals with the scenario in which the biological absorption of carbon takes place into the plants by means of photosynthesis. This consideration will make biological air pollution emission ($CO_{2,\ biological}$) that released during the combustion of biomass as carbon neutral to the environment. System boundary, represented as *B*, deals with the biological absorption of carbon into the plants by photosynthesis along with the permanent absorption of the carbon within the system boundary by way of carbon sequestration of char. Both the system boundaries are shown in Figure 13.9. Considering LAIP and ISIP process chains when applied for these two different routes of association of carbon into the system makes four possible routes of straw utilization for the intermediate pyrolysis process. The routes of intermediate pyrolysis process that correspond to the use of char

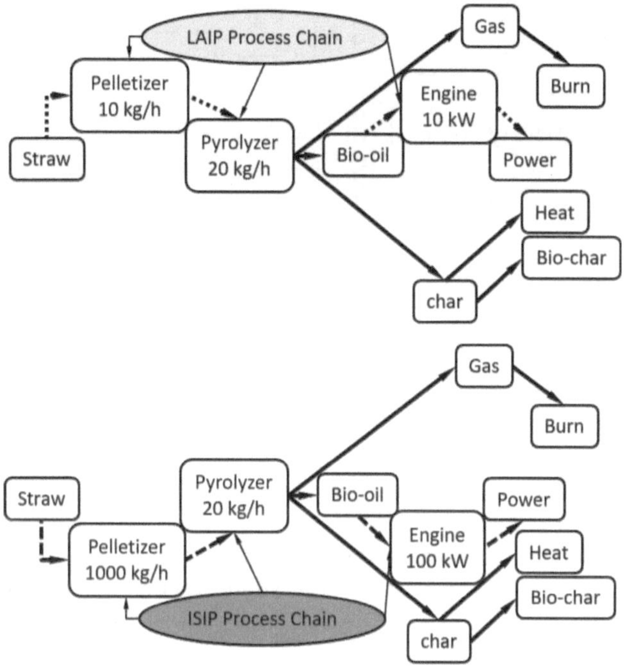


**FIGURE 13.7** Pyrolysis process chains showing the capacity of different units under LAIP and ISIP processes.

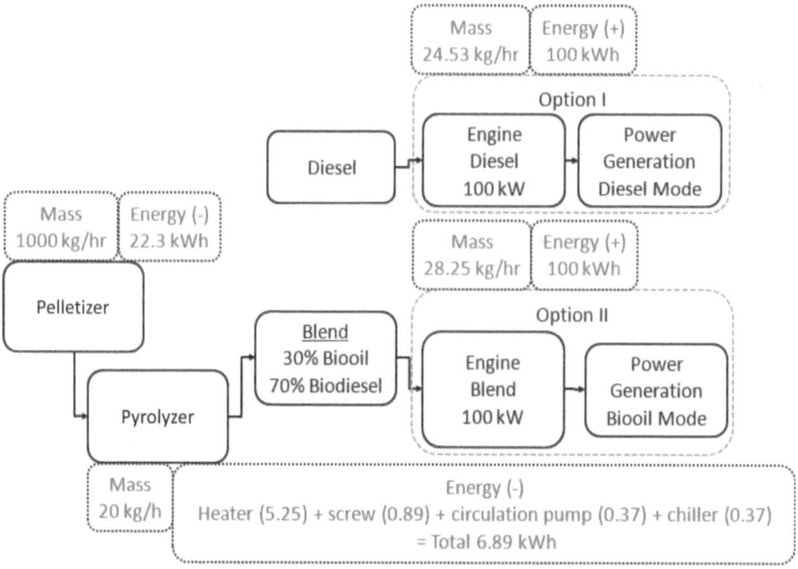


**FIGURE 13.8** Capacity, mass balance, and energy consumed/generated at different stages for LAIP intermediate pyrolysis process.

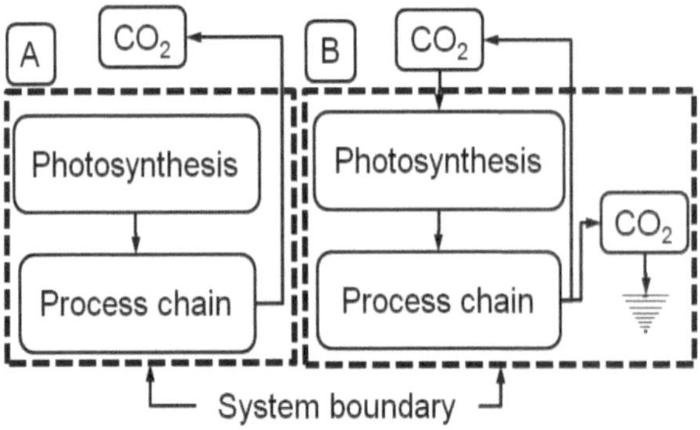


**FIGURE 13.9** System boundary and end-use scenarios under two routes for wheat straw using intermediate pyrolysis process.

for thermal energy are LAIP$^A$ and ISIP$^A$ scenarios, while the routes of intermediate pyrolysis process that correspond to the use of the char in the field as fertilizer, this route will also serve a low-cost carbon sequestration option for biochar, are LAIP$^B$ and ISIP$^B$.

### 13.2.6 Assumptions of the Study

The following assumptions are made in the LCA study:

1. No air pollution emission due to the burning of straw in the open field except for the production of $CO_2$. The effect of trace gases and particulate matter is neglected.
2. Results of embedded energy and embedded emission are derived based on the borrowed data (Ashby 2012). Embedded energy and embedded emission vary widely for the same material and even for the same process. It depends on the production, location, and method chosen for manufacturing. However, uncertainty and sensitivity analysis showed that the deviation in the embedded emission and embedded energy is well within the limit. A benchmark of 10% is considered as maximum limiting value for the deviation (kindly refer Annexure I for greater details).
3. It is assumed that engines release no other gases except $CO_2$. The effect of unburned hydrocarbon in the environment and its associated health risk is omitted from the scope of this study.
4. Effect of an increase in wear and tear and the corrosive effect of biooil on the engine parts are assumed to be not affecting the engine performance significantly. In an extreme case, this may affect the operational life of the engine, affecting the durability of the product under consideration (Chiaramonti, Oasmaa, and Solantausta 2007; Q. Zhang et al. 2007).
5. The use of the non-condensable gases, derived from the pyrolysis process, into an engine for the power generation is precluded from the scope of the

present study, presently all the gases are flared under static flame after proper combustion.

6. The effect of aerosol on human health and on the environment in terms of global warming potential (GWP) is not considered.

### 13.2.7 Embedded Energy and Embedded Emission

Embedded energy and embedded emissions are the quantities that are already consumed, in case of energy or released, in case of emission during the process of establishing the plant and machinery. Therefore, the determination of embedded energy and embedded emission will give estimates of the energy and emission that is already incurred in the process. Table 13.1 gives the embedded energy and emission data considered for the different materials consumed in the plant and machinery, while Table 13.2 gives information about the type of materials used and its corresponding weight of different parts at the various processing units. Calculation of embedded energy and emission is carried out as per the methodology suggested in the literature (Ashby 2012), the calculation procedure is as shown in (A. Patel et al. 2016). Embedded energy obtained in MJ is converted into the high-grade energy, i.e., in kWh by assuming a Carnot cycle efficiency factor of 40% (Çengel and Boles 2011), the detailed procedure is as suggested in Chr and Stefan (2005).

The ratio of the weight of an industrial unit and that of a laboratory counterpart, in case of pellet-making machine and engine, was obtained, which is approximately 1.6 for both the cases. Based on this, the embedded energy and embedded emission for a scaled-up pellet-making machine and engine unit are assumed to increase by 60% compared to its laboratory counterparts. The data obtained for this is based on the telephonic consultation with the manufacturer of each machine, the assumption is made in order to make calculations simple.

**TABLE 13.1**
**Eco-Properties of Materials**

| Embedded Energy for Different Materials and Processes [MJ/kg] [min–max] | | | | | |
|---|---|---|---|---|---|
| **Processes** | **CI** | **MS** | **SS** | **Al alloy** | **Cu alloy** |
| Material processing | 16–18 | 29–35 | 77–85 | 200–240 | 68–74 |
| Manufacturing by casting | 3.2–3.7 | 4.2–4.6 | 3.9–4.4 | 2.4–2.9 | 2.5–2.8 |
| Manufacturing by deformation | 2.5–3.1 | 2.2–2.6 | 3.0–3.7 | 2.4–2.9 | 1.8–2.2 |
| Recycle of material | 4.9–5.5 | 8.1–9.8 | 22–24 | 18–21 | 17.0–18.5 |
| **Embedded Emission for Different Materials and Processes [kg of $CO_2$/kg] [min–max]** | | | | | |
| Material processing | 1.0–1.1 | 2.2–2.8 | 4.6–5.4 | 11–13 | 4.9–5.6 |
| Manufacturing by casting | 0.19–0.22 | 0.25–0.28 | 0.24–0.29 | 0.14–0.17 | 0.15–0.17 |
| Manufacturing by deformation | 0.20–0.24 | 0.18–0.21 | 0.26–0.31 | 0.19–0.23 | 0.14–0.16 |
| Recycle of material | 0.29–0.32 | 0.6–0.8 | 1.4–1.5 | 1.1–1.2 | 1.2–1.4 |

*Source:* Ashby (2012).
*Note:* CI, MS, SS, Al, and Cu stand for cast iron, mild steel, stainless steel, aluminum, and copper.

### TABLE 13.2
### Details of the Machine Parts, Its Type, and Weight of the Material of Respective Units Considered for the Analysis

| Chopping Machine [kg] | | | Pelletizer [kg] | | |
|---|---|---|---|---|---|
| **Parts** | **Material** | **Wt.** | **Parts** | **Material** | **Wt.** |
| Body | MS | 25 | Die | SS | 8 |
| Wheel | CI | 12 | Roller | SS | 4 |
| Cutter | SS | 4 | Body | MS | 30 |
| | | | Motor | CI+Cu | 5 |
| **Pyrolyzer Machine [kg]** | | | **Engine with Generator Set [kg]** | | |
| **Parts** | **Material** | **Wt** | **Parts** | **Material** | **Wt** |
| Screw in | MS | 20 | Flywheel | CI | 50 |
| Screw out | MS | 30 | Engine | Al alloy | 3 |
| Cover | MS | 50 | Engine | MS | 30 |
| End shell | MS | 60 | Gen. | MS, Cu | 30 |
| Roller | Al Alloy | 20 | Base | MS | 10 |
| Motor | CI, Cu | 10 | | | |
| Condenser | MS | 30 | | | |

*Note:* CI, MS, SS, Al, and Cu stand for cast iron, mild steel, stainless steel, aluminum, and copper.

It is observed that Table 13.3 gives the embedded energy and embedded emission data for LAIP and ISIP processes. It is important to note that LAIP results in embedded energy and emission liability of approximately four times that of a lab-scale diesel-operated process. The increase is due to the inclusion of an intermediate pyrolysis process that requires the addition of a pelletizer and a pyrolyzer in the process. The increase shows the increase in the process liability, in terms of the addition of the machine, increased infrastructure, and other operating facilities.

### TABLE 13.3
### Embedded Energy and Embedded Emission Data

| # | Equipment | Embedded Energy (kWh) | Embedded Emission (kg of $CO_2$) |
|---|---|---|---|
| 1. | Chopping machine | 136 | 86 |
| 2. | Pelletizer, 10 kg/h | 205 | 128 |
| 3. | Pelletizer, 1 t/h[a] | 328 | 205 |
| 4. | Pyrolyzer, 20 kg/h | 902 | 574 |
| 5. | Engine, 10 kWh | 432 | 267 |
| 6. | Engine, 100 kWh[a] | 691 | 427 |
| 7. | LAIP | 1,675 | 1,055 |
| 8. | ISIP | 2,058 | 1,292 |

[a] Quantity obtained based on the corresponding laboratory-scale unit and a multiplying factor of 1.6.

### 13.2.8 Data Used in the Study

The present section summarizes the different data collected and used in the study. The refining energy and emission data used for the diesel are given in Table 13.4 (Safa and Samarasinghe 2012), while the transport emission is data is as shown in Table 13.5. It is assumed that the diesel is transported to the processing unit using a light goods vehicle, having the transportation emission as 0.097 kg of $CO_2$/t-km (Ashby 2012). Distance between local petrol pump and the nearest refinery is assumed as 200 km, and the distance of the petrol pump from the processing unit is considered as 10 km. The grid emission of 0.84 kg of $CO_2$/kWh is considered in the study, this is as per the grid emission data published by the Government of India for the year 2011, to account the effect of indirect emission generated due to the use of electricity when consumed from the grid (GoI 2011).

Table 13.6 shows the data relevant to the property of fuel and corresponding emission resulting due to the combustion of diesel (Indian Standard (IS-1460) 2005) and that of straw (Hornung, Apfelbacher, and Sagi 2011). Table 13.7 provides the emission data for a transport vehicle under a stationary test conducted using diesel and biodiesel as fuel referred to in the literature (Beer et al. 2000). The result shows that, although there is a significant variation in the emission from the individual test, the GWP from these emissions does not differ significantly. Since the emission from the biodiesel is environment neutral, the GWP of emission resulting from the use of biodiesel is in the negligible range.

Table 13.8 provides the data for the mass balance for various parts of a single wheat straw sample when it is ready to harvest. The first and the second columns under mass distribution are the percentage distribution and the weight of the different items. From the data, one can infer that for each kg of grain, 416 kg of

**TABLE 13.4**
**Embedded Energy and Embedded Emission Released from Fuel**

| Particulars | Embedded Energy and Emission | |
|---|---|---|
| Well-to-pump energy use | 8.83 (0.98) | MJ/Kg of diesel (kWh/kg of diesel) |
| Well-to-pump emission use | 0.75 | Kg of $CO_2$/kg of diesel |

*Source:* Safa and Samarasinghe (2012).

**TABLE 13.5**
**Transport Emission**

| Fuel and Vehicle Type | Energy [MJ/t-km] | $CO_2$ Emission [Kg/t-km] |
|---|---|---|
| Diesel—light goods vehicle | 1.36 | 0.097 |

*Source:* Ashby (2012).

**TABLE 13.6**
**Data Related to the Fuel from the Operation**

| | | |
|---|---|---|
| Density of diesel | 0.84 | kg/l (Indian Standard (IS-1460) 2005) |
| Calorific value | 44.17 | MJ/kg (Indian Standard (IS-1460) 2005) |
| Carbon percentage in diesel | 87% | Carbon/kg of diesel (Indian Standard (IS-1460) 2005) |
| $CO_2$ production by diesel | 3.19 | kg/kg of diesel (Indian Standard (IS-1460) 2005) |
| Energy contained in the straw | 17 | MJ/kg (Sagi et al. 2014; A. R. Patel 2018) |
| Energy contained in the char | 20 | MJ/kg (Sagi et al. 2014; A. R. Patel 2018) |
| Energy contained in the biooil | 23 | MJ/kg (Sagi et al. 2014; A. R. Patel 2018) |
| Carbon percentage in straw | 47% | Sagi et al. (2014); A. R. Patel (2018) |
| Carbon percentage in biooil | 40% | Sagi et al. (2014); A. R. Patel (2018) |
| Carbon percentage in biochar | 64% | Sagi et al. (2014); A. R. Patel (2018) |

*Source:* A. R. Patel (2018).

**TABLE 13.7**
**Emission from Engine Using Diesel and Biodiesel**

| Emission from Engine Using Diesel | | | | | |
|---|---|---|---|---|---|
| | **g/km** | | **kg/kg** | **GWP** | **$CO_2$ Equivalent** |
| $CO_2$ | 1,736 | 98.24% | 0.982 | 1 | 0.982 |
| CO | 7.72 | 0.44% | 0.004 | 1 | 0.004 |
| $NO_x$ | 21.26 | 1.20% | 0.012 | 7 | 0.084 |
| VOC | 1.3 | 0.07% | 0.001 | 0 | 0.000 |
| PM | 0.79 | 0.04% | 0.000 | 0 | 0.000 |
| Total | 1,768 | | 1.00 | | 1.0710 |
| **Emission from Engine Using Biodiesel** | | | | | |
| | **g/km** | | **kg/kg** | **GWP** | **$CO_2$ Equivalent** |
| $CO_2$ | 1,948 | 98.03% | 0.980 | 1 | 0.980 |
| CO | 11.41 | 0.57% | 0.006 | 1 | 0.006 |
| $NO_x$ | 25.66 | 1.29% | 0.013 | 7 | 0.090 |
| VOC | 1.21 | 0.06% | 0.001 | 0 | 0.000 |
| PM | 0.9 | 0.05% | 0.000 | 0 | 0.000 |
| Total | 1,987 | | 1.00 | | 1.0764 |

*Source:* Beer et al. (2000).

mass is produced by farming. It is noted that excluding grain and the reaches, almost everything else can be used in the pyrolysis process. Table 13.9 gives information regarding the indirect air pollution emission generated during the growth of wheat. Indirect air pollution emission produced during the growth is from the fertilizer and agrochemical consumed during the growth of the plant. Table 13.10 provides

**TABLE 13.8**
**Mass Balance of Wheat Straw**

| # | Particulars | Mass Distribution [%] | [Kg] | [Per kg of Wheat] |
|---|---|---|---|---|
| 1. | Internodes | 61.54 | 256.42 | 36.391% |
| 2. | Leaf | 25.92 | 108.00 | 15.328% |
| 3. | Node | 8.56 | 35.67 | 5.062% |
| 4. | Rachis | 2.09 | 8.71 | 1.236% |
| 5. | Grain | 0.24 | 1.00 | 0.142% |
| 6. | Other plants | 0.64 | 2.67 | 0.378% |
| 7. | Non-wheat spikes | 1.02 | 4.25 | 0.603% |
| | Total | 100 | 416.71 | 59.14% |

*Source:* Atik and Ates (2012) (Emission from 1 kg of wheat = 0.5914 kg of $CO_2$); Rajaniemi, Mikkola, and Ahokas (2011)

**TABLE 13.9**
**$CO_2$ Emissions from Input during Wheat Production (kg of $CO_2$/t)**

| # | Particulars | Emission from Wheat Production kg of $CO_2$/tonne | [%] |
|---|---|---|---|
| 1. | Fertilizer | 53.39 | 52.25% |
| 2. | Agrichemical | 5.41 | 5.30% |

*Source:* Safa and Samarasinghe (2012).

**TABLE 13.10**
**Emission from Open-Field Burning of Straw in the Fields**

| Particulars | kg of $CO_2$/kg of Straw | GWP $CO_2$ Equivalent | GWP [$CO_2$ Equivalent] | % |
|---|---|---|---|---|
| $CO_2$ | 1.47 | 1 | 1.47 | 84.50 |
| CO | 0.06 | 1 | 0.06 | 3.45 |
| $CH_4$ | 0.0034 | 26 | 0.0884 | 5.08 |
| NMHC | 0.0075 | 11 | 0.0825 | 4.74 |
| $N_2O$ | 0.00007 | 206 | 0.01442 | 0.83 |
| $NH_3$ | 0.00037 | 1 | 0.00037 | 0.02 |
| $SO_2$ | 0.00085 | 1 | 0.00085 | 0.05 |
| $NO_x$ | 0.0033 | 7 | 0.0231 | 1.33 |
| Total | 1.545 | Total | 1.740 | 100 |

*Source:* Li et al. (2007).

**TABLE 13.11**
**Energy Allocation in a Blend of Biooil (Based on the Results from Hornung, Apfelbacher, and Sagi, 2011 and A. R. Patel, 2018)**

| | Vol. | Density | Mass | | CV | Energy | |
|---|---|---|---|---|---|---|---|
| | % | kg/L | kg | % | MJ/kg | MJ | % |
| Biodiesel | 70 | 0.88 | 0.62 | 65 | 42 | 27.5 | 78 |
| Biooil | 30 | 1.10 | 0.33 | 35 | 23 | 8.0 | 22 |
| Blend | | 0.95 | 0.95 | | 36 | 35.6 | |

the data for the air pollution emission resulting due to the open-field burning of straw and GWP of $CO_2$ and equivalent warming potential for gases released during burning.

Based on the results obtained from the testing of the blend of biooil and biodiesel, the data on the energy content of biooil and biodiesel is worked out and is provided in Table 13.11. It is important to note that biodiesel is used to condense the biooil in the condenser. A blend of biooil (biooil 30% vol.% and biodiesel as 70% vol) is obtained from the process. The blend is supplied to the power generation unit and power is obtained. Energy content, calorific value of each fuel, and the density value for each fuel are obtained, and using this data, the energy distribution of each fuel in the blend being consumed for the power generation is obtained. The calculation shows that the energy content of biooil in the overall blend is 22% of the total content of the blend.

## 13.3 ASSESSMENT OF ENVIRONMENTAL SUSTAINABILITY

There has not been any evaluation of intermediate pyrolysis technology option with regard to its LCA against the practice of open-field burning of the straw in the fields or for the off-grid power production using diesel as fuel. In this context, the functional unit for the system is defined as the generation of electrical energy with effective use of agricultural waste and at a lower environmental burden (i.e., kWh/kg of $CO_2$) (Weidema et al. 2004). Considering this, the processing of 1,000 kg of straw as a functional unit is considered in the study. The embedded energy and embedded emission are determined for the elements of processes. The allocation of power produced is considered based on the energy content of the fuel, i.e., for the biooil and biodiesel present in the blend. A detailed inventory is worked out and energy and emission generated for different processes are obtained. The methodology followed is as given by *International Reference Life Cycle Data (ILCD) Handbook* (ILCD Handbook 2010). The functional unit for the LCA considered is air pollution emission released in kg of $CO_2$ for each 1,000 kg of straw processed.

Table 13.12 shows the estimated air pollution emission resulting from the open-field burning of straw, while Table 13.13 shows the energy and emission flow due to use of diesel from the laboratory as well as an industrial scale unit. Similarly, Table 13.14 shows the energy and emission flow from the pyrolysis-assisted unit opting for two different routes of biochar.

**TABLE 13.12**
**Emission from Open-Field Burning of Straw in the Fields due to the Burning of 500 kg of Straw**

| # | For 500 kg of Straw | Emission<br>kg of $CO_2$ |
|---|---|---|
| 1 | Growth of the straw | 332.40 |
| 2 | Open-field burning | 869.82 |

**TABLE 13.13**
**Energy and Emission Flow for Power Generation Using Diesel**

Operation with Diesel (Industrial-Scale Unit—192 kg of Diesel)

| # | Particulars | Embedded | | Process | |
|---|---|---|---|---|---|
| 1 | Refining of fuel | | | −188.89 | 144.50 |
| 2 | Transportation of fuel | | | | 3.92 |
| 3 | Engine operation | 691 | 427 | 784.52 | 206.08 |
| | | 691.23 | 426.92 | 595.63 | 354.50 |

**TABLE 13.14**
**Energy and Emission for Industrial-Scale Intermediate Pyrolysis Option per 1,000 kg of Straw**

| | | Embedded | | Process | Use of Char | |
|---|---|---|---|---|---|---|
| | | | | | Heat | Biochar |
| # | Particulars | kWh | kg of $CO_2$ | kWh | kg of $CO_2$ | |
| 1 | Growth of straw | | | | 333 | 333 |
| 2 | Chopping machine | 136 | 86 | | | |
| 3 | Transp. to lab | | | | 0.97 | 0.97 |
| 4 | Pellet making | 328 | 205 | −15 | 13 | 13 |
| 5 | Pyrolysis of straw | 902 | 574 | −334 | 281 | 281 |
| 6 | Treatment to biochar | | | | | −751 |
| 7 | Transp. of biochar | | | | | 1.55 |
| 8 | Engine ind. scale | 691 | 427 | 733 | 36 | 36 |
| | Total | 2,058 | 1,292 | 384 | 663 | −87 |

It is observed from Table 13.15 that it is possible to achieve a negative EF using intermediate pyrolysis options under biochar route ($ISIP^B$). Intermediate pyrolysis process, when followed using biochar route, can absorb carbon from the environment of the order of 87 kg of $CO_2$/1,000 kg of straw. Processing of 14 tonnes of straw

**TABLE 13.15**
**Energy and Emission Summary for Industrial-Scale Intermediate Pyrolysis Process per 1,000 kg of Straw Processing**

| | | Embedded Quantities | | Process Quantities/1,000 kg of Straw | |
|---|---|---|---|---|---|
| # | Scenario | kWh | kg of $CO_2$ | kWh | kg of $CO_2$ |
| 1 | BaU | 691 | 427 | 596 | 354.50 |
| 2 | ISIP[A] | 2,058 | 1,292 | 384 | 663 |
| 3 | ISIP[B] | 2,058 | 1,292 | 384 | −87 |

can make the intermediate pyrolysis process a carbon-neutral. Intermediate pyrolysis process also generates 384 kWh of electricity for each 1,000 kg of straw processed.

## 13.4 ASSESSMENT OF SOCIAL SUSTAINABILITY

An sLCA model for the processing of the straw under the PAS is presented in Figure 13.10. The model calls for a centralized unit that receives straw from different farms for the processing. The processing unit can run as a pyrolysis cooperative. The cooperative may distribute biooil and biochar generated from the input of straw.

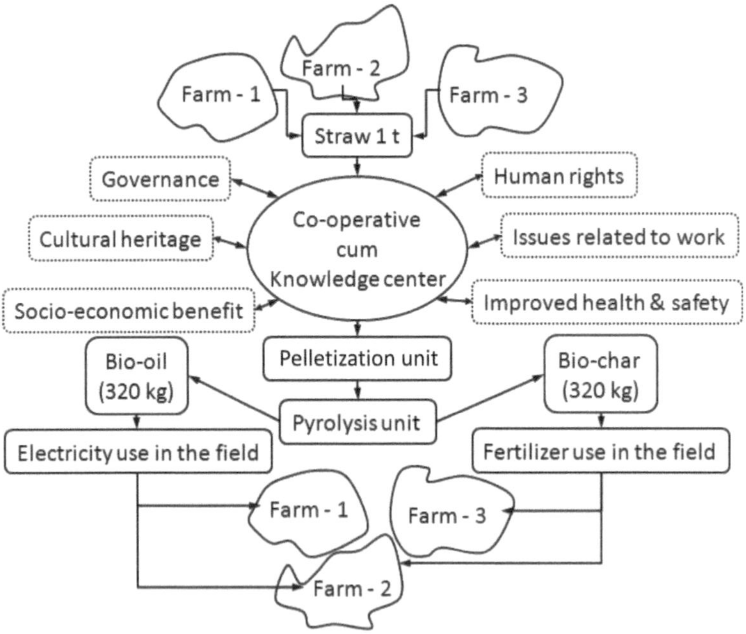


**FIGURE 13.10** Overview of sLCA model showing different units (bold lines) and important element considered for the analysis (dot lines).

Efficient management of such co-operative can serve as a hub or knowledge center (KC) and can support the people by providing required resources and train them for this. In the present study, a scale of +3 to –3 is considered to appreciate the positive and the negative effect of subcategory on the impact category.

## 13.4.1 Important Elements of sLCA Model

### 13.4.1.1 Cooperatives

Cooperatives are professionally managed, institutional models. It holds social benefits (SBs) such as empowerment of weaker section and women, employment of labor, the opportunity for landless people of society, rural employment, improvement in living standard, and community development (Banerjee 2007). Cooperative experiments showed that this can increase members' incomes, decrease the risks, and provide them with an opportunity to participate in societal functions (Murray-Prior 2008), provide a forum for discussion and mobilize local resource without increasing risk (Birchall 2003). Some examples are women's agricultural cooperatives in Greece, cooperative care for elderly in Wisconsin (Birchall 2003), while "Anand Model" for milk producers' cooperative company and contract farming arrangement are good examples to cite success in India. Cooperative experiments also achieve socio-economic development for violence embedded state during the conflict and in the failing economy. It helped failed and recovering states such as Bosnia, East Timor, Lebanon, Mozambique, Rwanda, Macedonia, and Nepal (Weihe 2004).

### 13.4.1.2 Knowledge Center

A village KC can impart SBs, one good example to cite this is the village KC program carried out in Pondicherry, India. The KC is having computer facility and is equipped with the internet. It helps to empower people through improved access to information and helps in creating community assets and empowering the users (Senthilkumaran and Arunachalam 2002). For instance, women started manufacturing their own brand of incense with the help of such KC, a school teacher using KC for taking tuition, some used KC as a place to share job vacancies, a high number of women managing KC leads to an increase in the number of female users help to resolve health-related problems of female visitors, the user additionally gets information regarding vegetable prices, pre- and post-natal care was obtained, the KC was given special protection during communal clashes, the idea of building KCs in public libraries and schools is also mooted (Senthilkumaran and Arunachalam 2002).

## 13.4.2 Goal and Scope of sLCA Study

The goal of the present study is to compare the sLCA impact for two end-use scenarios during the management of the farm and utilizing straw residue in the field of Punjab, India. The social dimension of this study involves farmer, community, society, and state. The system boundary of the study is the same as that of LCA. Under the present scenario the exploration, processing and use of diesel at the farm for the power generation are considered along with the end-use scenario of straw by way of

**TABLE 13.16**
**Stockholder Categories and Subcategories, Impact Category, and Scale**

| | |
|---|---|
| 1) Farm and farmworkers | Forced labor (FL), child labor (CL), FA and collective bargaining, SBs/social security, WH, FS, EO/discrimination, HS. |
| 2) Local communities | CI, access to material resources, access to immaterial resources, DM, CH, SHLCs, RIR, CE, LE, secure living conditions (SLCs), CO, ER. |
| 3) Society | PCSI, CED, PMACs, corruption. |
| 4) Value chain actor | FC, PSR, SR, RIPR. |
| 5) Consumers | SCP, FM, HS, TR, EoLR. |
| 6) State | PCS, IRP, ARU. |
| Impact category | Governance (G), human rights (Hu), working conditions (W), socioeconomic benefit (S), health and safety (He), culture heritage (C). |
| Scenario | Existing one or present scenario—BaU (E) and proposed scenario—PAS (P) |
| Scale: effect ranges from +3 to −3 | 3 Strongly positive 2 Highly positive 1 Moderately positive<br>0 No effect<br>−1 More or less negative −2 Very negative −3 Strongly negative |

open-field burning. The proposed scenario consists of collection and use of 1,000 kg of straw under the PAS. The theme of interest of the present study is how farmers manage their farm activity and utilize waste straw and agricultural residues in order to generate electricity in a more sustainable way.

Data considered for the study are from Punjab, India, and also application specific. The data source is based on personal visits to farms and discussion with farmers, farm works, and people involved in different activities in the form at village Khawaspur, near IIT Ropar in Punjab, India. Appropriate stakeholder categories and subcategories are considered for the study as per the guideline suggested under Benoît et al. (2009) and is as stated in Table 13.16.

### 13.4.3 Impact Assessment for Different Subcategory

#### 13.4.3.1 Farm and Farmworkers

##### *13.4.3.1.1 Freedom of Association (FA) and Collective Bargaining*

The farm work is not of high skill work, replacement is easy, and workers are suffered by hire and fire system of work. The choosing cooperative system gives them a chance to form associations, under this job skills increase, security is greater, and workers have greater bargaining power and, hence, give protection to their basic human rights (Birchall 2003). They may expect facility at work, e.g., clean drinking facility and gloves for spreading fertilizer, reduction in poverty due to increase in yield, and improvement in their living standard. The association will help the group to resolve social issues and improve social cohesion (Weihe 2004). The associative group can organize social function that preserves the culture and promote a sense of pride to the members.

#### *13.4.3.1.2 Social Benefit/Social Security*

India hosts largest number of poor, illiterates, and unemployed in the world (Kurian 2007). It is essential to focus on social development of the people since it is observed that social and economic developments are mutually reinforcing (Kurian 2007). However, in the case of a farmworker, current situation is not encouraging, the missing record of farmworker's availing provident funds benefit is only one such example. There is also gross negligence observed for intangible SB, e.g., accident compensation, child care at work, education, insurance, and accommodation.

Farmer suicides are also a burning social issue for government, for instance, Vidarbha in India, reached to around 2 persons/day in 2006 (Aggarwal 2008). Farmer suicide behavior is a major public and mental health problem (Kahn and Lester 2013). Farmers in India undergo stress mostly due to social issues such as illiteracy, the bonded labor system, family, and government corruption except for debts from local money lenders (Aggarwal 2008). Suicides are also a major problem in certain flourished states like Punjab, India (Kaur 2010). It is reported that the productivity of farmer was 8.7% higher if he had completed four years of schooling (Schultz 1988).

Developing human capital is a good alternative to counter such social issues. This can be achieved by formal training, education, and unconsciously through experience (Ostrom and Ahn 2003). Such skill helps in better decision-making, e.g., understanding the cropping pattern and investment strategy (Parthasarathy and Chopde 2001). KC at pyrolysis cooperative can provide a good alternative for this problem.

#### *13.4.3.1.3 Working Hours (WH)*

It is common in India to start work on time, but no defined time for calling off the day. Arranging for a siren in the cooperative for declaring the completion of work hour is a simple solution that helps illiterate farmworker not too dependent on the owner to declare the day. The cooperative can introduce flexible hours for childcare and may allow people to attend a school or other social function. Cooperative can take care of the proper working facility at night time in farm and reduce the risk of working in the night hours. The issue of risk from animal and unpleasant working condition is a concern for workers working at night time. Flexible hours and arranged holiday can increase participation in cultural events.

#### *13.4.3.1.4 Fair Salary (FS)*

The salary paid should be competitive with the best in the industry. Due to poor record and book-keeping, it is difficult to observe the payment of FS to the farmworker. As stated earlier, this should be maintained by keeping a strict record. This will help achieve worker basic rights like better housing and sanitation and societal uprightness. If the fair wage is ensured, worker may opt for good transportation facility and can reduce the risk of accident and promote safety to his and her family members. Pyrolysis cooperative can ensure FS and help increase awareness among workers about this.

#### *13.4.3.1.5 Equal Opportunity (EO)/Discrimination*

The Indian labor force is highly biased, and widespread discrimination is observed, the contribution of female in the workforce is only 25%, and women receive about half salary than male (Bhalla and Kaur 2011). The ironical gender discrimination is most pronounced in states of Punjab and Haryana (most prosperous states) (Pal and Ghosh 2007). Several incidences of discrimination are observed in the case of farmworkers, for instance, many times women are allowed to work even if she is pregnant. In afternoon break, women sleep on the floor while man sleeps on a cot. Giving respect to women is a culture and asking them to work is stigma. Giving an EO to work is a good alternate and goes well with the culture. SEWA (Self-Employed Women's Organization) of India works to empower poor women. It works by way of organizing a union, cooperatives, banking, production of milk, home-based and artisan items, petty trade, agriculture and forestry, land development, water harvesting, child care, housing, and others (Sen 1997). Cooperative is a good alternative when it comes to empowering women. Therefore, pyrolysis cooperative run by women can be a good socioeconomic tool.

#### *13.4.3.1.6 Health and Safety (HS)*

Burning of agricultural residue in open field and rampant use of chemicals and fertilizers in the farm are currently carried out causing the violation of HS standards. Currently, there is no systematic record of death and accident on the farm which is required and be made public. Carrying out programs to educate, counsel, and train farmworkers can prevent and control risk to members, their families, or community members. Lack of awareness many times causes deterioration in health, for instance, improper use of fertilizer without basic protection can risk human life and safety.

Currently, fertilizer bags are stored in the house under the bed, similar ignorance is observed in its use and disposal that are also not with adequate safety provision. Human health of farmworker concerned due to health risks from pesticides, emissions from burning fields, and accidents from occupational risks (Fritsche et al. 2012). Common symptoms are muscle weakness, difficulty in breathing, vomiting, headache, respiratory distress, leukemia, brain and liver damage, and even death in case of ingestion of high dose (Amar, Prahcee, and Mahesh 2010). Toxic chemicals from pesticides can cause a variety of adverse health effects, such as polychlorinated biphenyl (PCB), dioxins, and dichloro diphenyl trichloroethane (DDT), may cause endocrine disruption, and reduce disease resistance (Corvalán et al. 2005). Inappropriate and unsafe handling, by an illiterate farmer, of these agricultural chemicals is a major cause of harm to human health (Amar, Prahcee, and Mahesh 2010). Certain pesticides banned in advanced countries are still being used in India. Out of 50 poisoning cases, 50% are due to inappropriate use, only 7% users were aware of the health hazard. Agriculture Canada stressed to avoid crop residue open-field burning citing reasons such as impacts on emissions, safety risks, and fire management; reduced visibility on roads; and respiratory health impact and on waste management ("Heating Methods and Open Burning" 2013; Garg 2008).

Straw burning produces fine particles that can travel long distances in the wind and can enter breathing system and can affect people with lung diseases, such as asthma and chronic obstructive pulmonary disease (COPD) ("Heating Methods and Open Burning" 2013). Environment and Health Canada have declared that inhalation of PM10 can be leading to serious respiratory problems; CO may cause fatigue, headaches, dizziness, nausea, and in extreme case lead to unconsciousness and death; $NO_x$ can cause shortness of breath and irritate patients of emphysema and asthma; and the volatile organic compounds (VOCs) are carcinogenic and can cause respiratory irritation and illness, prolong exposure of polycyclic aromatic hydrocarbons (PAHs) pose a cancer risk ("Heating Methods and Open Burning" 2013). Poor health may lead to more leave from work and the cyclic increase in poverty and a decrease in poor social status. As opposed to it, a healthy person can participate in cultural activity well and can uphold the culture well. In poor countries, in rural areas, the health of people depends highly on the services of local productive ecosystems for food (Corvalán et al. 2005).

#### 13.4.3.2 Local Community

##### *13.4.3.2.1 Community Involvement (CI)*

Community centers can help aware community members about required information such as a basic right of wage, food security, health, and government schemes and can be instrumental in bringing social change (Senthilkumaran and Arunachalam 2002). Such centers can help empower weaker section of society and provide a forum for discussion of the problem and evident example to carry out the task. Such KCs can host community prayers or religious gathering and can strengthen the culture.

##### *13.4.3.2.2 Access to Material Resources (AR)*

It is observed that there is an acute shortage of water and electricity observed in rural areas (Parikh 2011). The basic facility for electricity and water can help students to study longer hours at night. Also, this will help social interaction among members, this will help greater interaction and help in greater bonding and discussion of social issues. Health improved due to the use of electricity in gadgets that save strenuous activity (especially for women) washing of cloths, flour mill, etc.; electricity allows the use of media that help improve cultural ties.

##### *13.4.3.2.3 Delocalization and Migration (DM)*

Recently, most regions of the world have experienced significant human migration due to industrialization, the green revolution, draught, and issues ranging from social, racial, and even religion in certain cases (Corvalán et al. 2005). Large-scale mass migration is reported in rural India in search of work, especially in states like Punjab and Gujarat (Ghosh 2005). No systematic record of migrants is observed in countries like India (Ghosh 2005). Migrants cause a host of socioeconomic problems in the region, increase the crime rates in host countries, and bring various diseases (Wickramasekera 2002). Migrant workers steal jobs from the local population and pose a risk to the maintenance of social-cultural homogeneity of the host country population (Wickramasekera 2002). The health of migrant is more likely to affect due to poor quality of food intake and staying in insecure and in shabby conditions. Such conditions may influence negatively on his culture and significantly weaken their heritage. However, by virtue of the remittance, they provide good living conditions, proper housing accommodation, better schooling, and lifestyle back home.

#### *13.4.3.2.4 Cultural Heritage (CH)*

The cooperative helps organize social functions, community development, and encourages members to take part in it (Banerjee 2007). This may help maintain their lost culture and forgotten ties. The cooperative can play an instrumental role in this aspect. This will reduce the socio-cultural divide and racial discrimination, which might result due to migration and help offer equitable societal human rights.

#### *13.4.3.2.5 Safe and Healthy Living Conditions (SHLC)*

The administration must agree that expecting clean air is a basic right of every community member (Diesendorf 2000). The stringent law must be enacted and enforced critically against open-field burning. However, there is no specific law to date to stop the burning of straw in the field. It is essential to keep records of people who opt for open-field burning and inspecting fields, especially in cropping seasons. The government can carry out a social campaign against the burning of straw. Similarly, at the consumer level, pesticides can contaminate not only food but also reservoir and water well, leading to health problems. The problem of contamination is so serious that DDT residue was found even in mother's milk. This is enough to state that the price for contamination has far-reaching consequences (Jhamtani 2010). Use of biochar can reduce the use of fertilizer and hazardous chemicals (Lehmann, Gaunt, and Rondon 2006). The PAS improves living conditions.

#### *13.4.3.2.6 Respecting Indigenous Right (RIR)*

It is observed that indigenous people are generally detached from mainstream activity. It is essential to prevent racial discrimination by providing record and acknowledge their contribution and equal rights to them. Involving indigenous people in the cooperative can improve their social status and financial condition and can uphold their culture better.

#### *13.4.3.2.7 Community Engagement (CE)*

KC may provide basic facilities for community interaction and talk to deliver knowledge and educate members by building social capital. Building social capital essentially deals with learning skills and establishing regularized patterns of relationships with others. Investing in social capital may act like "glue," and that can "gear" collective democratic action (Nakagawa and Shaw 2004). This can be achieved by bonding (ties between members), bridging (ties among the member of different background), and linking (ties between community members and those in influencing positions) (Nakagawa and Shaw 2004). It is observed that schooling and training are effective tools for increasing socioeconomic returns or building social capital (Hansen, University of Wisconsin, and Conference on Research in Income and Wealth 1970).

##### *13.4.3.2.8 Local Employment (LE)*

Increase the local people in employment also offers positive social-culture effect. Reduction in migrant worker can be considered as positive effect on the society and also reduction in a conflict-like situation. Mandatory display of the number of workers required at the farmhouse is an option to give the best chance for the local worker. The local worker may exhibit good skills to deal with issues dealing with local system and community. They can work in adverse condition due to better health and better quality of life and can participate in a community program and can strengthen the culture.

##### *13.4.3.2.9 Cohabitation (CO)*

Utilization of renewable sources has a positive consequence on the environment and hence helps people cohabit. For instance, the use of solar cell can reduce mining activity, and hydropower offers improved flood control and access to irrigation water (A. Evans, Strezov, and Evans 2009). Use of straw by the PAS may reduce exploration of crude oil, import to country, reduction in emission, and better waste management by avoiding open-field burning. The farmer may inform in advance to the people about certain activities that cause the disturbances in the society, due to noise, dust, and smell, pollution, for instance, stopping open-field burning may improve CO. It is everybody's right to protect from dust, noise, and smell due to various agricultural activities. Fertilizers and chemicals have several long-term effects on the community; avoiding the use of excess fertilizer and use of bio-fertilizers like mycorrhizal technology (Gulati 2009) and biochar (Lehmann and Joseph 2009) can increase CO with the community.

##### *13.4.3.2.10 Esthetic Responsibility (ER)*

To promote culture and improve the esthetic of the town, the statue of freedom fighters, etc., can be planned.

#### 13.4.3.3 Society

##### *13.4.3.3.1 Public Commitment to Sustainability Issues (PCSI)*

Open-field burning of agricultural residue is a burning problem. Indian National Biomass Cookstoves Initiative suggested that efficient cookstoves alone could avoid 5,70,000 premature deaths and a reduction in 4% of India's greenhouse gas emissions (Bhattacharyya 2012). In order to improve public commitment over this issue, direct appeal to people and farmers using media and newspaper is a good strategy. This will build social opinion in the public and positive approach towards stopping this practice. Such an appeal should be linked with a culture so that it has a positive effect.

##### *13.4.3.3.2 Contribution to Economic Development (CED)*

Use of biooil could reduce imports of crude oil. It will also result in a push for indigenous technology, use of the indigenous product, rural employment, self-employment, and economic welfare. Culture of using the indigenous product will be strengthened.

##### *13.4.3.3.3 Prevention and Mitigation of Armed Conflict (PMAC)*

Use of PAS may bring de-escalation of conflict as different types of groups can come close and work under the cooperative model. This will increase the safety of human and the protection of human rights. Reduction in conflict can also improve the socio-cultural ties and help preserve heritage. For instance, in the initial phase of biofuels, technology witnessed protests, some extreme violence in the streets, and a backlash against the use of biofuels since its use was competing for the food (Nuffield Council on Bioethics 2011). However, biooil generated using agricultural residue is a welcome step since it will reduce emission due to open-field burning and does not influence by "food-for-oil" crisis.

##### *13.4.3.3.4 Corruption (CR)*

Corruption is the exploitation of basic rights due to lack of awareness and education and results in the diminishing value in the society. Common examples prevailing in the farms are allowing smoking at the workplace or liquor to farmworkers, etc.; asking farmworkers to come and work on holidays and asking women to help men (her husband) employed for farm activity without pay; supply of farmworker through an agent or middleman and allowing them to charge money from the worker's pay; not informing the harmful effects of fertilizer and chemicals to illiterate farmworkers or not giving proper safety precautions. Strategy for government is enacting laws that allow information about basic rights, register, and investigate such incidences. Social and volunteer groups may appeal to people to quit smoking and liquor so that they could save more and uplift their socioeconomic status.

#### 13.4.3.4 Value Chain Actor

##### *13.4.3.4 1 Fair Competition (FC)*

Government can significantly alter the scenario; it must facilitate needed infrastructure so that farmer may get the best price without the need of the middleman. Concept of cooperative organization to sale production and purchase of seeds and fertilizers will improve the competition and ensure best returns. Encouraging the role of self-governance can improve cohesion and support cultural values.

##### *13.4.3.4 2 Promoting Social Responsibility (PSR)*

Cooperative can take lead in taking adequate steps towards social responsibility. For instance, by building a sanitation facility for poor people, mandating code of conduct for suppliers (farmers) regarding, use and disposal of chemical, fertilizer, and agricultural waste that harms human health, and code of conduct also can be issued for promoting human rights. Ensuring food security as social responsibility, the culture of distributing free meals at religious places is an example of the same, this will help better health and increased social security.

##### *13.4.3.4 3 Respect of Intellectual Property Rights (IPR) (RIPR)*

The PAS uses certain items protected under patent law (Hornung 2011). A record of such product must be mandated and see that the IPR is respected. Stringent laws to observe IPR is a government responsibility. Protecting IPR must be considered as a fundamental right.

### 13.4.3.5 Consumers

#### *13.4.3.5.1 Supplier Relationship (SR)*

Under the cooperative model, it is possible to preserve the basic right of supplier, payment on time, and avoid coercive communication between members and supplier. In return, members can expect good word of mouth about the community.

#### *13.4.3.5.2 Satisfaction/Consumer Privacy (SCP)*

Use of bio-fertilizer in the agriculture gives satisfaction to the consumer that they are not directly or indirectly promoting a practice of global pollution and their satisfaction increases. Displaying a declaration about the use of bio-fertilizer can reduce complaints from consumers, and it is possible to promote such brand in the long term.

#### *13.4.3.5.3 Feedback Mechanism (FM)*

The cooperative may organize farmer program and recognize best practice in framing by awards. This will encourage awareness and competition among farmers.

#### *13.4.3.5.4 Health and Safety (HS)*

Improvement in the health of farmworker can be ensured by avoiding the use of fertilizers and chemicals and use of biochar instead. Health is an important determinant of the quality of life of people (Murthy and Mundial 1990). For instance, infant mortality rate decrease by 22% (1991–1985), and it is expected to reach 20% by 2020 for India. However, the poor have not benefitted much, there is a lot of scope for the state in this front (Murthy and Mundial 1990).

#### *13.4.3.5.5 Transparency (TR)*

Limiting the fertilizer used per kg of grain production and declaring such quantity can bring transparency (TR) in the farming practice. Declaring message on the pack that end products like agricultural residues are processed in a sustainable manner can also bring TR and can help in stopping open-field burning of residue. This will promote organic fertilizer and reduce the cost and preserve the environment.

#### *13.4.3.5.6 End-of-Life Responsibility (EoLR)*

Currently, there is no mandatory requirement for disposal of bags of fertilizer and bottles or can use for poisonous chemicals and agricultural waste (Amar, Prahcee, and Mahesh 2010). Proper disposal can avoid not only contamination of water and land but also accidents. End-of-life scenario for agricultural residue using biochar route can give the effect of carbon sequestration and help in protecting the environment (Gaunt and Lehmann 2008). The cooperative can observe such end-of-life responsibility (EoLR) better and can further impose it on the farm suppliers.

### 13.4.3.6 State

#### *13.4.3.6.1 Policy and Commitment of State (PCS)*

It is reported that 65% of Indian agriculture is heavily dependent on natural factors, particularly rainfall, while rainfall variance accounts for more than 50% of the variability in crop yields (Bhise, Ambhore, and Jagdale 2007). The result of two decades of effort could result into the insurance of only 20% of farmers (Desai, Mitra, and Netaji Institute for Asian Studies 2011). In a country like India, only 1% of losses are insured compared to 30% in advanced countries (Linnerooth-Bayer and Mechler 2009). Agriculture subsidy of Rs. 227 billion during 1992–2000 is released on social grounds; however, it has failed to achieve any substantial social objective (Audinet 2002). These subsidies result in poor revenue generation for the state of the public utility and increase financial burden and result into public debt. Such a solution fails to fulfil the aim of social welfare in the long term (Bhattacharyya 2012).

It is argued that instead of opting for subsidy route, if the same resources are invested in improvement in irrigation and infrastructure to supply water in the long run, results would be far better (Sinha 2007). Social overhead plays an important role in the development of a country, many countries, including India, had implemented the importance of education, health, nutrition, housing, social security, and social welfare (Murthy and Mundial 1990).

The poor contribution of farmers in GDP (2001–2002) shows policy paralysis for the farming sector (Shiva and Jalees 2005; Sidhu 1974). Failure of institutional credits for small and marginal farmers is one of the reason for and involvement of gram panchayats to evolve a mechanism to overcome farmer crises (Shiva and Jalees 2005). The state can influence the existing structure by way of framing policy, organizing public debate, and building public opinion for and against a particular technology or aspect of technology, providing financial assistance, legal back-up, etc. In the present case, the state can ensure best practice and take measures that discourage open-field burning, avoid excess use of fertilizer, ensure equal wages, avoid pollution (inclusion of the right of clean air), avoid the use of harmful chemicals (right to clean water and land), pushing the use of biochar, better waste disposal in a more sustainable manner, and discouraging farmer suicides.

##### *13.4.3.6.2 Infrastructure Promotion and Publicity (IRP)*

State can be instrumental in pushing finance for opening up of pyrolysis processing unit may serve a need of community center; it can also play a role in technology dissemination programs. It is reported that making a person more informed will give strength to coping up with the losses due to failure of the crop (Desai, Mitra, and Netaji Institute for Asian Studies 2011) and making the person more social will reduce suicide rates (Aggarwal 2008). The state can act as a center of information exchange. Need to promote measures like contract farming is a forward agreement between a farmer and agencies to purchase produce at predetermined prices, it reduces risk of farmer initial success is observed in Punjab, India (Sharma 2008).

##### *13.4.3.6.3 Availability of Resource and Utilization (ARU)*

The current trend of financing farm activity is a policy failure, growth in agricultural subsidies of India is 12%/year between 1993–1994 and 2002–2003, while the contribution of farm output in GDP actually decreased in the same period (Mohanty 2013). On the other hand, India observed very low level of public health expenditure (5.1% of GDP), which is lowest in the world (Pal and Ghosh 2007). In Punjab, India, successive government has given free electricity to the farmer and very low canal water charges; both have encouraged inefficient use (Sharma 2008). High-fertilized subsidy encourages the rampant use and even beyond the prescribed limits, e.g., NPK (N, $P_2O_5$, and $K_2O$) consumption, which was less than 1 kg/ha in 1950–1951, jumped to 113 kg/ha in 2006–2007. Nitrogenous fertilizer in Punjab, India, found to be 20:6:1 against recommended 4:2:1 (Sharma 2008).

It is suggested that there is a need for effective utilization of resources. The state can take a policy decision regarding reallocating or abolishing all the subsidy on electricity, the supply of water, fertilizer, seeds, etc. Issue of availing easy finance for new technology like PAS at a lower interest rate can also be tried. This will encourage healthy farming practice and make farming compete in the global world. It is also seen that the wider usability of electricity among poor has the strongest social effect (Bhattacharyya 2012). PAS can discourage the use of fossil fuel, which already has a negative social impact (Ekener-Petersen, Höglund, and Finnveden 2013).

### 13.4.4 Impact Assessment Summary

Impact assessment is carried out by allocating points in the range from +3 to −3 to the present (BaU) as well as the proposed scenario (PAS). The relation between subcategory and impact category effects is rated from strongly positive (+3) to strongly negative (−3). Social impact assessment approaches are labeling of processes with the important attribute, inclusion or exclusion, of a particular element (e.g., child labor). The use of statistical value representing quantity as an indicator (e.g., number of WH), considering the risk to process (e.g., risk of armed conflict). Allocation procedure is based on personal judgment and interpreted based on total score in a given subcategory. The detailed discussion of each of these subcategories for the present and proposed scenario is presented hereunder.

Table 13.17 shows the social life cycle impact assessment table. The overall social impact under each subcategory is shown visually by the radar chart as Figure 13.11. The numerical implication of the score has less relevance, and information remains indicative because different category or added value does, by themselves, not say anything about the importance of social impacts. From the radar chart, it is concluded that the pyrolysis-assisted scenario has an all-round strong positive impact on the different aspects and has an overall high social impact.

## 13.5 ASSESSMENT OF ECONOMIC SUSTAINABILITY

LCC was developed and used first by the US military in the 1960s (Gluch and Baumann 2004) to assess the total cost of its long living goods such as tanks. LCC is an analysis technique which assesses and account all costs related to a product, over its entire life cycle, starting from its production to use, and including its maintenance and disposal (Real 2010; Benoît et al. 2009). LCC allows detailed viability assessment of the projects considering all the costs right from acquisition to operate and to disposal and allows optimizing the economic performances of enterprises. International Organization for Standardization defines LCC as a methodology for the systematic economic appraisal of products/processes by classifying costs into construction, management, maintenance, and end-of-life cost categories. Although the LCC model evaluates the monetary values of processes and flows associated

**TABLE 13.17**
**Impact Assessment Summary**

| | | G | | Hu | | W | | S | | He | | C | |
|---|---|---|---|---|---|---|---|---|---|---|---|---|---|
| **Stakeholder** | **Issue of Concern** | **E** | **P** | **E** | **P** | **E** | **P** | **E** | **P** | **E** | **P** | **E** | **P** |
| Farm and farmworkers | FL | −1 | 2 | −1 | 2 | | | −1 | 2 | −1 | 2 | −1 | 2 |
| | CL | −1 | 2 | −1 | 2 | −1 | 1 | −1 | 2 | −2 | 2 | −1 | 1 |
| | FA | −1 | 1 | −1 | 1 | −1 | 1 | −1 | 1 | | | −1 | 1 |
| | SB | −1 | 1 | −1 | 1 | −1 | 1 | −1 | 1 | −1 | 1 | −1 | 1 |
| | WH | −1 | 1 | −1 | 1 | −1 | 1 | | | −1 | 1 | −1 | 2 |
| | FS | −1 | 1 | −1 | 1 | | | −1 | 1 | −1 | 1 | −1 | 1 |
| | EO | −1 | 1 | −1 | 1 | −1 | 1 | −1 | 1 | −1 | 1 | −1 | 1 |
| | HS | −1 | 1 | −1 | 1 | −1 | 1 | −1 | 1 | | | −1 | 1 |
| Local communities | CI | −1 | 1 | −1 | 3 | | | −1 | 1 | −1 | 1 | −1 | 1 |
| | AR | −1 | 1 | −1 | 1 | −1 | 1 | −1 | 1 | 0 | 2 | 0 | 2 |
| | Access to immaterial resources | −1 | 1 | | | −1 | 1 | | | −1 | 1 | | |
| | DM | −1 | 1 | −1 | 1 | −1 | 1 | −1 | 1 | −1 | 1 | 1 | −2 |
| | CH | −1 | 2 | −1 | 1 | | | −1 | 2 | | | | |
| | SHLC | −1 | 1 | −1 | 2 | −1 | 2 | −1 | 1 | −1 | 1 | | |
| | RIR | −1 | 1 | | | | | −1 | 1 | | | −1 | 2 |
| | CE | −1 | 2 | | | | | −1 | 1 | | | −1 | 1 |
| | LE | −1 | 1 | −1 | 2 | | | −1 | 2 | −1 | 1 | −1 | 1 |
| | SLC | −1 | 1 | | | −1 | 1 | | | −1 | 1 | −1 | 1 |
| | CO | −1 | 1 | −1 | 1 | | | −1 | 2 | −1 | 1 | −1 | 1 |
| | ER | | | | | | | | | | | −1 | 1 |
| Society | PCSI | −1 | 1 | | | | | −1 | 1 | | | −1 | 1 |
| | CED | −1 | 1 | | | | | −1 | 1 | | | −1 | 1 |
| | PMAC | −1 | 1 | | | | | −1 | 1 | −1 | 1 | 0 | 1 |
| | CR | −1 | 1 | −1 | 2 | | | −1 | 1 | | | −1 | 1 |
| Value chain actor | FC | −1 | 2 | | | | | | | | | −1 | 2 |
| | PSR | −1 | 1 | −1 | 1 | −1 | 1 | | | −1 | 1 | −1 | 1 |
| | SR | | | −1 | 1 | | | | | | | | |
| | RIPR | −1 | 1 | −1 | 1 | | | 1 | −1 | | | | |
| Consumers | SCP | −1 | 1 | −1 | 1 | −1 | 1 | −1 | 1 | | | | |
| | FM | −1 | 2 | | | | | | | | | | |
| | H&S | | | | | | | −1 | 1 | | | | |
| | TR | −2 | 2 | | | | | −2 | 2 | | | | |
| | EoLR | −2 | 3 | | | −2 | 3 | | | −2 | 2 | | |
| State | PCS | −2 | 3 | −1 | 2 | | | −1 | 1 | | | −1 | 2 |
| | IRP | −2 | 3 | | | | | −1 | 3 | | | −1 | 3 |
| | ARU | −2 | 3 | | | | | −1 | 3 | | | | |

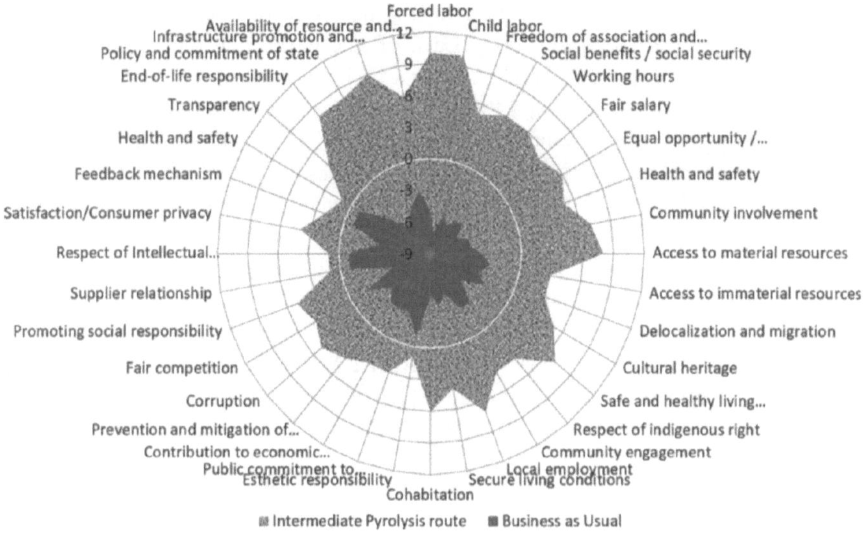


**FIGURE 13.11** Multi-axis radar chart showing the result of sLCA for BaU and PAS scenarios considered for the study.

with a product or the system, its scope and depth varies significantly (Biondini et al. 2008). In spite of its widespread use, a comprehensive methodology and detailed guidelines is yet to publish (Benoît et al. 2009). LCC is a useful technique while comparing competing alternatives for projects with different costs and benefits and having different spans of time (Kaan Ozbay et al. 2003).

In principle, LCC and Net Present Value (NPV) are similar, both deal with discounting cash flows. Scope of NPV is limited to strategic decisions and business planning, while the scope of LCC is wider and it takes into account a wide range of technical data spanning the entire lifecycle, e.g., energy or fuel consumption in use, maintenance operations, and end-of-life scenarios and can influence in decision-making concerned to design, production, operation, performance, maintenance, logistic, and disposal (Favi et al. 2012). In particular, energy conservation projects are excellent examples of the application of LCC (Favi et al. 2012). LCC is in direct contrast to the payback method of economic analysis. The latter focus on how quickly the investment can be recovered without much concerned with long-term economic performance. It ignores all costs and risk beyond the recovery of invested capital. It even compares projects with different life span and many times do not account time value for money (Fuller and Petersen 1996).

LCC deals with the future, and since the future is uncertain, LCC has a high element of uncertainty. LCC has several parameters that are inherently uncertain, e.g., discount rate, the life of the machine, inflation, and maintenance cost for operation under new technology. In order to conduct LCC in a reliable and trustworthy manner, a thorough understanding of the theoretical engineering and economics background must be acquired (Ozbay et al. 2003). Lack of acknowledging method is the biggest limitation of the application of LCC for a problem at hand (Oduyemi, Okoroh, and Dean 2014; Pulakka and Sarja 1999). The approach of the user also affects the LLC results, e.g., building manufacturer is many times not the building user and this lacks long-term interest in it; therefore, cost considerations are different. Due to uncertainty in the market condition, the maintenance, fuel costs, rate of interests, inflation, and the material cost may vary from time to time and from place to place (to be precise economy to economy) (Oduyemi, Okoroh, and Dean 2014). LCC fails to handle uncertainty, irreversible decisions; neglects environment concern, future environmental calamity, availability, and reliability of input inventory; and the influence of decision maker's personal values (Real 2010). LCC analysis also fails to consider cost added due to the intangible aspects of product, e.g., user's comfort or cost of greener building, recyclable or natural material (Fuller and Petersen 1996).

### 13.5.1 Determination of Cost Component

The first step in the LCC study is to determine the life of the project. It is recommended that the duration of the project considered, different competing options should be the same, and it should be equal to the longest life of the component of the system (Ozbay et al. 2003). Next is the determination of cost. Certainly, cost is occurring once in the lifetime or occurring in a non-annual interval, e.g., initial investment costs, replacement costs, residual values, break-down costs, and repair costs. Investment in capital cost of the machine may occur more than once if the life span of study is more than the life of the machine; under these circumstances, it is essential to consider salvaging cost accounting at the end of the project (Ozbay et al. 2003). Conversely, there is some cost that occurs annually recurring during the project period, e.g., energy costs, annual maintenance cost, fuel cost, and labor cost (Fuller and Petersen 1996).

The discount rate in LCC is the most sensitive parameter, a typical discount rate ranges from 4% to 10% (Ozbay et al. 2003). Uncertainty in the discount rate is an issue, especially when the project is evaluated in developing countries; in such circumstance, the IRR format is a preferred indicator (Ozbay et al. 2003). Lower discount rate favors project with higher capital cost (building, infrastructure project, hydro project). Conversely, a high discount rate favors project that has higher future costs (mining, transportation project) (Ozbay et al. 2003). Inflation reduces the purchasing power of the money and also the effect of the discounting rate. Choosing the correct inflation rate is the key in the LCC study and depends on the economic condition of a project undertaken, etc. To measure costs with inflated or deflated money is meaningless, is like measuring building's dimensions with an elastic tape (Favi et al. 2012), inflation is considered in the range of 4%–6% (Ozbay et al. 2003). The real discount rate ($d$) which is diluted due to the effect of inflation ($i$) for a nominal discount rate ($D$) can be obtained as under

$$d = \left( \frac{(1+D)}{(1+i)} - 1 \right) \tag{13.1}$$

Rise in the price of a commodity due to factors other than inflation can be considered in the analysis using escalation rate. The actual escalation in price ($e$) can be obtained from nominal price escalation ($E$) under the current rate of inflation ($i$) as under

$$e = \left( \frac{(1+E)}{(1+i)} - 1 \right) \tag{13.2}$$

Inflation will weaken the value of money in the long term. If this effect is not considered in the analysis, the accounting of money under this situation is called as constant dollar accounting. However, if the effect of inflation is considered, there will be a distinction in the value of money between two time periods. Accounting under this consideration is called current dollar accounting; under this, the value of today's money will reduce in tomorrow's scenario and vice versa. The concept is known as the time value of money or present value (PV) of money. For instance, predicting the single present value (SPV) or amount ($A_0$) that is required to be invested today at a given discount rate ($d$) so that after ($n$) years the amount would be ($A_n$) can be given by the following relationship.

$$A_0 = \left( \frac{A_n}{(1+d)^n} \right) \tag{13.3}$$

If the number of such non-uniform amount (multiple non-uniform present value—MUPV) is generated from 1 to $n$ years, it is possible to convert this amount into a single equivalent amount ($A_0$) under discounting rate ($d$) using the following relationship.

$$A_0 = \sum_{t=1}^{n} \left( A_n \left( \frac{1}{1+d} \right)^n \right) \tag{13.4}$$

LCC analysis is carried out in literature, and the outcome is presented in terms of the benefit–cost ratio method or savings–investment ratio method, same is followed in the present case also. It is the ratio of the present worth of positive cash flows to the present worth of negative cash flows. Besides this, the comparison of the cost of power generation under discounted and the non-discounted method is also presented. In order to show the reliability of the result, the result is presented by varying such individual parameters. Sensitivity analysis is accepted as a risk assessment technique (Real 2010) and warn users against assumptions in the study, discounting rate, inflation rate, and different cost. Lack of sensitivity analysis devaluates the reliability of the entire study (Korpi and Ala-Risku 2008).

Figures 13.12 and 13.13 show the cost component goes into the analysis of two scenarios, BaU and PAS. Number of times of occurrence of a cost corresponds to the number of time it is displayed in the figure. Note that capital cost can occur more than once in the life of the project, this is possible when the life of the individual component is lower than project life. Maintenance cost, depreciation cost, labor cost, and fuel cost occur every year. However, salvage cost occurs only in the last year of the project. Tables 13.18–13.21 show input cost of the study, data pertaining to different units, experimental data, and rate considered for the analysis, respectively.

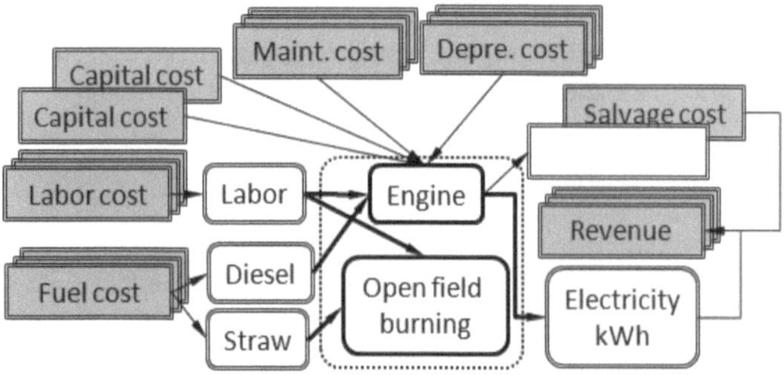


FIGURE 13.12 The cost component occurring in different years of operations under BaU scenario.

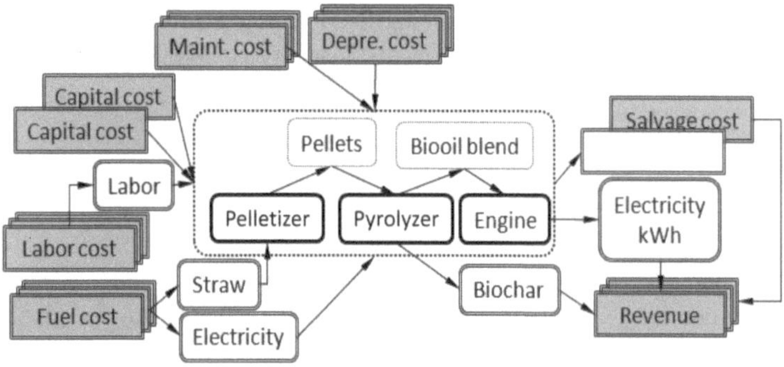


FIGURE 13.13 The cost component occurring in different years of operations under PAS scenario.

**TABLE 13.18**
**Input Cost Considered under BaU and PAS Scenario for the Study**

| Input Cost | Value | Input Cost | Value |
|---|---|---|---|
| Cost of straw | 3 Rs/kg | Electricity cost | 4 Rs/kWh |
| Cost of diesel | 45 Rs/l | Manpower | 10 Rs/h |
| Electric load | 5 HP | O&M charges | 20% |
| Duty factor | 70% | Cost of char | 3 Rs/kg |

Since the life of the project is greater than the life of an individual machine, the capital cost of the pelletizer will occur in 1st year, 9th year, and 18th year. Similarly, the cost of pyrolyzer will appear in 1st year and 16th year and cost of engine will occur in 1st year and 11th year. The replacement cost of the machine is obtained based on the straight-line depreciation method amount adjusted to the present value accrued for the period. Salvage value of a machine is considered as the present value of the book value of a machine at the end of life of the project.

**TABLE 13.19**
**Data Pertaining to Different Units under PAS Scenario**

| Machine | Cost (Rs) | Life (y) | Capacity |
|---|---|---|---|
| Pelletizer | 1,50,000 | 8 | 1,000 kg/h |
| Pyrolyzer | 10,00,000 | 15 | 20 kg/h |
| Engine | 1,50,000 | 10 | 100 kWh |

**TABLE 13.20**
**Experimental Data Obtained during Engine Trial under BaU and PAS Scenario**

| Fuel Consumption | From Experiment (A. R. Patel 2018) |
|---|---|
| Diesel consumed | 1.17 kg/h |
| Blend consumed | 1.35 kg/h |
| SFC—diesel | 0.355 kg/kWh for 10 kW engine |
| SFC—blend | 0.390 kg/kWh for 10 kW engine |
| Fuel spent—diesel | 29.2 l/h for 100 kW engine[a] |
| Fuel spent—blend | l/h for 100 kW engine[b] |

[a] Data supplied by manufacturer.
[b] Data based on the test on a 10 kW engine.

**TABLE 13.21**
**Rate Considered for the Analysis**

| Rate | Value | Description |
|---|---|---|
| $i$ | 4% | Inflation rate |
| $d$ | 8.2% | Discount rate |
| $E_f$ | 8.0% | Fuel price escalation rate |
| $E_m$ | 6.0% | Maintenance escalation rate |
| $E$ | 6.0% | General escalation |

The fuel cost for the pelletizer is calculated based on the cost of straw and the number of hours of operation. Since the capacity of the pyrolyzer is only 20 kg/h compared to the pelletizer which is 1,000 kg/h, the cost of fuel at pelletizer is considered as 2%. While considering the cost of pyrolyzer, the cost of pellets is considered as zero since pellets are an internal product, only electricity cost is considered in case of pyrolyzer as fuel cost. Operation and maintenance charges are considered as 20% of the capital cost for all units. In the case of a pelletizer, the cost of maintenance and labor is considered as 2% since its utilization is 2% compared to the pyrolyzer capacity. There will be three components of revenue; they are income from the sale

**TABLE 13.22**
**Power Generation Costs Using BaU and PAS Scenario**

| Particulars | BaU | % Cost | PAS | % Cost |
|---|---|---|---|---|
| Capital cost (Rs/y) | 150,000 | 0.28 | 5,300,000 | 23.22 |
| Fuel cost (Rs/y) | 52,060,376 | 97.67 | 9,450,574 | 41.41 |
| Maint. cost (Rs/y) | 49,710 | 0.09 | 382,350 | 1.68 |
| Labor cost (Rs/y) | 795,354 | 1.49 | 1,606,615 | 7.04 |
| Replacement cost (Rs/y) | 248,548 | 0.47 | 6,082,525 | 26.65 |
| Total (Rs/y) | 53,303,988 | 100 | 22,822,064 | 100 |
| Revenue (Rs/y) | 18,929,363 | | 24,057,913 | |
| Net surplus (+) / loss (−) | −34,374,625 | | 1,235,849 | |
| Benefit/cost ratio | −0.64 | | 0.054 | |
| Cost of elect. (Rs/kWh) | 11.10 | | 4.75 | |

of electricity, income from the sale of biochar, and income from salvage cost of the machine at the end of the project. The difference in the present value of the total cost and total revenue will be net surplus or loss.

### 13.5.2 Life Cycle Cost of Electricity Generation

Table 13.22 summarizes the result of power generation using diesel and PAS. It is observed that there is a net loss due to the operation in case of the diesel-assisted system. This suggests that the BaU scenario is not financially viable and results in negative benefit-to-cost ratio, whereas pyrolysis-assisted scenario offers surplus revenue and therefore financially viable and leads to positive benefit-to-cost ratio (0.054). The cost of power generation under the BaU scenario is 11.10 Rs/kWh, whereas the same for pyrolysis-assisted scenario it is 4.75 Rs/kWh.

From the table, it is concluded that the life cycle cost of electricity production under PAS scenario can be reduced to Rs. 4.75 per kWh from Rs. 11.10 per kwh under BaU scenario.

## 13.6 CONCLUDING REMARKS

This work discusses the sustainability studies of an intermediate pyrolysis process considering a triple bottom line analysis approach for the two scenarios of farm management. It is observed that intermediate pyrolysis process when carried out at a scaled-up unit and following biochar route is attractive from the environment point of view. It is possible to achieve a negative EF using intermediate pyrolysis options under biochar route, and therefore, such an option provides an alternative for low-cost carbon sequestration. It is inferred that a LAIP is technically viable but not a feasible alternative. The present analysis uses a borrowed set of embedded data and has high uncertainty. The uncertainty of embedded energy in "all max" case is 14%, which is high (as given in Appendix I). The sensitivity of embedded energy and emission data is found to be well under 10%, and hence, overall uncertainty is

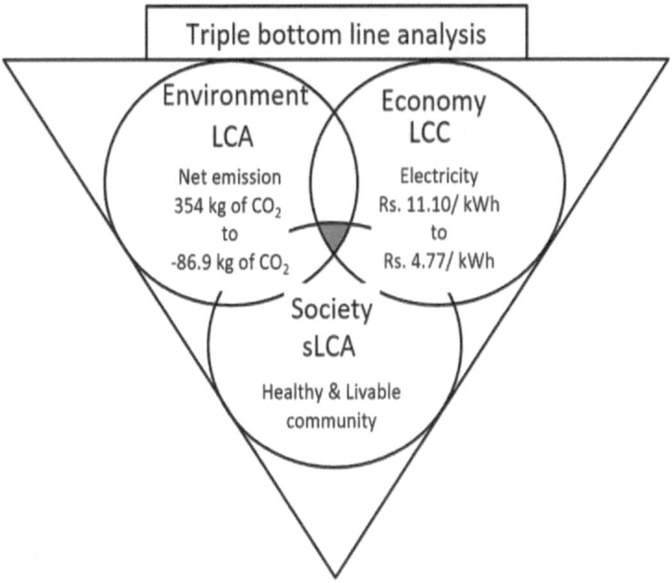


**FIGURE 13.14** Result of tipple bottom line analysis across its three pillars, for BaU and PAS scenarios.

well within the control, and data can be considered as to satisfactory quality. Highest sensitivity is observed in the case of the material phase of MS.

From the study, it is observed that the PAS route has the capability of positive impact on the social life cycle analysis on every front. An attractive feature of this analysis is a cooperative model with the KC in it and the generation and use of biochar. The effect of sLCA recommendation is greatest on the local community, followed by the state, while farm and farmworkers will be third largest beneficiaries of social change due to the proposed scenario. It can be concluded that the pyrolysis-assisted scenario has a strong positive impact on the social aspects, and implementation of this technology can certainly make the realization of a **Lively Society** and a **Happy** and **Healthy Lifestyle** for People. From the result of LCC study, it is observed that pyrolysis-assisted scenario results in positive NPV (with IRR as 9.7%) and with a benefit–cost ratio of 0.054. Electricity generation cost improves from 11.10 Rs/kWh to 4.75 Rs/kWh. The result of all three sustainability pillars is summarized as given in Figure 13.14. It is easy to conclude that intermediate pyrolysis technology influences all the three pillars of sustainability assessment. The net effect of the technology is the improvement in all three pillars of sustainability.

## ACKNOWLEDGMENT

Authors are thankful to Oglesby Charitable Trust (http://www.oglesbycharitabletrust.co.uk/) for its generous grant to support the program. The support received from School of Mechanical, Materials & Energy Engineering at IIT Ropar is gratefully acknowledged.

## APPENDIX 13.A1: UNCERTAINTY AND SENSITIVITY OF THE DATA

The usage of borrowed data is common in LCA studies due to its, ease in analysis, common practice, broad availability, lack of relevant data, or no data available. However, borrowed data is also inherent uncertainty with it, few of the reasons could be different sources of origin, geographical location, market conditions, prevailing technology, legal infrastructure, climatic conditions, cultural-oriented habits, and preferences, to mention a few (ILCD Handbook 2010). It is essential to quantify the error and ensure the reliability of the results since the error in the data is reflected in the result, and in extreme case, this may also lead to the selection of the wrong alternative (Ashby 2012). An uncertainty analysis is carried out for this reason. Under this, the least and the worst effect of uncertainty in the input data is considered to predict the two extreme outcomes. The range of uncertainty is obtained by comparing the deviation with the actual result (ILCD Handbook 2010). In the present case, the data used to determine the embedded energy and embedded emission is borrowed one. The data considered for the material is as recommended by Ashby (2012). The data is given in terms of a range (as shown in Table 13.1). Predicting the embedded energy for the process, considering all lower bound values of respective use of material is termed as "all low" and, similarly considering all higher bound values of respective use of material is termed as "all high" values. Results are summarized in Table 13.A1.1. It is observed from the table that uncertainty in case of embedded energy has a bound off −11% and +14%, for all lower value use to all higher value use in a given range. This suggests a high uncertainty value due to the quality of input data, it is also observed that there is a skewed distribution and higher uncertainty is observed in "all high" case. The high uncertainty in the embedded energy data for the use of material emphasizes a strong need for evolving specific data relevant to the study. This will effectively bring down the overall uncertainty. Similar analysis when carried out for embedded emission data, the uncertainty found out to be satisfactory (< 10%) and is within the control.

Sensitivity analysis is needed to identify the most sensitive inventory data influencing the outcome. It is obtained by considering a finite change in a singular input parameter by 10% (Table 13.A1.1) (ILCD Handbook 2010). Table 13.A1.2 shows the effect of the increase in a single input parameter on the embedded energy and

**TABLE 13.A1.1**
**Uncertainty in Embedded Energy and Emission**

| | Base | All Min | | All Max | |
|---|---|---|---|---|---|
| Energy (MJ) | 1,675 | 1,491 | *−11%* | 1,909 | *14%* |
| Emission (kg of $CO_2$/t) | 1,055 | 955 | −9% | 1,154 | 9% |

emission values, keeping values of other input variables as constant. As a rule of thumb, sensitivity below 10% is considered as a sign of good reliability. However, the certain study advocates allowable sensitivity as high as 15%–20% (Taylor et al. 1997), suggesting that the choice of sensitivity depends on the criticality of the result wherein these data are applied. The result of sensitivity is shown in the table. It is observed from the table that sensitivity is less than 10% of all the cases. Among all material phases, mild steel (MS) has a high sensitivity for embedded energy and embedded emission. This is because, MS is the most largely used material and has a high contribution to the material phase of embedded data (66%). The present analysis strongly recommends the requirement of a country-specific original embedded energy and emission data for the material processing phase for material MS. Such data may improve the reliability of the result in the case of a similar study.

Emission generated from the production of each kWh of energy at power plants or in the energy mix of the grid increases with the increase in the utilization of emission intensive fuels (e.g., coal). As per the emission data of the year 2011, EF of the Indian grid is 0.84 kg of $CO_2$/kWh (GoI 2011). The sensitivity of EF on the embedded emission data is worked out, as 5.59%, this is low and within the control. Refining burden is another important input parameter. Consumption of the crude oil at the refinery varies from 4% to 15% of the crude input. Hydro-cracking technology is largely used in India. EF under this refining route ranges from 0.2 to 0.35 (UNIDO 2010). Considering EF as 0.22 kg of $CO_2$/kg of diesel, the sensitivity of EF on the embedded energy and emission is found to be low and well under control.

**TABLE 13.A1.2**
**Result of Sensitivity Analysis**

| | **Sensitivity in Embodied Energy [%]** | | | | |
|---|---|---|---|---|---|
| | **CI** | **MS** | **SS** | **Al Alloy** | **Cu Alloy** |
| Material | 0.84 | *6.69* | 0.90 | 1.91 | 0.78 |
| Casting | 0.18 | 0.48 | – | – | – |
| Deformation | 0.18 | 0.84 | 0.12 | 0.06 | 0.06 |
| Recycle | 0.24 | 1.85 | 0.24 | 0.12 | 0.18 |
| | **Sensitivity in Embodied Emission [%]** | | | | |
| Material | 0.66 | *7.58* | 0.76 | 1.52 | 0.76 |
| Casting | 0.09 | 0.38 | – | – | – |
| Deformation | 0.09 | 0.66 | 0.09 | – | – |
| Recycle | 0.19 | 2.09 | 0.19 | 0.19 | 0.19 |

## REFERENCES


Adam, J. 2013. *Alternatives to Open-Field Burning on Paddy Farms*, Vol. 18. Malaysia: Agricultural and Food Policy Studies Institute.

Aggarwal, N. 2008. "Editorial: Farmer Suicides in India: The Role of Psychiatry and Anthropology." *International Journal of Social Psychiatry* 54 (4): 291–292. doi: 10.1177/0020764007088396.

Amar, D. M, J. Prahcee, and J. Mahesh. 2010. "Modern Agricultural Practices: A Dilemma of Farmers and Farm Workers' Health in the Cash Crop Zone in the Maharashtra State." *Bhatter College Journal of Multidisciplinary Studies* 1 (1): 80–88.

Andrews, E., P. Lesage, C. Benoît, J. Parent, G. Norris, and J-P. Revéret. 2009. "Life Cycle Attribute Assessment: Case Study of Quebec Greenhouse Tomatoes." *Journal of Industrial Ecology* 13 (4): 565–578. doi: 10.1111/j.1530-9290.2009.00142.x.

Antal, M. J., and M. Grønli. 2003. "The Art, Science, and Technology of Charcoal Production." *Industrial & Engineering Chemistry Research* 42 (8): 1619–1640. doi: 10.1021/ie0207919.

Ashby, M. F. 2012. *Materials and the Environment: Eco-Informed Material Choice*. Oxford, UK: Butterworth-Heinemann.

Atik, C., and S. Ates. 2012. "Mass Balance in Silica in Straw from the Prospective of Silica Reduction in Straw Pulp" *BioResource* 7 (3): 3274–3282.

Atkinson, C. J., J. D. Fitzgerald, and N. A. Hipps. 2010. "Potential Mechanisms for Achieving Agricultural Benefits from Biochar Application to Temperate Soils: A Review." *Plant and Soil* 337 (1–2): 1–18. doi: 10.1007/s11104-010-0464-5.

Audinet, P. 2002. "Electricity Prices in India." *Energy Prices and Taxes* 2 (2): 11–21. http://mahadiscom.com/emagazine/jan06/india1%5B1%5D.pdf.

Banerjee, A. 2007. "Lessons Learned Studies: India." *Report of the Animal Production and Health Commission for Asia and the Pacific*. Food and Agriculture Organization of the United Nations (FAO), New Delhi, India. http://www.csa-be.org/IMG/pdf_LLS_India_revised_20Dec07.pdf.

Beer, T., T. Grant, R. Brown, J. Edwards, P. Nelson, H. Wattson, and D. Williams. 2000. "Life-Cycle Emissions Analysis of Alternative Fuels for Heavy Vehicles." Stage 1 C/0411/1.1/F2. To the Australian Greenhouse Office: CSIRO Atmospheric Research.

Benoît, C., B. Mazijn, E. S. Andrews, United Nations Environment Programme, and Sustainable Consumption and Production Branch. 2009. *Guidelines for Social Life Cycle Assessment of Products: Social and Socio-Economic LCA Guidelines Complementing Environmental LCA and Life Cycle Costing, Contributing to the Full Assessment of Goods and Services within the Context of Sustainable Development*. Paris: United Nations Environment Programme.

Bhalla, S., and R. Kaur. 2011. "Labour Force Participation of Women in India: Some Facts, Some Queries." *Asia Research Centre Working Paper* 40 (LSE India Observatory Research Project). http://eprints.lse.ac.uk/38367/1/ARCWP40-BhallaKaur.pdf.

Bhattacharyya, S. C. 2012. "Energy Access Programmes and Sustainable Development: A Critical Review and Analysis." *Energy for Sustainable Development* 16 (3): 260–271. doi: 10.1016/j.esd.2012.05.002.

Bhise, V. B., S. S. Ambhore, and S. H. Jagdale. 2007. "Performance of Agriculture Insurance Schemes in India." *101st EAAE Seminar 'Management of Climate Risks in Agriculture'*, Berlin, Germany, July, 5–6. http://ageconsearch.umn.edu/bitstream/9245/1/sp07bh01.pdf.

Biondini, F., D. M. Frangopol, International Symposium on Life Cycle Civil Engineering, and International Association for Life Cycle Civil Engineering, eds. 2008. "Life-Cycle Civil Engineering." *Proceedings of the First International Symposium on Life-Cycle Civil Engineering*, Varenna, Lake Como, Italy, June 10–14, 2008. Boca Raton, FL: CRC Press/Balkema.

Birchall, J. 2003. *Rediscovering the Cooperative Advantage: Poverty Reduction through Self-Help*. Geneva: International Labour Organisation.

Bramley, G., N. Dempsey, S. Power, and C. Brown. 2006. "What Is 'Social Sustainability', and How Do Our Existing Urban Forms Perform in Nurturing It?" *Sustainable Communities and Green Futures' Conference*, Bartlett School of Planning, University College London, London

Bridgwater, A. V., and G. V. C. Peacocke. 2000. "Fast Pyrolysis Processes for Biomass." *Renewable and Sustainable Energy Reviews* 4 (1): 1–73.

Bridgwater, A.V. 2012a. "Review of Fast Pyrolysis of Biomass and Product Upgrading." *Biomass and Bioenergy* 38 (March): 68–94. doi: 10.1016/j.biombioe.2011.01.048.

Brova, G. 2008. "The Emergence of Environmental and Social Sustainability." *Student Reviewers*, 53–59.

Çengel, Y. A., and M. A. Boles. 2011. *Thermodynamics: An Engineering Approach*. New York: McGraw-Hill.

Chiaramonti, D., A. Oasmaa, and Y. Solantausta. 2007. "Power Generation Using Fast Pyrolysis Liquids from Biomass." *Renewable and Sustainable Energy Reviews* 11 (6): 1056–1086. doi: 10.1016/j.rser.2005.07.008.

Chr, S. H., and N. Stefan. 2005. "Life Cycle Assessment of the Wave Energy Converter: Wave Dragon." http://orbit.dtu.dk/services/downloadRegister/3711218/WaveDragon.pdf.

Corvalán, C., S. Hales, A. J. McMichael, Millennium Ecosystem Assessment (Program), and World Health Organization, eds. 2005. *Ecosystems and Human Well-Being: Health Synthesis*. Millennium Ecosystem Assessment. Geneva: World Health Organization.

Desai, M., S. Mitra, and Netaji Institute for Asian Studies, eds. 2011. *Cloud, Stone, and the Mind: The People and Environment of Darjeeling Hill Area*. Kolkata: K.P. Bagchi & Company.

Diesendorf, M. 2000. "Sustainability and Sustainable Development." *Sustainability: The Corporate Challenge of the 21st Century* 2: 19–37. http://www.ies.unsw.edu.au/sites/all/files/CorpSust2000.pdf.

Ekener-Petersen, E., J. Höglund, and G. Finnveden. 2013. "Social and Socioeconomic Impacts from Vehicle Fuels." http://www.diva-portal.org/smash/record.jsf?pid=diva2:721761.

Elkington, J. 1999. *Cannibals with Forks: The Triple Bottom Line of 21st Century Business*. Oxford, UK: Capstone.

Evans, A., V. Strezov, and T. J. Evans. 2009. "Assessment of Sustainability Indicators for Renewable Energy Technologies." *Renewable and Sustainable Energy Reviews* 13 (5): 1082–1088. doi: 10.1016/j.rser.2008.03.008.

Evans, R. J., and T. A. Milne. 1987. "Molecular Characterization of the Pyrolysis of Biomass. 2. Applications." *Energy & Fuels* 1 (4): 311–319.

Favi, C., M. Peruzzini, M. Germani. 2012. "A Lifecycle Design Approach to Analyse the Eco-Sustainability of Industrial Products and Product-Service Systems." *Proceedings of International Design Conference DESIGN*, 879–888. http://www.researchgate.net/profile/Margherita_Peruzzini/publication/258862230_A_Lifecycle_design_approach_to_analyze_the_Eco-sustainability_of_industrial_products_and_product-service_systems/links/02e7e53a440c683262000000.pdf.

Finkbeiner, M., E. M. Schau, A. Lehmann, and M. Traverso. 2010. "Towards life cycle sustainability assessment." *Sustainability* 2 (10): 3309–3322. doi: 10.3390/su2103309.

Fritsche, U. R., K. J. Hennenberg, K. Hünecke, R. Herrera, and K. Wiegmann. 2012. Sustainable Bioenergy: Key Criteria and Indicators. *Final D* 4, 1–29. http://www.biomassfutures.eu/public_docs/final_deliverables/WP4/D4.1%20Sustainable%20Bioenergy%20-%20criteria%20and%20indicators.pdf.

Fuller, S. K., and S. R. Petersen. 1996. "Life Cycle Costing Manual." *NIST Handbook* 135. http://aesl.hyu.ac.kr/resource/blcc/blcc-handbk1.pdf.

Garg, S. 2008. "Trace Gases Emission from Field Burning of Crop Residues." *Indian Journal of Air Pollution Control* 1: 76–86.

Gaunt, J. L., and J. Lehmann. 2008. "Energy Balance and Emissions Associated with Biochar Sequestration and Pyrolysis Bioenergy Production." *Environmental Science & Technology* 42 (11): 4152–4158. doi: 10.1021/es071361i.

Ghosh, J. 2005. "The Political Economy of Farmer Suicides in India." *Freedom from Hunger Lecture Series, New Delhi*. http://www.macroscan.com/fet/dec05/pdf/freedom_hunger.pdf.

Gluch, P., and H. Baumann. 2004. "The Life Cycle Costing (LCC) Approach: A Conceptual Discussion of Its Usefulness for Environmental Decision-Making." *Building and Environment* 39 (5): 571–580. doi: 10.1016/j.buildenv.2003.10.008.

GoI, Min of Power. 2011. *$CO_2$ Baseline Database for the Indian Power Sector, User Guide, Version 6.0*. New Delhi: Government of India, Ministry of Power.

Gulati, A. 2009. "Emerging Trends in Indian Agriculture: What Can We Learn from These." *Agricultural Economics Research Review* 22 (2): 171–184. http://core.ac.uk/download/pdf/6689677.pdf.

Gupta, P., S. Sahai, N. Singh, C. K. Dixit, D. P. Singh, C. Sharma, M. K. Tiwari, R. Gupta, and S. C. Garg. 2004. "Residue Burning in Rice-Wheat Cropping System: Causes and Implications." *Current Science*, 87 (12): 1713–1717.

Hansen, W. L., University of Wisconsin, and Conference on Research in Income and Wealth, eds. 1970. "Education, Income, and Human Capital." *Studies in Income and Wealth by the Conference on Research in Income and Wealth*, v. 35. New York: National Bureau of Economic Research; distributed by Columbia University Press.

Hassan, M. N. A., P. Jaramillo, and W. M. Griffin. 2011. "Life Cycle GHG Emissions from Malaysian Oil Palm Bioenergy Development: The Impact on Transportation Sector's Energy Security." *Energy Policy* 39 (5): 2615–25. doi: 10.1016/j.enpol.2011.02.030.

"Heating Methods and Open Burning." 2013. Canadian Lung Association.

Heijungs, R., G. Huppes, and J. Guinée. 2009. "A Scientific Framework for LCA." *Deliverable (D15) of Work Package 2*. http://www.leidenuniv.nl/cml/ssp/publications/calcas_report_d15.pdf.

Hohnen, P., and J. Potts. 2007. *Corporate Social Responsibility an Implementation Guide for Business*. Vol. 1. Manitoba: International Institute for Sustainable Development. www.iisd.org/pdf/2007/csr_guide.pdf.

Hornung, A. 2011. "Patent Application, Thermal Treatment of Biomass, Publication No. US 2011/0067991 A1."

Hornung, A, A. Apfelbacher, and S Sagi. 2011. "Intermediate Pyrolysis. A Sustainable Biomass-to-Energy Concept - Biothermal Valorisation of Biomass (BtVB) Process." *Journal of Scientific & Industrial Research* 70: 664–667.

ILCD Handbook. 2010. European Commission - Institute for Environment and Sustainability: International Reference Life Cycle Data System (ILCD) Handbook. EUR 24708 EN. Luxembourg, EU.

Indian Standard (IS-1460). 2005. "Automotive Diesel Fuel - Specification (Vth Revision)." Bureau of Indian Standards.

ISO, 14044. 2006. "ISO 14044-2006." International Standard ISO 14044:2006(E). Environmental Management — Life Cycle Assessment — Requirements and Guidelines. Switzerland. www.ISO.org.

Jacobsen, M. L. 2006. "Communicating Corporate Social Responsibility". MA - International Marketing, Netherlands: Aarhus School of Business.

Janeiro, L. 2011. "Towards a Methodology for the Sustainability Assessment of Technologies." *Master Thesis*, Netherlands: Faculty of Geoscience, Universiteit Utrecht. http://www.prosuite.org/c/document_library/get_file?uuid=268c9330-a02f-496c-8fa6-2d2608bc7352&groupId=12772.

Jhamtani, H. 2010. "The Green Revolution in Asia: Lessons for Africa." *FAO Paper.* http://www.icosgroup.net/static/foodsec/text/fao_the_green_revolution_in_Asia.pdf.

Kahn, D. L., and D. Lester. 2013. "Efforts to Decriminalize Suicide in Ghana, India and Singapore." *Suicidology Online* 4: 96–104. http://www.suicidology-online.com/pdf/SOL-2013-4-96-104.pdf.

Kaur, M. 2010. "The Paradox of India's Bread Basket: Farmer Suicides in Punjab." *PRAXIS: The Fletcher School of Human Security* 25: 39–60. http://fletcher.tufts.edu/praxis/archives/xxv/~/media/DD8E2DF1EA6C47028EB789DCD7ED5D37.pdf.

Kloepffer, W. 2008. "Life Cycle Sustainability Assessment of Products: (With Comments by Helias A. Udo de Haes, p. 95)." *The International Journal of Life Cycle Assessment* 13 (2): 89–95. doi: 10.1065/lca2008.02.376.

Kneifel, J. 2010. "Life-Cycle Carbon and Cost Analysis of Energy Efficiency Measures in New Commercial Buildings." *Energy and Buildings* 42 (3): 333–340. doi: 10.1016/j.enbuild.2009.09.011.

Korpi, E., and T. Ala-Risku. 2008. "Life Cycle Costing: A Review of Published Case Studies." *Managerial Auditing Journal* 23 (3): 240–261. http://www.emeraldinsight.com/doi/abs/10.1108/02686900810857703.

Kumar, P., and S. Kumar. 2010. "Valuing the Health Effects of Air Pollution from Agricultural Residue Burning." In *Invited Paper to the ACIAR Workshop "Policy Options to Reduce Rice Stubble Burning"*, Punjab State Marketing (Mandi) Board, Chandigarh, 13–15. http://www.cerdi.org/uploads/sfCmsContent/html/323/kumar.pdf.

Kurian, N. J. 2007. "Widening Economic & Social Disparities: Implications for India." *Indian Journal of Medical Research* 126 (4): 374. http://www.academia.edu/download/30569472/1014.pdf.

Lehmann, J., J. Gaunt, and M. Rondon. 2006. "Bio-Char Sequestration in Terrestrial Ecosystems – A Review." *Mitigation and Adaptation Strategies for Global Change* 11 (2): 395–419. doi: 10.1007/s11027-005-9006-5.

Lehmann, J., and S. Joseph. 2009. "Biochar for Environmental Management: An Introduction." In Biochar for Environmental Management: Science and Technology, edited by Johannes Lehmann and Stephen Joseph, p. cm. Includes bibliographical references and index. ISBN 978-1-84407-658-1 (hardback) 1. Charcoal. 2. Soil amendments. 3. Environmental management. I. Lehmann, Johannes, Dr. II. Joseph, Stephen, 1950-TP331.B56 2009 631.4'22–dc22, 1–12.

Li, X., S. Wang, L. Duan, J. Hao, C. Li, Y. Chen, and L. Yang. 2007. "Particulate and Trace Gas Emissions from Open Burning of Wheat Straw and Corn Stover in China." *Environmental Science & Technology* 41 (17): 6052–6058. doi: 10.1021/es0705137.

Linnerooth-Bayer, J., and R. Mechler. 2009. "Insurance against Losses from Natural Disasters in Developing Countries." New York. http://www.ruralfinance.org/fileadmin/templates/rflc/documents/Insurance__against_pdf.pdf.

Major, J. 2009. "A Guide to Conducting Biochar Trials." *International Biochar Initiative.* http://www.biochar-international.org/images/IBI_Biochar_Trial_Guidev1.1.pdf.

McKenzie, S. 2004. *Social Sustainability: Towards Some Definitions.* Hawke Research Institute, University of South Australia. http://unisa.edu.au/hawkeinstitute/publications/downloads/wp27.pdf.

Mohanty, B. B. 2013. "Farmer Suicides in India." *Economic & Political Weekly* 48 (21): 45. http://www.im4change.org.previewdns.com/siteadmin/tinymce/uploaded/Farmer_Suicides_in_India.pdf.

Morrison, M. L., and K. Sinclair. 1998. *Avian Risk and Fatality Protocol.* Springfield, VA: National Renewable Energy Laboratory. http://www.nrel.gov/wind/pdfs/24997.pdf. http://www.doe.gov/bridge/home.html.

Murakami, T. 1999. "GRI Symposium and Workshop." Symposium & workshop Japan Environment Agency. GRI (Global Reporting Initiative). Chuo Univ. Surugadai Kinenkan, Kanda-Surugadai, Chiyoda-ku, Tokyo: EARG (Environmental Auditing Research Group), GRI (Global Reporting Initiative).

Murray-Prior, R. 2008. "Are Farmers in Transitional Economies Likely to Benefit from Forming Collaborative Marketing Groups." *Banwa Management* 5 (2): 10–21. http://ageconsearch.umn.edu/bitstream/145093/2/Banwa%20vol.8%20no.2_Murray-Prior_pp10-21.pdf.

Murthy, N., and B. Mundial. 1990. *How Well Do India's Social Service Programs Serve the Poor?* Vol. 491. Public Sector Management and Private Sector Development Division, Country Economics Department, World Bank. http://www-wds.worldbank.org/external/default/WDSContentServer/IW3P/IB/1990/08/01/000009265_3960929182420/Rendered/PDF/multi_page.pdf.

Nakagawa, Y., and R. Shaw. 2004. "Social Capital: A Missing Link to Disaster Recovery." *International Journal of Mass Emergencies and Disasters* 22 (1): 5–34. http://www.alnap.org/pool/files/socialcapital2004.pdf.

Norman, W., and C. MacDonald. 2004. "Getting to the Bottom of 'Triple Bottom Line." Philosophy Documentation Center. *Business Ethics Quarterly*, 14 (2): 243–262. http://www.jstor.org/stable/3857909.

Novak, J. M., W. J. Busscher, D. L. Laird, M. Ahmedna, D. W. Watts, and M. A. S. Niandou. 2009. "Impact of Biochar Amendment on Fertility of a Southeastern Coastal Plain Soil." *Soil Science* 174 (2): 105–112. doi: 10.1097/SS.0b013e3181981d9a.

Nuffield Council on Bioethics. 2011. *Biofuels: Ethical Issues*. London: Nuffield Council on Bioethics.

O'Connell, M. 2005. "Regulatory Framework Development: Understanding Current Sustainability Regulations and Policy." *Report PR101 for Beacon Pathway. Auckland: Beacon Pathway*. http://beaconpathway.co.nz/images/uploads/Final_Report_PR101_Regulatory_Framework_Development.pdf.

Oduyemi, O., M. Okoroh, and A. Dean. 2014. "Barrier to Life Cycle Costing Usage." *30th Annual ARCOM Conference*, Portsmouth, UK.

Ostrom, E., and T. K. Ahn. 2003. "A Social Science Perspective on Social Capital: Social Capital and Collective Action." *Revista Mexicana De Sociologia* 65 (1): 155–233. http://www.researchgate.net/profile/T_Ahn/publication/242569787_A_SOCIAL_SCIENCE_PERSPECTIVE_ON_SOCIAL_CAPITAL_SOCIAL_CAPITAL_AND_COLLECTIVE_ACTION/links/00b495314b33f5220f000000.pdf.

Ouadi, M., J.G. Brammer, A. Hornung, and M. Kay. 2012. "Waste to Power." *TAPPI Journal of Bioenergy* 11 (2) (February): 55–64.

Ozbay, K., N. A. Parker, D. Jawad, and S. Hussain. 2003. "Guidelines for Life Cycle Cost Analysis." *Technical Report FHWA-NJ-2003-012*. NJ, USA: New Jersey Department of Transportation.

Pal, P., and J. Ghosh. 2007. "Inequality in India: A Survey of Recent Trends." *Economic and Social Affairs Working Paper* 45. http://www.chereum.umontreal.ca/activites_pdf/Session%201/Inequality%20in%20India%20(1).pdf.

Pant, D., A. Singh, G. Van Bogaert, Y. A. Gallego, L. Diels, and K. Vanbroekhoven. 2011. "An Introduction to the Life Cycle Assessment (LCA) of Bioelectrochemical Systems (BES) for Sustainable Energy and Product Generation: Relevance and Key Aspects." *Renewable and Sustainable Energy Reviews* 15 (2): 1305–1313. doi: 10.1016/j.rser.2010.10.005.

Parikh, K. 2011. "Low Carbon Strategies for Inclusive Growth - An Interim Report." Planning Commission Government of India.

Parthasarathy, D., and V. K. Chopde. 2001. *Building Social Capital: Collective Action, Adoption of Agricultural Innovations, and Poverty Reduction in the Indian Semi-Arid Tropics*. International Crops Research Institute for the Semi-Arid Tropics (ICRISAT). http://sites.google.com/site/sustoilfg/India.pdf. https://www.eldis.org/document/A28628.

Patel, A., P. Sarkar, H. Tyagi, and H. Singh. 2016. "Time Value of Emission and Technology Discounting Rate for Off-Grid Electricity Generation in India Using Intermediate Pyrolysis." *Environmental Impact Assessment Review* 59: 10–26. doi: 10.1016/j.eiar.2016.03.001.

Patel, A. R. 2018. "Decentralised Off-Grid Electricity Generation from Biooil Produced Using Intermediate Pyrolysis of Agricultural Residual Waste." *Ph D Thesis*, Ropar, Punjab, India: Indian Institute of Technology Ropar. doi: 10.13140/RG.2.2.12089.62567.

Pulakka, S., and A. Sarja. 1999. "Life Cycle Cost Design Methods and Tools." *Durability of Building Materials and Components* 8: 2715. http://citeseerx.ist.psu.edu/viewdoc/download?doi=10.1.1.515.4990&rep=rep1&type=pdf.

Rajaniemi, H. M., and J. Ahokas. 2011. "Greenhouse Gas Emissions from Oats, Barley, Wheat and Rye Production." *Agronomy Research Biosystem Engineering* 1 (Special issue): 189–195.

Ravindranath, N., P. Balachandra, S. Dasappa, and K. Usharao. 2006. "Bioenergy Technologies for Carbon Abatement." *Biomass and Bioenergy* 30 (10): 826–837. doi: 10.1016/j.biombioe.2006.02.003.

Real, S. F. 2010. "Contribution of Life Cycle Cost Analysis to Design Sustainability in Construction." *Masters Dissertation in Civil Engineering*, Department of Civil Engineering and Architecture-Instituto Superior Técnico-Technical University of Lisbon, Lisbon, Portugal. 153 p.

Roberts, K. G., B. A. Gloy, S. Joseph, N. R. Scott, and J. Lehmann. 2010. "Life Cycle Assessment of Biochar Systems: Estimating the Energetic, Economic, and Climate Change Potential." *Environmental Science & Technology* 44 (2): 827–833. doi: 10.1021/es902266r.

Safa, M., and S. Samarasinghe. 2012. "$CO_2$ Emissions from Farm Inputs 'Case Study of Wheat Production in Canterbury, New Zealand.'" *Environmental Pollution* 171 (December): 126–132. doi: 10.1016/j.envpol.2012.07.032.

Sagi, S., R. Berry, A. Patel, H. Singh, and T., Himanshu. 2014. "Decentralised Off-Grid Electricity Generation in India Using Intermediate Pyrolysis of Residue Straws." Overall Report of Phase I and Phase II. IIT Ropar, India and Aston University, UK.

Schultz, T. P. 1988. "Education Investments and Returns." *Handbook of Development Economics* 1 (1): 543–630. http://www.researchgate.net/profile/Paul_Schultz4/publication/4825955_Education_investments_and_returns/links/0deec52c454b8e0ed7000000.pdf.

Sen, G. 1997. "Empowerment as an Approach to Poverty." *Background Paper to Human Development Report*. http://ieham.org/html/docs/Empowerment_as_an_approach_to_Poverty.pdf.

Senthilkumaran, S., and S. Arunachalam. 2002. "Expanding the Village Knowledge Centres in Pondicherry." *Regional Development Dialogue* 23 (2; SEAS AUT): 65–84. http://104.238.81.58/~support/sites/default/files/RDD-IVRP.pdf.

Sharma, V. P. 2008. "India's Agrarian Crisis and Corporate-Led Contract Farming: Socio-Economic Implications for Smallholder Producers." *International Food and Agribusiness Management Review* 11 (4): 25–48. http://www.iimahd.ernet.in:8181/assets/upload/faculty/vijaypdf2.pdf.

Shiva, V., and K. Jalees. 2005. *Farmers Suicides in India*. Research Foundation for Science, Technology and Ecology. Hauz Khas, New Delhi: Research Foundation for Science, Technology and Ecology A-60. http://www.navdanya.org/attachments/Organic_Farming10.pdf.

Sidhu, S. S. 1974. "Relative Efficiency in Wheat Production in the Indian Punjab." *The American Economic Review*, 742–751. http://www.jstor.org/stable/1813329.

Singh, A., D. Pant, N. E. Korres, A-S. Nizami, S. Prasad, and J. D. Murphy. 2010. "Key Issues in Life Cycle Assessment of Ethanol Production from Lignocellulosic Biomass: Challenges and Perspectives." *Special Issue on Lignocellulosic Bioethanol: Current Status and Perspectives* 101(13): 5003–5012.

Sinha, S. 2007. "Agriculture Insurance in India." *CIRM Working Paper Series* (Chennai, India, CIRM). http://ifmr.co.in/wp-content/uploads/2011/03/cirm_sinha_paper_june2007.pdf.

Swarr, T. E., D. Hunkeler, W. Klöpffer, H-L. Pesonen, A. Ciroth, A C. Brent, and Robert Pagan. 2011. "Environmental Life-Cycle Costing: A Code of Practice." *The International Journal of Life Cycle Assessment* 16 (5): 389–391. doi: 10.1007/s11367-011-0287-5.

Taylor, G. D., T. S. Meinert, R. C. Killian, and G. L. Whicker. 1997. *Development and Analysis of Efficient Delivery Lanes and Zones in Truckload Trucking*. University of Arkansas, Mack-Blackwell National Rural Transportation Study Center. http://ww2.mackblackwell.org/web/research/ALL_RESEARCH_PROJECTS/1000s/1061-taylor/MBTC1061.pdf.

UNIDO. 2010. "Global Technology Roadmap for CCS in Industry Sectoral Assessment: Refineries, for United Nations Industrial Development Organisations, Report No./DNV Reg No.:/12P5TPP-9, Draft Rev 3, 2010-08-25."

Weidema, B., H. Wenzel, C. Petersen, and K. Hansen. 2004. "The Product, Functional Unit and Reference Flows in LCA." *Danish Environmental Protection Agency, Environmental News*, no. 70. http://gfc.force.dk/resources/777.pdf.

Weihe, T. 2004. *Cooperatives in Conflict & Failed States*. US Overseas Cooperative Development Council. http://www.uwcc.wisc.edu/info/intl/weihe.pdf.

Wickramasekera, P. 2002. *Asian Labour Migration: Issues and Challenges in an Era of Globalization*. International Migration Programme, International Labour Office. http://web.waseda.jp/gsaps/eaui/educational_program/PDF_WS2015/Lecture5_Reading1_Honda.pdf.

Williams, A. S. 2009. "Life Cycle Analysis: A Step by Step Approach." http://www.ideals.illinois.edu/handle/2142/14450.

Woolf, D., J. E. Amonette, F. Alayne Street-Perrott, J. Lehmann, and S. Joseph. 2010. "Sustainable Biochar to Mitigate Global Climate Change." *Nature Communications* 1 (5): 1–9. doi: 10.1038/ncomms1053.

Zhang, A., L. Cui, G. Pan, L. Li, Q. Hussain, X. Zhang, J. Zheng, and D. Crowley. 2010. "Effect of Biochar Amendment on Yield and Methane and Nitrous Oxide Emissions from a Rice Paddy from Tai Lake Plain, China." *Agriculture, Ecosystems & Environment* 139 (4): 469–475. https://doi.org/10.1016/j.agee.2010.09.003.

Zhang, Q., J. Chang, T. Wang, and Y. Xu. 2007. "Review of Biomass Pyrolysis Oil Properties and Upgrading Research." Energy Conversion and Management 48 (1): 87–92. doi: 10.1016/j.enconman.2006.05.010.

# Index